Pitman Research Notes in Mathematics Series

Submission of proposals for consideration
Suggestions for publication, in the form of outlines and representative samples, are invited by the Editorial Board for assessment. Intending authors should approach one of the main editors or another member of the Editorial Board, citing the relevant AMS subject classifications. Alternatively, outlines may be sent directly to the publisher's offices. Refereeing is by members of the board and other mathematical authorities in the topic concerned, throughout the world.

Preparation of accepted manuscripts
On acceptance of a proposal, the publisher will supply full instructions for the preparation of manuscripts in a form suitable for direct photo-lithographic reproduction. Specially printed grid sheets can be provided and a contribution is offered by the publisher towards the cost of typing. Word processor output, subject to the publisher's approval, is also acceptable.

Illustrations should be prepared by the authors, ready for direct reproduction without further improvement. The use of hand-drawn symbols should be avoided wherever possible, in order to maintain maximum clarity of the text.

The publisher will be pleased to give any guidance necessary during the preparation of a typescript, and will be happy to answer any queries.

Important note
In order to avoid later retyping, intending authors are strongly urged not to begin final preparation of a typescript before receiving the publisher's guidelines. In this way it is hoped to preserve the uniform appearance of the series.

Addison Wesley Longman Ltd
Edinburgh Gate
Harlow, Essex, CM20 2JE
UK
(Telephone (0) 1279 623623)

Titles in this series. A full list is available from the publisher on request.

151 A stochastic maximum principle for optimal control of diffusions
U G Haussmann
152 Semigroups, theory and applications. Volume II
H Brezis, M G Crandall and F Kappel
153 A general theory of integration in function spaces
P Muldowney
154 Oakland Conference on partial differential equations and applied mathematics
L R Bragg and J W Dettman
155 Contributions to nonlinear partial differential equations. Volume II
J I Díaz and P L Lions
156 Semigroups of linear operators: an introduction
A C McBride
157 Ordinary and partial differential equations
B D Sleeman and R J Jarvis
158 Hyperbolic equations
F Colombini and M K V Murthy
159 Linear topologies on a ring: an overview
J S Golan
160 Dynamical systems and bifurcation theory
M I Camacho, M J Pacifico and F Takens
161 Branched coverings and algebraic functions
M Namba
162 Perturbation bounds for matrix eigenvalues
R Bhatia
163 Defect minimization in operator equations: theory and applications
R Reemtsen
164 Multidimensional Brownian excursions and potential theory
K Burdzy
165 Viscosity solutions and optimal control
R J Elliott
166 Nonlinear partial differential equations and their applications: Collège de France Seminar. Volume VIII
H Brezis and J L Lions
167 Theory and applications of inverse problems
H Haario
168 Energy stability and convection
G P Galdi and B Straughan
169 Additive groups of rings. Volume II
S Feigelstock
170 Numerical analysis 1987
D F Griffiths and G A Watson
171 Surveys of some recent results in operator theory. Volume I
J B Conway and B B Morrel
172 Amenable Banach algebras
J-P Pier
173 Pseudo-orbits of contact forms
A Bahri
174 Poisson algebras and Poisson manifolds
K H Bhaskara and K Viswanath
175 Maximum principles and eigenvalue problems in partial differential equations
P W Schaefer
176 Mathematical analysis of nonlinear, dynamic processes
K U Grusa
177 Cordes' two-parameter spectral representation theory
D F McGhee and R H Picard
178 Equivariant K-theory for proper actions
N C Phillips
179 Elliptic operators, topology and asymptotic methods
J Roe
180 Nonlinear evolution equations
J K Engelbrecht, V E Fridman and E N Pelinovski
181 Nonlinear partial differential equations and their applications: Collège de France Seminar. Volume IX
H Brezis and J L Lions
182 Critical points at infinity in some variational problems
A Bahri
183 Recent developments in hyperbolic equations
L Cattabriga, F Colombini, M K V Murthy and S Spagnolo
184 Optimization and identification of systems governed by evolution equations on Banach space
N U Ahmed
185 Free boundary problems: theory and applications. Volume I
K H Hoffmann and J Sprekels
186 Free boundary problems: theory and applications. Volume II
K H Hoffmann and J Sprekels
187 An introduction to intersection homology theory
F Kirwan
188 Derivatives, nuclei and dimensions on the frame of torsion theories
J S Golan and H Simmons
189 Theory of reproducing kernels and its applications
S Saitoh
190 Volterra integrodifferential equations in Banach spaces and applications
G Da Prato and M Iannelli
191 Nest algebras
K R Davidson
192 Surveys of some recent results in operator theory. Volume II
J B Conway and B B Morrel
193 Nonlinear variational problems. Volume II
A Marino and M K V Murthy
194 Stochastic processes with multidimensional parameter
M E Dozzi
195 Prestressed bodies
D Iesan
196 Hilbert space approach to some classical transforms
R H Picard
197 Stochastic calculus in application
J R Norris
198 Radical theory
B J Gardner
199 The C^*-algebras of a class of solvable Lie groups
X Wang
200 Stochastic analysis, path integration and dynamics
K D Elworthy and J C Zambrini

201 Riemannian geometry and holonomy groups
S Salamon
202 Strong asymptotics for extremal errors and polynomials associated with Erdös type weights
D S Lubinsky
203 Optimal control of diffusion processes
V S Borkar
204 Rings, modules and radicals
B J Gardner
205 Two-parameter eigenvalue problems in ordinary differential equations
M Faierman
206 Distributions and analytic functions
R D Carmichael and D Mitrovic
207 Semicontinuity, relaxation and integral representation in the calculus of variations
G Buttazzo
208 Recent advances in nonlinear elliptic and parabolic problems
P Bénilan, M Chipot, L Evans and M Pierre
209 Model completions, ring representations and the topology of the Pierce sheaf
A Carson
210 Retarded dynamical systems
G Stepan
211 Function spaces, differential operators and nonlinear analysis
L Paivarinta
212 Analytic function theory of one complex variable
C C Yang, Y Komatu and K Niino
213 Elements of stability of visco-elastic fluids
J Dunwoody
214 Jordan decomposition of generalized vector measures
K D Schmidt
215 A mathematical analysis of bending of plates with transverse shear deformation
C Constanda
216 Ordinary and partial differential equations. Volume II
B D Sleeman and R J Jarvis
217 Hilbert modules over function algebras
R G Douglas and V I Paulsen
218 Graph colourings
R Wilson and R Nelson
219 Hardy-type inequalities
A Kufner and B Opic
220 Nonlinear partial differential equations and their applications: Collège de France Seminar. Volume X
H Brezis and J L Lions
221 Workshop on dynamical systems
E Shiels and Z Coelho
222 Geometry and analysis in nonlinear dynamics
H W Broer and F Takens
223 Fluid dynamical aspects of combustion theory
M Onofri and A Tesei
224 Approximation of Hilbert space operators. Volume I. 2nd edition
D Herrero
225 Operator theory: proceedings of the 1988 GPOTS–Wabash conference
J B Conway and B B Morrel
226 Local cohomology and localization
J L Bueso Montero, B Torrecillas Jover and A Verschoren
227 Nonlinear waves and dissipative effects
D Fusco and A Jeffrey
228 Numerical analysis 1989
D F Griffiths and G A Watson
229 Recent developments in structured continua. Volume II
D De Kee and P Kaloni
230 Boolean methods in interpolation and approximation
F J Delvos and W Schempp
231 Further advances in twistor theory. Volume I
L J Mason and L P Hughston
232 Further advances in twistor theory. Volume II
L J Mason, L P Hughston and P Z Kobak
233 Geometry in the neighborhood of invariant manifolds of maps and flows and linearization
U Kirchgraber and K Palmer
234 Quantales and their applications
K I Rosenthal
235 Integral equations and inverse problems
V Petkov and R Lazarov
236 Pseudo-differential operators
S R Simanca
237 A functional analytic approach to statistical experiments
I M Bomze
238 Quantum mechanics, algebras and distributions
D Dubin and M Hennings
239 Hamilton flows and evolution semigroups
J Gzyl
240 Topics in controlled Markov chains
V S Borkar
241 Invariant manifold theory for hydrodynamic transition
S Sritharan
242 Lectures on the spectrum of $L^2(\Gamma\backslash G)$
F L Williams
243 Progress in variational methods in Hamiltonian systems and elliptic equations
M Girardi, M Matzeu and F Pacella
244 Optimization and nonlinear analysis
A Ioffe, M Marcus and S Reich
245 Inverse problems and imaging
G F Roach
246 Semigroup theory with applications to systems and control
N U Ahmed
247 Periodic-parabolic boundary value problems and positivity
P Hess
248 Distributions and pseudo-differential operators
S Zaidman
249 Progress in partial differential equations: the Metz surveys
M Chipot and J Saint Jean Paulin
250 Differential equations and control theory
V Barbu

251 Stability of stochastic differential equations with respect to semimartingales
X Mao
252 Fixed point theory and applications
J Baillon and M Théra
253 Nonlinear hyperbolic equations and field theory
M K V Murthy and S Spagnolo
254 Ordinary and partial differential equations. Volume III
B D Sleeman and R J Jarvis
255 Harmonic maps into homogeneous spaces
M Black
256 Boundary value and initial value problems in complex analysis: studies in complex analysis and its applications to PDEs 1
R Kühnau and W Tutschke
257 Geometric function theory and applications of complex analysis in mechanics: studies in complex analysis and its applications to PDEs 2
R Kühnau and W Tutschke
258 The development of statistics: recent contributions from China
X R Chen, K T Fang and C C Yang
259 Multiplication of distributions and applications to partial differential equations
M Oberguggenberger
260 Numerical analysis 1991
D F Griffiths and G A Watson
261 Schur's algorithm and several applications
M Bakonyi and T Constantinescu
262 Partial differential equations with complex analysis
H Begehr and A Jeffrey
263 Partial differential equations with real analysis
H Begehr and A Jeffrey
264 Solvability and bifurcations of nonlinear equations
P Drábek
265 Orientational averaging in mechanics of solids
A Lagzdins, V Tamuzs, G Teters and A Kregers
266 Progress in partial differential equations: elliptic and parabolic problems
C Bandle, J Bemelmans, M Chipot, M Grüter and J Saint Jean Paulin
267 Progress in partial differential equations: calculus of variations, applications
C Bandle, J Bemelmans, M Chipot, M Grüter and J Saint Jean Paulin
268 Stochastic partial differential equations and applications
G Da Prato and L Tubaro
269 Partial differential equations and related subjects
M Miranda
270 Operator algebras and topology
W B Arveson, A S Mishchenko, M Putinar, M A Rieffel and S Stratila
271 Operator algebras and operator theory
W B Arveson, A S Mishchenko, M Putinar, M A Rieffel and S Stratila
272 Ordinary and delay differential equations
J Wiener and J K Hale
273 Partial differential equations
J Wiener and J K Hale
274 Mathematical topics in fluid mechanics
J F Rodrigues and A Sequeira
275 Green functions for second order parabolic integro-differential problems
M G Garroni and J F Menaldi
276 Riemann waves and their applications
M W Kalinowski
277 Banach C(K)-modules and operators preserving disjointness
Y A Abramovich, E L Arenson and A K Kitover
278 Limit algebras: an introduction to subalgebras of C*-algebras
S C Power
279 Abstract evolution equations, periodic problems and applications
D Daners and P Koch Medina
280 Emerging applications in free boundary problems
J Chadam and H Rasmussen
281 Free boundary problems involving solids
J Chadam and H Rasmussen
282 Free boundary problems in fluid flow with applications
J Chadam and H Rasmussen
283 Asymptotic problems in probability theory: stochastic models and diffusions on fractals
K D Elworthy and N Ikeda
284 Asymptotic problems in probability theory: Wiener functionals and asymptotics
K D Elworthy and N Ikeda
285 Dynamical systems
R Bamon, R Labarca, J Lewowicz and J Palis
286 Models of hysteresis
A Visintin
287 Moments in probability and approximation theory
G A Anastassiou
288 Mathematical aspects of penetrative convection
B Straughan
289 Ordinary and partial differential equations. Volume IV
B D Sleeman and R J Jarvis
290 *K*-theory for real *C**-algebras
H Schröder
291 Recent developments in theoretical fluid mechanics
G P Galdi and J Necas
292 Propagation of a curved shock and nonlinear ray theory
P Prasad
293 Non-classical elastic solids
M Ciarletta and D Ieşan
294 Multigrid methods
J Bramble
295 Entropy and partial differential equations
W A Day
296 Progress in partial differential equations: the Metz surveys 2
M Chipot
297 Nonstandard methods in the calculus of variations
C Tuckey
298 Barrelledness, Baire-like- and (LF)-spaces
M Kunzinger
299 Nonlinear partial differential equations and their applications. Collège de France Seminar. Volume XI
H Brezis and J L Lions

300 Introduction to operator theory
T Yoshino
301 Generalized fractional calculus and applications
V Kiryakova
302 Nonlinear partial differential equations and their applications. Collège de France Seminar Volume XII
H Brezis and J I Lions
303 Numerical analysis 1993
D F Griffiths and G A Watson
304 Topics in abstract differential equations
S Zaidman
305 Complex analysis and its applications
C C Yang, G C Wen, K Y Li and Y M Chiang
306 Computational methods for fluid-structure interaction
J M Crolet and R Ohayon
307 Random geometrically graph directed self-similar multifractals
L Olsen
308 Progress in theoretical and computational fluid mechanics
G P Galdi, J Málek and J Necas
309 Variational methods in Lorentzian geometry
A Masiello
310 Stochastic analysis on infinite dimensional spaces
H Kunita and H-H Kuo
311 Representations of Lie groups and quantum groups
V Baldoni and M Picardello
312 Common zeros of polynomials in several variables and higher dimensional quadrature
Y Xu
313 Extending modules
N V Dung, D van Huynh, P F Smith and R Wisbauer
314 Progress in partial differential equations: the Metz surveys 3
M Chipot, J Saint Jean Paulin and I Shafrir
315 Refined large deviation limit theorems
V Vinogradov
316 Topological vector spaces, algebras and related areas
A Lau and I Tweddle
317 Integral methods in science and engineering
C Constanda
318 A method for computing unsteady flows in porous media
R Raghavan and E Ozkan
319 Asymptotic theories for plates and shells
R P Gilbert and K Hackl
320 Nonlinear variational problems and partial differential equations
A Marino and M K V Murthy
321 Topics in abstract differential equations II
S Zaidman
322 Diffraction by wedges
B Budaev
323 Free boundary problems: theory and applications
J I Diaz, M A Herrero, A Liñan and J L Vazquez
324 Recent developments in evolution equations
A C McBride and G F Roach
325 Elliptic and parabolic problems: Pont-à-Mousson 1994
C Bandle, J Bemelmans, M Chipot, J Saint Jean Paulin and I Shafrir
326 Calculus of variations, applications and computations: Pont-à-Mousson 1994
C Bandle, J Bemelmans, M Chipot, J Saint Jean Paulin and I Shafrir
327 Conjugate gradient type methods for ill-posed problems
M Hanke
328 A survey of preconditioned iterative methods
A M Bruaset
329 A generalized Taylor's formula for functions of several variables and certain of its applications
J-A Riestra
330 Semigroups of operators and spectral theory
S Kantorovitz
331 Boundary-field equation methods for a class of nonlinear problems
G N Gatica and G C Hsiao
332 Metrizable barrelled spaces
J C Ferrando, M López Pellicer and L M Sánchez Ruiz
333 Real and complex singularities
W L Marar
334 Hyperbolic sets, shadowing and persistence for noninvertible mappings in Banach spaces
B Lani-Wayda
335 Nonlinear dynamics and pattern formation in the natural environment
A Doelman and A van Harten
336 Developments in nonstandard mathematics
N J Cutland, V Neves, F Oliveira and J Sousa-Pinto
337 Topological circle planes and topological quadrangles
A E Schroth
338 Graph dynamics
E Prisner
339 Localization and sheaves: a relative point of view
P Jara, A Verschoren and C Vidal
340 Mathematical problems in semiconductor physics
P Marcati, P A Markowich and R Natalini
341 Surveying a dynamical system: a study of the Gray–Scott reaction in a two-phase reactor
K Alhumaizi and R Aris
342 Solution sets of differential equations in abstract spaces
R Dragoni, J W Macki, P Nistri and P Zecca
343 Nonlinear partial differential equations
A Benkirane and J-P Gossez
344 Numerical analysis 1995
D F Griffiths and G A Watson
345 Progress in partial differential equations: the Metz surveys 4
M Chipot and I Shafrir
346 Rings and radicals
B J Gardner, Liu Shaoxue and R Wiegandt
347 Complex analysis, harmonic analysis and applications
R Deville, J Esterle, V Petkov, A Sebbar and A Yger
348 The theory of quantaloids
K I Rosenthal
349 General theory of partial differential equations and microlocal analysis
Qi Min-you and L Rodino

350 Progress in elliptic and parabolic partial differential equations
A Alvino, P Buonocore, V Ferone, E Giarrusso, S Matarasso, R Toscano and G Trombetti
351 Integral representations for spatial models of mathematical physics
V V Kravchenko and M V Shapiro
352 Dynamics of nonlinear waves in dissipative systems: reduction, bifurcation and stability
G Dangelmayr, B Fiedler, K Kirchgässner and A Mielke
353 Singularities of solutions of second order quasilinear equations
L Véron
354 Mathematical theory in fluid mechanics
G P Galdi, J Málek and J Necas
355 Eigenfunction expansions, operator algebras and symmetric spaces
R M Kauffman
356 Lectures on bifurcations, dynamics and symmetry
M Field
357 Noncoercive variational problems and related results
D Goeleven
358 Generalised optimal stopping problems and financial markets
D Wong
359 Topics in pseudo-differential operators
S Zaidman
360 The Dirichlet problem for the Laplacian in bounded and unbounded domains
C G Simader and H Sohr
361 Direct and inverse electromagnetic scattering
A H Serbest and S R Cloude
362 International conference on dynamical systems
F Ledrappier, J Lewowicz and S Newhouse
363 Free boundary problems, theory and applications
M Niezgódka and P Strzelecki
364 Backward stochastic differential equations
N El Karoui and L Mazliak
365 Topological and variational methods for nonlinear boundary value problems
P Drábek
366 Complex analysis and geometry
V Ancona, E Ballico, R M Mirò-Roig and A Silva
367 Integral expansions related to Mehler–Fock type transforms
B N Mandal and N Mandal
368 Elliptic boundary value problems with indefinite weights: variational formulations of the principal eigenvalue and applications
F Belgacem
369 Integral transforms, reproducing kernels and their applications
S Saitoh
370 Ordinary and partial differential equations. Volume V
P D Smith and R J Jarvis
371 Numerical methods in mechanics
C Conca and G N Gatica
372 Generalized manifolds
K-G Schlesinger
373 Independent axioms for Minkowski space-time
J W Schutz
374 Integral methods in science and engineering Volume one: analytic methods
C Constanda, J Saranen and S Seikkala
375 Integral methods in science and engineering Volume two: approximation methods
C Constanda, J Saranen and S Seikkala
376 Inner product spaces and applications
T M Rassias
377 Functional analysis with current applications in science, technology and industry
M Brokate and A H Siddiqi

Martin Brokate

University of Kiel, Germany

and

Abul Hasan Siddiqi

Aligarh Muslim University, India

(Editors)

Functional analysis with current applications in science, technology and industry

CRC Press
Taylor & Francis Group
Boca Raton London • New York

CRC Press is an imprint of and
Taylor & Francis Group, an Informa business

FIRST INDIAN REPRINT, 2014

Addison Wesley Longman Limited
Edinburgh Gate, Harlow
Essex CM20 2JE, England
and Associated Companies throughout the world.

Published in the United States of America
by Addison Wesley Longman Inc.

First published 1998

AMS Subject Classifications (Main) 46, 47, 49
(Subsidiary) 35, 65, 68

ISSN 0269-3674

ISBN 0 582 31260 4

British Library Cataloguing in Publication Data

A catalogue record for this book is
available from the British Library

Printed and bound in India by Bhavish Graphics.

Table of Contents

Part C: Applications in Science and Industry

Preface

As science and technology become more refined and interconnected, the demand for advanced mathematics beyond basic vector algebra and differential and integral calculus increases. Functional analysis, besides being an area of independent innermathematical interest, provides many fundamental notions relevant for the description, analysis, numerical approximation and computer simulation of processes which evolve in a continuum of time or space. Awareness of that fact continues to spread over the world – indeed, in several respects functional analysis has already become part of the common language of science.

During the week December 16 – 19, 1996, the conference "Functional Analysis and Applications" took place at Aligarh Muslim University, Aligarh, India. Invited speakers were W. Żelasko (Poland), S. Saxon (USA) and L. Sánchez Ruiz (Spain) for structural results concerning topological algebras and topological vector spaces, S. Kesavan (India), P.N. Srikanth (India), Adimurthi (India), W. Oettli (Germany) and H. Ben-El-Mechaiekh (Canada) on variational problems, and H. Neunzert (Germany), M. Brokate (Germany), G. Evans (U.K.) and A.K. Pani (India) on applications in science and industry. Their as well as some other selected contributions form the contents of this volume. It is dedicated to Dr. Taqdir Husain, Professor Emeritus of McMaster University, Hamilton, Canada, a distinguished alumnus of Aligarh Muslim University who played a significant role in promoting teaching and research of functional analysis, especially in the third world countries.

The conference was organized by a committee including Dr. M. Rahman (Vice Chancellor of Aligarh Muslim University) as chairman and professors A.H. Siddiqi, J.N. Kapur, H.P. Dixit, M. Mohsin, M.Z. Khan, M.A. Qadri, Z.U. Ahmad and M.A. Pathan as members.

As a guest of honor, Professor D.N. Misra whose efforts helped a lot in providing for mathematics its due place within the Indian scientific system, also addressed the delegates during the inaugural ceremony.

We join the sentiments Dr. Rahman expresses in his address to the publisher and to the organizations supporting the conference; in addition we want to thank the contributors for the timely submission of their manuscripts and the universities of Kaiserslautern and Kiel for providing their facilities to complete the present volume.

Martin Brokate　　　　Abul Hasan Siddiqi

Sitting Row (Left to Right):
S.Kesavan, V.Vardharajan, H.Ben-El-Mechaiekh, T.Husain, M.Brokate, A.H.Siddiqi, W.Żelasko, M.Rahman (VC), D.N.Misra, M.Shafi, M.A.Qadri, Zandra, M.Mohsin, M.A.Pathan, W.Oettli.

Standing First Row (Left to Right):
Adimurthi, A.Sofi, A.Mehdi, B.Moosvi, Asima, Sobuhi, Anupama Gupta, Pammy Manchanda, P.Singh, A.P.Singh, R.D.Giri, B.S.Komal, S.N.Purohit, S.N.Nanda, A.K.Pani, P.K.Jain, R.S.Pathak, P.K.Banerji, Q.H.Ansari.

Standing Second Row (Left to Right):
S.B.Joshi, P.N.Srikanth, R.Kumar, M.Rehman, S.C.Gupta, Z.Ehsan, N.Ahmad, S.Patnaik, Firdaus, K.R.Kazmi, A.Ahmad, A.Anguraj, R.Kumar, K.N.Rangnathan, R.P.Sinha, S.M.Zaidi, S.M.Rizvi.

Standing Third Row (Left to Right):
R.K.George, R.N.Mukherjee, Ghazi Salama, S.A.V.Manickam, R.Jones, Abdullah A.Asad, A.Mukheimer, M.J.S.Shahwan, Shadab, Tariq, Nabiullah, S.Husain, T.Jamil, P.K.Sengupta.

ALIGARH MUSLIM UNIVERSITY

ALIGARH-202 002

Vice Chancellor

M. RAHMAN

May 13, 1997

I am very happy to know that the Addison Wesley Longman Limited, United Kingdom, is publishing all important papers presented during the International Conference on "Functional Analysis with Applications" held during 16-19 December 1996 at the AMU with the title ` Functional Analysis with Current Applications to Science, Technology and Industry` as a volume in the Pitman Research Notes in Mathematics series. It was my privilege to have acted as Chairman of the Organising Committee. The holding of this Conference was possible because of the financial support of the ICTP, Trieste, Italy, DST, CSIR, UGC and NBHM, Government of India and NGOS like Nadim Tarin Educational Society, Duty Society and Al-Falah Group.

In my opinion this volume will provide inspiration and guidelines to our young researchers to work in the emerging areas of Mathematics which are of vital importance for Science, Technology and Industry. It was a sheer joy to have had the opportunity of interaction with experts from all over the globe. The conference has left an undying landmark in our academic life and we will cherish this memory for all time to come. We also wish that we will have another opportunity to host a similar conference not too distant a future.

(M. Rahman)
Vice-Chancellor

Phone : (O) (0571) 400994, (R) (0571) 400173 FAX : (O) (0571) 401815, (R) (0571) 400087
GRAM : MUSLIM UNIVERSITY

Part A

Structural Results

W ŻELAZKO

On some old problems and new questions in topological algebras

To Professor Taqdir Husain

The aim of this article is two-fold, firstly we report on the present status of the 33 problems posed in [27] (called in the sequel *Problems* and numbered with natural numbers). We mention also the additional problems posed in [29], called here *Problems* too, and numbered there and in this paper with the capital Roman letters. Secondly we pose several new problems (called in the sequel *Questions*). As we shall see, most of the problems in [27] are solved by now, we shall formulate the corresponding results and provide the readers with suitable references. The article is based upon the talk the author gave during the international conference described in this volume, here he wishes to express his gratitude to the organizers for the hospitality extended to him.

All algebras in this paper are associative, if not stated otherwise each considered algebra can be either real or complex as well. A *topological algebra* is an algebra A which is a (Hausdorff) topological vector space (shortly: t.v.s.) and the multiplication is a jointly continuous bilinear map of $A \times A$ into A. We say that A is a *semitopological algebra* if we assume that the multiplication is merely separately continuous. A semitopological algebra A is topological if and only if for each neighbourhood of the origin U there is such a neighbourhood V which satisfies

$$V^2 \subset U. \tag{1}$$

Every topological algebra A can be completed, in the sense that its multiplication uniquely extends to a jointly continuous multiplication on the completion of the t.v.s. A. Generally speaking, this is not longer true for a semitopological algebra. Also every complete and metrizable semitopological algebra (an F-*algebra*, or an algebra of *type* F) is topological. A topological, or semitopological algebra is said *locally convex*, if it is locally convex as a t.v.s. If A is a locally convex topological algebra, then its topology can be given by means of a family $(\|\cdot\|_\alpha)_{\alpha\in\mathfrak{a}}$ of seminorms such that for each α in $\mathfrak{a}$ there is a $\beta \in \mathfrak{a}$ with

(2) $$\|xy\|_\alpha \leq \|x\|_\beta \|y\|_\beta \text{ for all } x, y \in A, \text{ and } \|e\|_\alpha = 1 \text{ for all } \alpha,$$

in case when A is unital with a unity e (see [6]). For general information about topological algebras the reader is referred to [9], [14], [15], [21], [24], [25].

The first two problems in [27] concern the topologization of algebras, so first we describe the related questions and results.

Problem 1. Let A be a real or complex algebra, can it be given a topology, so that it becomes a topological algebra ?

Problem 2. Let A be a real or complex algebra, can it be given a topology so that it becomes a locally convex algebra ?

If we relax the requirements in Problems 1 and 2 and replace there a "topological algebra" by a "semitopological algebra" the answer to so modified Problems 1 and 2 is in positive:

Proposition 1. Let A be a real or complex algebra, then it can be topologized as a complete locally convex semitopological algebra.

The proof is almost obvious, we simply provide A with the maximal (strongest) locally convex topology τ_{max}^{LC}, given by means of all seminorms on A. It is known (see e.g. [19], example on p. 56) that it is a complete (Hausdorff) topology on each vector space, and under this topology all endomorphisms are continuous and all linear subspaces are closed. In particular, continuous are all operators of right and left multiplication by a fixed element in A, so that (A, τ_{max}^{LC}) is a semitopological algebra. There occurs a question (cf [35]) how many different topologies can make an infinite dimensional algebra a semitopological algebra. This is answered in [23] with the following result

Proposition 2. Let A be a real or complex infinite dimensional algebra, then A can be made a semitopological algebra in at least two different ways.

Here we use the topology τ_{max}^{LC} and the topology τ_{max}^{w} – the strongest weak topology given by means of all seminorms of the form $x \to |f(x)|$, where f is a linear functional on A. These topologies are different because first is complete and the second is never complete on an infinite dimensional vector space. We cannot, however answer the following

Question 1. Let A be an infinite dimensional real or complex algebra. Can it be topologized in at least two different ways as a complete semitopological algebra ?

This question is open only for countably dimensional algebras (= at most countably generated algebras), i.e. algebras having a countable Hamel basis. For uncountably dimensional algebras we have a stronger result [13]

Proposition 3. Let A be an uncountably dimensional real or complex algebra. Then A can be made a complete semitopological algebra in at least continuum different ways.

For the proof we can use the topologies τ^p_{max}, $0 < p \leq 1$, i.e. the maximal p-convex topologies given by means of all p-homogeneous seminorms. Again these topologies are complete and Hausdorff vector space topologies, with all endomorphisms continuous and all linear subspaces closed. Moreover (cf.[13]), these topologies are mutually different on a vector space X iff the dimension of X is uncountable. Otherwise they all coincide and then they are locally convex since $\tau^{LC}_{max} = \tau^1_{max}$. This implies, in particular, the following result in topological vector spaces, which is not connected with topological algebras (see[13]).

Proposition 4. There exists a complete non-locally convex t.v.s. with all separable subspaces locally convex.

For the proof we can take any uncountably dimensional vector space X provided with τ^p_{max} for $0 < p < 1$. Thus X in non-locally convex. If X_o is a separable subspace of X with a dense countable subset S, then the linear span of S is a dense subaspace of X_o, but all subapsces are closed, so it coincides with X_o. Consequently X_o has an at most countable Hamel basis. One can easily see that the topology of X restricted to X_o is again τ^p_{max}, but this time it coincides with τ^{LC}_{max}.

The Proposition 4 and further Proposition 13 suggest that the answer to the following question can be in positive

Question 2. Let A be an infinite dimensional algebra. Can it be topologized as a (complete) semitopological algebra in infinitely many ways ?

Maybe even a stronger question can be answered in positive

Question 3. Let A be as above. Can it be topologized as a (complete) semi-topological algebra in uncountaby many ways ?

Both questions can be understood in two ways, the weaker one asks about any topologies and the stronger one about the complete topologies.

We pass back to the topological algebras. The Problem 2 was answered in negative in [28]. The obtained result can be formulated as follows.

Proposition 5. Let X be a real or complex vector space, and denote by $L_{FD}(X)$ the algebra of all its finite dimensional endomorphisms. Then $L_{FD}(X)$ is topologizable as a locally convex topological algebra if and only if the dimension of X is finite. Consequently the algebra $L(X)$ of all endomorphisms of X is topologizable as a locally convex algebra if and only if X is finite dimensional.

Later Müller [17] answered in negative to the Problem 1. His result, which also gives the first example of a commutative non-topologizable algebra is as follows.

Proposition 6. There exists a commutative algebra A which is non-topologizable as a topological algebra. Moreover the algebra A can be imbedded into the algebra $L(X)$ for each infinite dimensional vector space X and consequently this algebra is non-topologizable as a topological algebra.

Further the author obtained a stronger version of the Proposition 5 (see [34]). To formulate it observe that if in the definition of a topological algebra we relax the condition that the topology is Hasdorff and merely assume the joint continuity of multiplication, equivalent to the relaion (1), then each algebra is topologizable (even as a locally convex algebra) with so called antidiscrete topology i.e. the minimal vector space topology τ_{min} for which the only neighbourhood of the origin is the whole space (see [22]). This follows that each algebra A has a maximal algebra topology τ^a_{max}, i.e. a maximal topology statysfying relation (1), and A is topologizable if and only if this topology is Hausdorff. Our result states that the maximal algebra topology can be as bad as possible, it reads as follows.

Proposition 7. Let X be an infinite dimensional vector space and put $A = L_{FD}(X)$. Then the maximal algebra topology for A coincides with the antidiscrete topology τ_{min}.

The above result gives also non-topologizability of $L(X)$ for an infinite dimensional vector space X and the proof of this fact is much simpler then the proof given by Proposition 6. We do not know whether the non-topologizability of $L(X)$ is due only to the fact that it contains (the two-sided ideal) $L_{FD}(X)$. So we pose the following two questions.

Question 4. Let X be an infinite dimensional (real or complex) vector space and put $A = L(X)/L_{FD}(X)$ (a Calkin type algebra). Is A topologizable as a topological (resp. topological locally convex) algebra ?

Question 5. Determine the topology τ^{a}_{max} for the algebra A of the question 4.

The above results could suggest that if an algebra is topologizable as a topological algebra then it is also topologizable as a topological locally convex algebra. Such a conjecture is disproved by the following result in [14].

Proposition 8. There exists a commutative topological algebra which is not topologizable as a locally convex topological algebra.

There exist, however, positive results concerning topologizability. In [31] we obtained the following results.

Proposition 9. Let A be an at most countably generated algebra. Then (A, τ^{LC}_{max}) is a topological algebra. Consequently each such an algebra is topologizable as a complete locally convex topological algebra.

The proof follows immediately from the following result. Recall that a free algebra is an algebra of polynomials in non-commutung variables and that any algebra A is a quotient of a free algebra (with number of variables equal to the number of generators of A).

Proposition 10. Let $\mathbf{t} = (t_\alpha)$ be an arbitrary family of variables and let $A = A_F(\mathbf{t})$ be a free algebra in these variables. Then (A, τ^{LC}_{max}) is a topological algebra if and only if the number of variables is at most countable.

Note that the above result is independent of the continuum hypothesis as well as the following result given in [7], which shows that the Proposition 10 gives the best possible result experessed in terms of the cardinality of a set of generators.

Proposition 11 For each cardinal number $\aleph > \aleph_o$ there exists an algebra with $\aleph$ generators which is not topologizable as a topological algebra.

The Proposition 10 implies also the following result given in [31].

Proposition 12. There exists a complete locally convex semitopological algebra which is not topological but all of whose commutative subalgebras are topological.

For the proof we can take a free algebra $A_F(\mathbf{t})$ in uncountably many variables equipped with the topology τ_{max}^{LC}. It is easy to see that each its commutative subalgebra can depend only upon a finitely many variables and so is topological under the topology τ_{max}^{LC} which is a restriction of the topology of $A_F(\mathbf{t})$ to this subalgebra.

Note that in Proposition 10 we can take instead of τ_{max}^{LC} any topology τ_{max}^{p} and the result remains true. Thus combining the result of Proposition 4 with that of Proposition 12 we can obtain a complete semitopological algebra which is non-locally convex and non-topological, but all of whose separable subalgebras are locally convex and topological (this seems to be a new observation).

The negative solutions to the Problems 1 and 2 show that the following problems (posed in [29]) make a sense.

Problem A. When a real or complex algebra can be topologized so that it becomes a topological algebra ?

Problem B. When a real or complex algebra can be topologized so that it becomes a locally convex algebra ?

Problem C. Under which conditions an algebra can be topologized as a normed topological algebra ?

We can also consider (harder) version of these Problems demanding the completeness in addition.

Not much is known about the uniqueness of a topological algebra topology. Note the following result obtained in [23].

Proposition 13. Let A be an algebra of polynomials, or a free algebra, both in arbitrarily many variables. Then A has at least continuum different complete locally convex topologies making it a topological algebra.

We pose the following question analogous to Questions 1-3.

Question 6. Let A be an infinite dimensional topological algebra, can it be made a topological algebra in at least two different ways (resp. ininitely, or uncountably many ways) ?

Question 7. Does there exist a topological algebra which is non-topologizable as a complete topological algebra ?

Question 8. Let A be an infinite dimensional complete topological algebra, can it be made a complete topological algebra in two different ways (resp. infinitely, or uncountably many ways) ?

Question 9. Let A be an infinitely dimensional locally convex topological algebra. Can it be made a locally convex topological algebra in two (resp. infinitely, or uncountably many) different ways ?

In spite of all mentioned results we are very far from full understanding the topologization problem. There are no necessary and sufficient conditions fo the topologization of an algebra, not mentioning the topologization it as a complete topological algebra, as a metrizable algebra etc. Since every semisimple Banach algebra has a unique Banach algebra topology (see [11] and [18]) this topology is hidden somewhere in the algebraic structure of the algebra in question–but we do not know how to obtain this topology. This leads us to the next problem posed in [27], we restricted there our attenion to the commutative algebras, which seems also to be a hard question.

Problem 3. Characterize commutative semisimple algebras which can be topologized as Banach algebras.

A locally convex algebra of type F is said a B_o-*algebra*, or an algebra of *type* B_o. A locally convex algebra is said *multiplicatively convex* (shortly m-convex), if its topology can be given by means of a family of submultiplicative seminorms.

A generalization of Johnson's result for m-convex algebras of type B_o was given in [5] for commutative algebras and in general case in [8] (even without assuming the m-convexity), however the latter proof is unavailable to the author. The question whether semisimple non-commutative B_o-algebras have a unique topology (as B_o-algebras) was posed as the Problem I in [29]. It is quite possible that this type of uniqueness can be proved for F-algebras, so we recall the following question posed in [29].

Problem K. Let A be a semisimple F-algebra. Is the topology of A a unique topology making it an F-algebra ?

The answer is not known, even if we assume additionally that A is commutative. It seems to be unreasonable to pass outside the metrizable case. For instance Michael [15] constructed on $C[0,1]$ a complete non metrizable topology making it an m-convex algebra. A weaker question was posed in [29]:

Problem J. Let A be a semisimple B_o-algebra. Is there only one topology making it an algebra of type B_o.

It seems that this problem is solved in [8].

The further problems in [27] concern relations of the topology of an algebra and its commutative subalgebras.

Problem 4. Suppose that there are two topologies τ_1 and τ_2 both making an algebra A a topological algebra. Suppose that for each commutative subalgebra $\mathcal{A} \in A$ we have $\tau_1|\mathcal{A} = \tau_2|\mathcal{A}$, where $\tau|\mathcal{A}$ is the restriction of the topology τ to $\mathcal{A}$. Does it follow $\tau_1 = \tau_2$?

A more general question is the following one.

Problem 5. Suppose that a homomorphism h of a topological algebra A into a topological algebra B is continuous when restricted to any commutative asubalgebra of A. Does it follow that h is continuos ?

The negative answer to both Problems 4 and 5 was given in [1]. It follows immediately from the following result.

Proposition 14. There exists a normed topological algebra with two non-equivalent norms, which are, however, equivalent on each commutative subalgebra.

The algebra of above example is incomplete in both norms, however all maximal commutative subalgebras are complete.

The next 20 Problems in [27] were given in a form of a table whose columns and rows are labelled with the following classes of topological algebras: FD – the class of all finitely dimensional algebras; B – the class of all Banach algebras; MB_o – the class of all multiplicatively convex algebras of type B_o; B_o – the class of all algebras of type B_o; F – the class of all algebras of type F; LC – the class of all complete locally convex topological algebras and T – the class of all complete topological algebras (in these problems we restrict our attention only to complete topological algebras, since every topological algebra can be completed we do not loose much of the generality).

	B	MB_0	B_0	F	LC	T
FD	6	7	8	9	10	11
B	–	12	13	14	15	16
MB_0	–	–	17	18	19	20
B_0	–	–	–	21	22	23
F	–	–	–	–	–	24
LC	–	–	–	–	–	25

In this table the k-th Problem occuring at the intersection of a column labelled with $\mathcal{K}_1$ and a row labelled with $\mathcal{K}_2$ makes a sense if for the classes $\mathcal{K}_1, \mathcal{K}_2$ of topological algebras we have $\mathcal{K}_2 \subset \mathcal{K}_1$, $\mathcal{K}_2 \neq \mathcal{K}_1$ and in this case it reads as follows.

Problem k.$(6 \leq k \leq 25)$. Let $A \in \mathcal{K}_1$ and suppose that all closed commutative subalgebras of A are in $\mathcal{K}_2$. Does it follow that A is in $\mathcal{K}_2$?

All the Problems 6-25 are solved by now and the Problems 6-11 were solved even before they were formulated in [27], but the author was not aware of this fact. The positive solutions of the Problems 6-11 is due to Thomas Laffey and follow form a purely algebraic result obtained in [12]:

Proposition 15. Let A be an infinitive dimensional algebra, then it has an infinite dimensional commutative subalgebra.

The result is true not only for the real or complex algebras.

The Problem 17 is connected with a result concerning action of entire functions on elements of topological algebras. Let $\phi(\zeta) = \sum_{k=0}^{\infty} a_k \zeta^k$ be an entire function of one complex variable. We say that this function operates on a topological algebra A if for every $x \in A$ the series $\sum a_k x^k$ is convergent in A (if A is real, qwe assume that all a_k are real numbers, if A is non-unital, we assume $a_0 = 0$). In [16] it was obtained the following result.

Proposition 16. Let A be a commutative algebra of type B_o. Then all entire functions operate on A if and only if A is m-convex.

It was left an open question, whether this result is true without assumption of commutativity and it is exactly the Problem 17. In fact, if A is a B_o-algebra on which operate all entire functions, then these functions operate also on all commutative subalgebras, which by above Proposition are m-convex and we ask about m-convexity of the whole of A. The Problem was solved in [30] with the following result (true both in real and in complex case). Reacall that A semitopological algebra is called a Q-algebra in case when the group of all its invertible (quasi-invertible in the non-unital case) elements is an open set.

Proposition 17. There exists a non unital m-convex algebra A of type B_o on which operate all entire functions. Moreover A is a Q-algebra and a radical algebra.

Since $B_o \subset F$ and $B_o \subset LC \subset T$, the above result solves also in negative the Problems 18-20.

Further Problems of this table were solved in [36], where we obtained the following two results.

Proposition 18. There exists an m-convex algebra of type B_o which is non-Banach, but all of whose closed commutative subalegbras are Banach Algebras.

It solves in negative the Problems 12-16 of and also the Problems D and E of [28] which ask resspectively whether a locally convex or a topological algebra with all commutative subalgebras normed must be itself a normed algebra.

Proposition 19. There exists a complete non-metrizable m-convex locally convex algebra all of whose commutative closed subalgebras are Banach.

Since $B \subset B_o$, it solves in negative the Problems 22-24. Later we have obtained ([37]) the following result.

Proposition 20. There is a complete non-locally convex topological algebra with all commutative subalgebras locally convex.

It solves the Problem 25. Finally we solved the remaining Problem 21, again in negative. Here we used a locally bounded algebra (see [24] or [25]), it is as a Banach algebra with the norm replaced by a (submultiplicative) p-homogeneous norm, $0 < p < 1$. We used also the fact that each Banach algebra is also an algebra of type B_o. The result obtained in [38] reads as follow.

Proposition 21. There exists a complete locally bounded algebra which is non-Banach, but all of whose closed commutative subalgebras are Banach algebras.

The common method in proving Propositions 18-21 lies in contructing an algebra with the property that it is a direct sum of its centre and a subspace, such that codimension of the centre in each maximal commutative subalgebra equals to one.

The Proposition 16 implies the following result (see [24] or [25]).

Proposition 22. Let A be a commutative complex algebra of type B_o which is a Q-algebra. Then A is an m-convex algebra.

The question whether this result is also true in the non-commutative case is the following

Problem 26. Suppose that A is a non-commutative Q-algebra of type B_o. Does it follow that A is m-convex ?

The negative answer to this question is given in the Proposition 17.

Turpin ([20]) generalized the Proposition 22 obtaining the following result.

Proposition 23. Let A be a complete complex locally convex commutative topological algebra with a continuous inverse. Then A is an m-convex algebra.

The next Problem asks whether the above result is true in the non-commutative case. Again a negative answer follows from the Proposition 16.

Problem 27. Suppose that A is a complete locally convex topological algebra with a continuous inverse. Does it follow that A is multiplicatively convex ?

The additional assumption of the continuity of inverse is necessary for the validity of the Proposition 23, because, for instance, one can construct a complete infinitely dimensional locally convex field which is a topological algebra. It is clearly a Q-algebra and it cannot be m-convex, because for these algebras there is true the Gelfand Mazur theorem (see [15], [24] or [25]). This assumption is unnecessary for metrizable complete topological algebras which are Q-algebras, because of the following result (see [24] or [25]), here $G(A)$ is the set (group) of all invertible elements of an algebra A.

Proposition 24. Let A be a unital algebra of type F. Then the operation of taking an inverse is continuous on $G(A)$ if and only if $G(A)$ is a G_δ-set.

We do not know whether the Propositions 22 and 23 are true in the real case, so we ask the following

Question 10. Let A be a complete real locally convex algebra with a continuous inverse. Does it follow that A is m-convex ?

We shall discuss now the possibility of a generalization of the Proposition 16.

As it stays it cannot be generalized onto non-metrizable algebras because of the following result ([26]).

Proposition 25. There exists a commutative complete non-m-convex locally convex topological algebra on which operate all entire functions.

However the operation of taking an inverse is discontinuous in the example of the above Proposition. This fact together with Proposition 23 supports the conjecure that the following question could be answered in positive, at least for the complex algebras.

Question 11. Suppose that A is a complete commutative locally convex topological algebra with a continuous inverse on which operate all entire functions. Does it follow that A is m-convex ?

The above Question has two versions: complex and (harder) real. The solution of the following problem would be useful for the theory of Q-algebras.

Question 12. Let A be a topological algebra and suppose that all maximal commutative subalgebras of A are Q-algebras. Does it follow that A is a Q-algebra ?

The remaining Problems in [27] concern ideals and subalgebras. The following Problem is the famous Michael-Mazur problem posed by Michael in [15] and considered also (independently) by Stanisław Mazur, it is, perhaps, the most challenging problem in the theory of topological algebras.

Problem 28. Let A be a complex m-convex algebra of type B_o. Does it follow that all multiplicative-linear functionals on A are continuous ?

For real algebras (and real valued multiplicative linear functionals) the problem is solved in affirmative, see [9].

The Problem is open also for commutative algebras. We do not know whether the general problem is equivalent to the commutative one. So we ask the following

Question 13. Suppose that A is a topological algebra with the property that all its multiplicative linear functionals are continuous on commutative subalgebras. Does it follow that all these functionals are continuous on A ?

A similar question can be asked for particular classes of topological algebras, such as algebras of type F, B_o, or MB_o.

The Problem 28 has many equivalent formulations. For details see [9] and [32].

A topological algebra A is said finitely generated if there is a finite subset S of A such that it coincides with the smallest closed subalgebra containg S. The strongest result concerning solution of the problem 28 seems to be the following result due to Richard Arens [3].

Proposition 26. Let A be a commutative complex finitely generated m-convex algebra of type B_o. Then all multiplicative linear functionals on A are continuous.

The following Problems are also open, even if the considered algebras are commutative and finitely generated (similarly as Problem 28 both of them are solved in affirmative for real algebras).

Problem 29. Let A be an algebra of type B_o. Is it true that all multiplicative-linear functionals in A are continuous ?

Problem 30. Let A be an algebra of type F. Is it tue that all multiplicative linear functionals on A are continuous ?

There are examples of commutative complete (non-metrizable) m-convex algebras with discontinuous multiplicative linear functionals. Such an example was already given in [15]. The algebra constructed in [26] has more discontinuous multiplicative linear functionals then the continuous ones. We ask the following question (posed in [32]).

Question 14. Does there exist a complete topological algebra with only discontinuous multiplicative linear functionals ?

There are commutative unital algebras of type B_o without multiplicative linear functionals. The example of such a complex algebra is the algebra L^ω constructed in [3] (see also [24] or [25]). All maximal ideals in this algebra are dense and of infinite codimension. If we consider the algebra $C[0,1]$ treated as a dense subalgebra of L^ω, then all its multiplicative linear functionals are discontinuous (otherwise they would extend by continuity to the whole of L^ω), this is the reason that we restrict the Question 14 only to complete algebras. However L^ω has many closed proper ideals. Not all complete algebras have this property. in [4] Atzmon obtained the following result.

Proposition 27. There exists a complete commutative locally convex topological algebra in which every proper ideal is dense.

We do not know, however, whether such a result can be true for metrizable algebras, so we asked in [27] the following problem.

Problem 31. Let A be a commutative complex infinite-dimensional unital algebra of type B_o. Does A has proper closed ideals ?

We did not ask a similar question for F-algebras, because in [21] Waelbroeck constructed an infinite dimensional field of type F, however this construction has some gaps and the author cannot say whether it is correct. So we ask the following two (not independent) questions.

Question 15. Does there exist an infinite-dimensional field of type F ?

Question 16. Let A be a commutative infinite-dimensional algebra of type F. Does it have a closed proper ideal ?

The answer is not known even if A has a unity.

The answer to the Question 16 is not known even for (non-unital) Banach algebras.

Problem 32. Does there exist an infinite dimensional commutative complex Banach algebra without closed proper ideals ?

Even a stronger question is not yet answered.

Problem 33. Does there exists an infinite dimensional commutative complex Banach algebra without closed proper subalgebras ?

Clearly, a semitopological algebra of dimension greater then one without proper closed subalgebras cannot be unital and it must be commutative. Recently the author obtained the following result [39].

Proposition 28. There exists an infinite-dimensional metrizable locally convex semitopological algebra without proper closed subalgebras.

But we donot know an answer to the following question (posed in [39]), it is a weaker version of the Problem 33.

Question 17. Does there exist an infinite dimensional topological algebra without proper closed subalgebras ?

One can ask also stronger versions of this question assuming in addition the completeness, local convexity, metrizability etc.

References

[1] B. Aniszczyk, R. Frankiewicz and C. Ryll-Nardzewski, Continuity of a homomorphism on commutative subalgebras is not sufficient for continuity, Studia Math. 98 (1991), 247-248.

[2] R. F. Arens, The space L^{ω} and convex topological rings, Bull. Amer. Math. Soc. 52 (1946), 931-935.

[3] — , Dense inverse limit rings, Michigan Math. J. 5 (1958), 169-182.

[4] A. Atzmon, An operator without invariant subaspeces on a nuclear Fréchet space, Ann. of Math. (2) 117 (1983), 669-684.

[5] R. L. Carpenter, Uniqueness of topology for commutative semisimple F-algebras, Proc. Amer. Math. Soc. 29 (1971), 113-117.

[6] A. Fernandez and V. Müller, Renormalizations of Banach and locally convex algebras, Studia Math. 96 (1990), 237-242.

[7] R. Frankiewicz and G. Plebanek, An example of a non-topologizable algebra, Studia Math. 116 (1995),85-87.

[8] Guan Bo, Uniqueness of topolgy an continuity of systems of derivations on locally convex F-algebras (Chinese), Acta Math. Sinica 31 (1988), No 5, 577-585.

[9] T. Husain, Multiplicative Functionals on Topological Algebras, Boston 1983.

[10] H. Jarchow, Locally Convex Spaces, Stuttgart 1981.

[11] B. E. Johnson, The uniqueness of the (complete) norm topology, Bull. Amer. Math. Soc. 73 (1967), 537-539.

[12] T. J. Laffey, Commutative subalgebras in infinite dimensional algebras, Bull. London Math. Soc. 5 (1973), 312-314.

[13] A. Kokk and W. Żelazko, On vector spaces and algebras with maximal locally pseudoconvex topologies, Studia Math. 12 (1995), 195-201.

[14] A. Mallios, Topological Algebras. Selected topics, Amsterdam 1986.

[15] E. Michael, Locally Multiplicatively-Convex Topological Algebras, Mem. Amer. Math. Soc. 11 (1952).

[16] B. S. Mityagin, S. Rolewicz and W. Żelazko, Entire functions in B_o-algebras, Studia Math. 21 (1962), 291-306.

[17] W. Müller, On topologizable algebras, Studia Math. 99, (1991), 149-153.

[18] T.J. Ransford, A short proof of Johnson's uniqueness theorem, Bull. London Math. Soc. 21 (1989), 487-488.

[19] H. H. Schaeffer, Topological Vector Spaces, Springer, New York 1971.

[20] Ph. Turpin, Un remarque sur les alg/'ebres a inverse continu, C. R. Acad. Sc. Paris 270 (1970), A1686-A1689.

[21] L. Waelbroeck, Topological Vector Spaces and Algebras, Lecture Notes in Math. 230, Springer Verlag 1971.

[22] A. Wilansky, Modern Methods in Topological Vector Spaces, New York 1978.

[23] M. Wojciechowski and W. Żelazko,Non-uniqueness of topology for algebras of polynomials, Coll. Math. 72 (1997), 111-121.

[24] W. Żelazko, Metric Generalizations of Banach Algebras, Dissert. Math. (Rozprawy Mat.) 47, 1965.

[25] — , Selected Topics in Topological Algebras, Aarhus Univ. Lecture Notes Series, No 31, 1971.

[26] — , A non-m-convex algebra on which operate all entire functions, Ann. Polon. Math. 46 (1985), 389-394.

[27] — , On certain open problems in topological algebras, Rend. Sem. Mat. Fis. Milano, 59 (1989), 1992,49-58.

[28] — , Example of an algebra which is non-topologizable as a locally convex algebra, Proc. Amer. Math. Soc. 110 (1990), 947-949.

[29] — , Some open problems and recent results in topological algebras, in Functional Analysis and Related Topics, World Scientific 1991, 58-65.

[30] — , Concerning entire functions in B_o-algebras, Studia Math. 110 (1994), 283-290.

[31] — , On topologization of countably generated algebras, Studia Math. 112 (1994), 83-88.

[32] — , What is known and what is not known about multiplicative linear functionals, in Topological Vector Spaces, Algebras and Related Areas, Pitman Res. Notes in Math. Sc. vol. 316 (1994), 102-115.

[33] — , Further examples of locally convex algebras, ibidem, 162-171.

[34] — , Concerning topologization of real or complex algebras, Coll. Math. 71 (1996), 111-113.

[35] — , Concerning topologization of algebras – the results and open problems, in Advances in Functional Analysis, New Age International Publishers, (in print).

[36] — , A non-Banach m-convex algebra all of whose closed commutative subalgebras are Banch algebras, Studia Math. 119 (1996), 195-198.

[37] — , A non-locally convex topological algebra with all commutative subalgebras locally convex, ibidem 120 (1996), 89-94.

[38] — , A locally bounded algebra all of whose closed commutative subalgebras are Banch algebras, Comment. Math. (Prace Mat.), in print.

[39] — , A semitopological algebra without proper closed subalgebras, Coll. Math. (in print).

MATHEMATICAL INSTITUTE, POLISH ACADEMY OF SCIENCES

Sniadeckich 8, P.O. Box 137,

00-950 Warszawa (Poland).

E-mail: zelazko@impan.gov.pl

L M SÁNCHEZ RUIZ AND S A SAXON

Weak barrelledness conditions

1 Introduction

Barrelledness is involved in several theorems considered to be basic in functional analysis, such as the uniform boundedness principle, closed graph theorems and open mapping theorems. However, in a number of situations barrelledness means more than is actually needed. This prompted several researchers to look for weaker conditions (cf. [7, Chapter 2] and [11, Section 8.2]) that yield the same or similar results. For instance, this was the main reason dual locally complete (dlc) spaces were studied by Valdivia [36], who was interested in closed graph theorems. On other occasions researchers were interested in just a particular feature of barrelledness. The inheritance of barrelledness by countable-codimensional subspaces (the Saxon-Levin-Valdivia Theorem) implies that every countable-codimensional subspace of a barrelled space E is Mackey. However, there are even dense one-codimensional subspaces of Mackey spaces that are not Mackey [10, p.100]. And so in studying what feature of barrelledness was really needed to assure the inheritance of Mackeyness by countable- codimensional subspaces F of a Mackey space E, Levin-Saxon [10] reduced it in the F separable case to the fact that E should have *property (C)*, i.e., that every $\sigma(E', E)$-bounded subset of E' is relatively $\sigma(E', E)$-countably compact, and in the F closed case to the fact that, E should have *property (S)*, i.e., that E' is $\sigma(E', E)$-sequentially complete. This last condition was subsequently weakened by Valdivia [37, Theorem 3] to be dlc, and by Ferrando-Sánchez Ruiz [5, Theorem 3] to be *inductive* (in cit. op., *weakly barrelled*), i.e., the inductive limit of any increasing covering sequence of subspaces, a condition weaker than dlc under Mackeyness. Saxon-Tweddle's even weaker condition of being primitive (see below), equivalently, having a compatible inductive topology, is crucial in splitting a barrelled space into fit and flat components [31]. And the present authors substitute primitivity for barrelledness [29] to improve the main result of Robertson-Saxon-Robertson [12]. Decades ago, the likes of Valdivia, De Wilde, Houet, Saxon collected weak barrelledness conditions that implied inductivity. Suppose F is a dense subspace of E; Valdivia proved that if F is barrelled and E is Baire-like, then F also enjoys the stronger property. What is the "weakest" barrelledness condition on F such that if E enjoys either a stronger weak barrelledness condition or, stronger yet, is quasi-Baire or Baire-like, then F also enjoys the same stronger property? (Cf. [29].) Or, if F is closed and barrelled, what is the "weakest" barrelledness condition on E/F that will ensure E is primitive?

Such lines of research and a profusion of terminology cry for the context of a systematic structural study of weak barrelledness. Years of evolution and quite recent developments now permit comprehension and simplicity unavailable to Ruess and other structural pioneers. To demonstrate, the authors offer their reorganization [29]. Even though there is now a wider range of concepts, clarity is achieved via a simple descriptive dichotomy and the discovery of precisely which concepts coincide under the Mackey topology [27], which answers the Levin-Saxon and Ferrando-López Pellicer questions. This paper derives from two talks, 'A review of weak barrelledness' by S.A. Saxon, and 'New focus of weak barrelledness conditions' by L.M. Sánchez Ruiz, and from the supporting joint research [26]-[30].

2 Weak barrelledness conditions

Throughout this paper the word *space* will stand for any Hausdorff locally convex topological vector space over the real or complex field $\mathbb{K}$. Let us recall the definitions of the weak barrelledness conditions we will deal with that have not been given in the introduction.

A space E is *barrelled* if every barrel in E is a neighborhood of 0 in E. An $\aleph_0$-*barrel* of a space E is a barrel U in E which is the intersection of a sequence $\{U_n\}_n$ of closed absolutely convex neighborhoods of 0 in E. A space E is $\aleph_0$-*barrelled* [6] if every $\aleph_0$-barrel of E is a neighborhood of 0 in E, and E is *C-barrelled* [11, p. 237] if every $\aleph_0$-barrel $U = \bigcap_n U_n$ in E, such that for each $x \in E$ there is some positive integer m with $x \in U_n$ for $n \geq m$, is a neighborhood of 0 in E.

The following two weakenings of barrelledness evoke Grothendieck's characterization of barrelled spaces as those spaces in which every $\sigma(E', E)$-bounded subset of E' is equicontinuous. A space E is ℓ^∞-*barrelled* [10] if every countable $\sigma(E', E)$-bounded subset of E' is equicontinuous, and E is c_0-*barrelled* [40] if every $\sigma(E', E)$-null sequence of E' is equicontinuous.

A non-decreasing sequence of (closed) absolutely convex sets $\Delta = \{A_n\}_n$ in a space E such that each $x \in E$ is absorbed by some A_n is called a (*closed*) *absorbing sequence*. Given an absorbing sequence Δ of the space $E(\tau)$, we will denote by τ_Δ the finest locally convex topology on E inducing the same neighborhoods of 0 as τ on each A_n. A space $E(\tau)$ has *property* (L) [15] if, given any absorbing sequence Δ of the space $E(\tau)$, $\tau_\Delta = \tau$, and $E(\tau)$ has the weaker *property* (LC) [15] if, given any absorbing sequence Δ of the space $E(\tau)$, τ_Δ happens to be compatible with the dual pairing (E, E'). The Final Remarks of Ruess [16] imply that property (LC) in a space E is equivalent to dlc, i.e., every $\sigma(E', E)$-bounded closed absolutely convex subset of E' is a Banach disk. New characterizations of dlc recently obtained by the authors [26] will be exposed in Section 5.

3 Seminorms versus linear forms

It is easy to realize that some of the following properties characterize a space E satisfying some of the aforementioned barrelledness conditions. We will say that E has property:

- $\langle(p_\alpha)\mathrm{b}\rangle$ $[(f_\alpha)\mathrm{b}]$ if each family of continuous $\langle$seminorms $\{p_\alpha\}_{\alpha\in I}\rangle$ [linear forms $\{f_\alpha\}_{\alpha\in I}$] on E which is pointwise bounded is equicontinuous.

- $\langle(p_n)\mathrm{b}\rangle$ $[(f_n)\mathrm{b}]$ if each sequence of continuous $\langle$seminorms $\{p_n\}_n\rangle$ [linear forms $\{f_n\}_n$] on E which is pointwise bounded is equicontinuous.

- $\langle(p_n)0\rangle$ $[(f_n)0]$ if each sequence of continuous $\langle$seminorms $\{p_n\}_n\rangle$ [linear forms $\{f_n\}_n$] on E which is pointwise convergent to 0 is equicontinuous.

- $\langle(p\mid_{A_n})\rangle$ $[(f\mid_{A_n})]$ if each $\langle$seminorm $p\rangle$ [linear form f] on E, for which there exists some absorbing sequence $\{A_n\}_n$ in E such that each restriction $\langle(p\mid_{A_n})\rangle$ $[(f\mid_{A_n})]$ is continuous, is itself continuous.

- $\langle(p\mid_{E_n})\rangle$ $[(f\mid_{E_n})]$ if each $\langle$seminorm $p\rangle$ [linear form f] on E, for which there exists some absorbing sequence $\{E_n\}_n$ in E formed by linear subspaces such that each restriction $\langle(p\mid_{E_n})\rangle$ $[(f\mid_{E_n})]$ is continuous, is itself continuous.

In fact the first of the above properties characterizes barrelled spaces; the second one, $\langle\aleph_0\rangle[\ell^\infty]$-barrelled spaces; the third one, $\langle C\rangle[c_0]$-barrelled spaces; the fourth one, property $\langle(L)\rangle[(LC)]$; and the fifth one, $\langle$inductive spaces$\rangle$ [*primitive* spaces].

4 The table

The properties of the previous section and the fact that $|f|$ is a seminorm whenever $f \in E^*$ make obvious most of the weak barrelledness relationships depicted in the following table, explained below.

barrelled; (p_α)b	$\Longleftrightarrow$	barrelled; (f_α)b
$\Downarrow$		$\Downarrow$
$\aleph_0$-barrelled; (p_n)b	$\Longrightarrow$	ℓ^∞-barrelled; (f_n)b
1) separable		
$\Downarrow$		$\Downarrow$
C-barrelled; $(p_n)0$	$\Longrightarrow$	c_0-barrelled; $(f_n)\,0$
$\Downarrow$		$\Downarrow$
property (L); $p\|_{A_n}$	$\Longrightarrow$	dlc; $f\|_{A_n}$
2) Mackey		
3) metrizable		
$\Downarrow$		$\Downarrow$
inductive; $p\|_{E_n}$	$\Longrightarrow$	primitive; $f\|_{E_n}$
4) Mackey		

This table in matrix form is not complete –for example properties (C) and (S) are not included– but may be considered as the basic skeleton to which other weak barrelledness conditions relate.

The matrix terms represent classes of spaces which are symmetrically characterized by seminorms and linear forms. Given any entry in the array, the weaker ones lie rightward and/or downward. Several of the classes represented in the matrix coincide under usual assumptions such as separability, metrizability, or Mackeyness. This is indicated by means of a numbered box in the table. So, for instance, Box 1) indicates that each separable ℓ^∞-barrelled space is barrelled [2, Corollay 4.a] and Box 3) that each metrizable dlc space is barrelled [15, Proposition 6.8]. Box 2) means that each Mackey dlc space is C-barrelled [27, Theorem 3.2]. It is our recently submitted answer to the question raised by Ferrando-López Pellicer [3], and contributes to the simplicity of the table. This result improves immediately those contained in [11, 5.1.33 and 8.1.29(i)], by means of which every Mackey dlc space is c_0-barrelled and has property (L), and that in [15, Proposition 5.6(3)], by means of which every Mackey space with property (C) has property (L). In [29] we develop a complete set of examples, extending the ones found in [5, 11], to show that the above arrows and boxes tell the entire interrelational story.

Properties (C) and (S) of Levin-Saxon fit inside Box 3) and outside Boxes 1) and 2), and clearly relate to the "basic skeleton" as follows:

$$\ell^\infty\text{-barrelled} \Rightarrow \text{property } (C) \Rightarrow \text{property } (S) \Rightarrow \text{dlc.}$$

In fact, it had been a long-standing question –since 1971 ([10])– whether there should be a box clustering the first two of these four classes of spaces under Mackeyness. Ferrando-Sánchez Ruiz [5, Proposition 1] answered positively under the additional hypothesis of separability. However, thanks to examples submitted in [27] we now know the answer is, in general, negative. Therefore, including properties (C) and (S), the picture simplifies to precisely seven different classes of spaces, linearly related, under Mackeyness:

$$\begin{aligned} \text{barrelled} &\Rightarrow \aleph_0\text{-barrelled} \Rightarrow \ell^\infty\text{-barrelled} \\ &\Rightarrow \text{prop. } (C) \Rightarrow \text{prop. } (S) \Rightarrow \text{Box 2)} \Rightarrow \text{Box 4).} \end{aligned}$$

Under metrizability only two different classes of spaces remain,

$$\text{barrelled} \Rightarrow \text{inductive,}$$

or, Box 3) $\Rightarrow$ Box 4), as is obvious from the fact that metrizable spaces must be Mackey.

The table's thematic swapping of p and f, one for the other, when applied to properties (C) and (S) generates no new classes of spaces, but new characterizations of $\aleph_0$-barrelledness instead, defined in the following theorem of [29]. We will denote by P_E the set of all seminorms on E endowed with the topology of pointwise convergence on E, which is a closed subset of the space of all real-valued functions on E with its usual product topology.

Theorem 4.1 *Let E be a space. The following assertions are equivalent:*

(1) *E is $\aleph_0$-barrelled.*

(2) *E has property (C_p), i.e., every pointwise bounded sequence of continuous seminorms on E has a limit point in P_E that is continuous.*

(3) *E has property* (S_p), *i.e., every limit of a sequence of continuous members of* P_E *is continuous.*

Proof. To see that $[(1) \Rightarrow (2)]$, note that Tychonoff's Theorem implies that any pointwise bounded sequence $\{p_n\}_n$ of continuous seminorms on E has a limit point $p \in P_E$ which must be continuous, since it is bounded on any neighborhood of 0 on which $\{p_n\}_n$ is uniformly bounded.

$[(2) \Rightarrow (3)]$. Obvious, since P_E is Hausdorff.

$[(3) \Rightarrow (1)]$. Given a pointwise bounded sequence $\{p_n\}_n$ of continuous seminorms on E, set $q_n(x) = \max_{k \leq n} p_k(x)$ for each n and each $x \in E$. Now $\{q_n\}_n$ is a pointwise convergent sequence of continuous seminorms on E whose continuous limit majorizes each p_n, so that $\{p_n\}_n$ is equicontinuous. ■

5 Local Completeness

Paradoxically, the authors' synthesis of weak barrelledness began with a recent article adding new characterizations of dual local completeness [26]. There we noted how in the Saxon-Levin-Valdivia Theorem the first two authors [21] proved that any countable-codimensional subspace of a barrelled space spanned by a closed absolutely convex subset is itself closed. Valdivia [35] got independently his proof following a different reasoning, and in his subsequent development he showed that any barrelled space satisfied what Ruess would later call property (L). Under Mackeyness, which was the case since in fact we have barrelledness, property (L) is equivalent to (LC) and therefore to dlc. And amazingly, it happens that dlc is equivalent to the above result obtained by Saxon-Levin, which in fact was given as a necessary condition for property (S) (see also [41] or [11, 8.2.17]). Before stating all the characterizations of dlc found in [26], we observe that each of them could be used to characterize local completeness in general, since it is duality invariant. A notable case in point involves ℓ^1-completeness.

Definition 5.1 *A space E is* ℓ^1-complete *if for each* $(\lambda_n)_n \in \ell^1$ *and bounded sequence* $\{x_n\}_n$ *in E, the series* $\sum_n \lambda_n x_n$ *converges in E. We say E is* dual ℓ^1-complete *if* $(E', \sigma(E', E))$ *is* ℓ^1-complete.

The proof of the equivalence of ℓ^1-completeness and local completeness, [26, Theorem 2.1], is based upon the fact that if E is locally complete, $(\lambda_n)_n \in \ell^1$ and $\{x_n\}_n$

is a bounded sequence in E, then the series $\sum_n \lambda_n x_n$ converges in the Banach space E_A, the linear span of the closed absolutely convex hull A of $\{x_n\}_n$ endowed with the gauge of A as norm, and therefore the series $\sum_n \lambda_n x_n$ converges in E. The converse shows that if E is ℓ^1-complete and A is a bounded closed absolutely convex subset of E, then A is a Banach disk, i.e., any Cauchy sequence $\{y_n\}_n$ in E_A converges in E_A. This is done by showing that $\{y_n\}_n$ has a subsequence $\{y_{n_k}\}_k$ which converges in E_A where the $\{n_k\}_k$ are determined in such a way that $\|y_p - y_q\| \leq 2^{-k}$, for $p, q \geq n_k$, $k \in \mathbb{N}$, and then use the fact that $(2^{-k})_k \in \ell^1$, $\{2^k(y_{n_{k+1}} - y_{n_k})\}_k$ is a bounded sequence contained in A and E is ℓ^1-complete.

The characterizations of dlc provided by [26, Theorems 2.3 and 2.10] are included in the following.

Theorem 5.1 *Let E be any space. The following assertions are equivalent:*

(1) *E is dual locally complete.*

(2) *E is dual ℓ^1-complete.*

(3) *E has property (LC).*

(4) *If A is an absorbing absolutely convex set and $f \in E^*$ with $f\,|_A$ continuous, then $f \in E'$.*

(5) *If $\{A_n\}_n$ is an absorbing sequence in a countable-codimensional subspace F of E and each A_n is closed in E, then F is closed.*

(6) *If $f \in E^*$ and A is a closed absolutely convex subset of E such that $f(A) = \{0\}$ and $\operatorname{sp}(A)$ is a countable-codimensional subspace of E, then $f \in E'$.*

(7) *If F and H are transverse subspaces of E with $F = \operatorname{sp}(A)$, where A is a closed absolutely convex subset of E and $\dim(E/F) \leq \aleph_0$, then both $F + H$ and $A + H$ are closed.*

(8) *If $F = \operatorname{sp}(A) \subseteq E$ with A a closed absolutely convex subset of E and $\dim(E/F)$ $\leq \aleph_0$, then F is closed.*

The above theorem provides common ground for the development found not only in [21] and [35], but also for other conditions some of which were known to be only necessary for barrelledness. So, for instance, [16, Proposition 3.5(1)] gave (5) just as a necessary condition for property (LC).

We can now easily prove the well known fact [11, 8.2.23(b)] that c_0-barrelled $\Rightarrow$ dlc. Indeed, if $\{f_n\}_n$ is $\sigma(E', E)$-bounded and $(\lambda_n)_n \in \ell^1$, choose a sequence $(\epsilon_n)_n$ of positive numbers converging to 0 so slowly that $(\epsilon_n^{-1}\lambda_n)_n \in \ell^1$. Because E is c_0-barrelled, $\{\epsilon_n f_n\}_n$ is equicontinuous, so that $\sum_n (\epsilon_n^{-1}\lambda_n)(\epsilon_n f_n) = \sum_n \lambda_n f_n$ is continuous, and E is dual ℓ^1-complete.

The dichotomy of seminorm versus linear form in the table and (2) of Theorem 5.1 suggest the following characterization of property (L) (cf. [26, Theorem 2.2]).

Theorem 5.2 *Given a space $E(\tau)$, the following assertions are equivalent:*

(1) *E has property (L).*

(2) *If $(\lambda_n)_n \in \ell^1$ and $\{p_n\}_n$ is a pointwise bounded sequence of continuous seminorms on E, then the seminorm defined by $\sum_n |\lambda_n| p_n$ is continuous on E.*

This result and the argument (analogous to) that we just gave prove immediately the table's other non-obvious implication, namely that C-barrelled $\Rightarrow$ property (L).

Now (7) of Theorem 5.1 provides a four-line proof [26] of the Saxon-Levin-Valdivia Theorem, while (1) and (8) similarly prove [26] the inheritance of dual local completeness by countable-codimensional subspaces (Valdivia's [37, Theorem 13]). In fact, (1) and (8) provide a quick uniform proof for the countable-codimensional subspace inheritance of each property at or above the dlc level, including properties (C) and (S) but excluding C- and c_0-barrelled spaces, where inheritance may fail [29]. Inheritance prevails at the lowest (inductive/primitive) level [5, 29].

E has the *Banach-Mackey property* if each $\sigma(E', E)$-bounded set is $\beta(E', E)$-bounded. There are many normed primitive spaces that are not barrelled [5, 28, 29] and therefore do not have the Banach-Mackey property. And any $\aleph_0$-dimensional space E with $\dim(E^*/E') = 1$ has the Banach-Mackey property but is not primitive. That is, the Banach-Mackey property, weaker than barrelledness, does not fit well into our table.

Yet, a clear view of the table would have facilitated the study of this property by a number of ranking experts. Indeed, as noted in [26], the following is a modern statement of the Banach-Mackey Theorem, with classical proof.

Theorem 5.3 *Every dlc space E has the Banach-Mackey property.*

We observe in [26, Theorem 2.4] that E has the Banach-Mackey property if and only if each barrel is bornivorous [42, Theorem 10-4-7], which is equivalent to the condition that some member of each closed absorbing sequence is bornivorous, and the latter is equivalent to Ruess' property (B) [16, Lemma1.2]. Thus [11, 5.1.34] and [15, Lemma 6.3] just restate the Banach-Mackey Theorem, while [35, Theorem 6], [2, Corollary 1], [40, Proposition 4.1(1)] say less, namely, that E has the Banach-Mackey property if, respectively, E is barrelled, is ℓ^∞-barrelled, is c_0-barrelled.

The above considerations and the fact that metrizable spaces are quasibarrelled prove metrizable dlc spaces are barrelled, the well-known Box 3). In *Mazur spaces* [42], i.e., those for which sequentially continuous linear forms are continuous, the converse of the Banach-Mackey Theorem holds [26, Theorem 2.6], and every c_0-barrelled Mazur space has property (S) [26, Theorem 2.7]. Hence Box 2) provides the following, which includes two results of Kalton [8].

Theorem 5.4 *For a Mackey Mazur space E the following assertions are equivalent:*

(1) *E is dual locally complete.*

(2) *E has the Banach-Mackey property.*

(3) *E has property (S).*

(4) *E is c_0-barrelled.*

(5) *E is C-barrelled.*

(6) *E has property (L).*

6 General patterns and countable enlargements

Properties (C) and (S), and dlc-ness and primitivity are duality invariant, the table's other seven conditions are not. Suppose F is a dense subspace of E. If F enjoys a given non-duality invariant condition, so does every subspace between F and E. This result fails spectacularly for all the duality invariant conditions, since there exists a nonprimitive space E with a dense hyperplane F that has property (C). If every subspace between F and E has a given one of the properties in the table's right column and E has the stronger property immediately to its left, then F also enjoys the stronger property. But in general, F cannot advance in this way from either the lowest or second lowest level to the next higher one. For example, there is a C-barrelled space E with a dense non-c_0-barrelled subspace F such that every subspace between F and E is dlc. (F must then also have property (L), but fails to reach the middle level.) But again, if every subspace between F and E is dlc and E enjoys one of the table's stronger properties not at the middle level or, stronger yet, is quasi-Baire or Baire-like, then F also enjoys the same stronger property [29]. This is an optimal improvement, inspired and certified by the table, of the Valdivia result that F barrelled and E Baire-like imply F is also Baire-like.

A similar pattern emerges in the study [30] of a *countable enlargement* (CE) η of a space (E, ξ), by which we mean the coarsest locally convex topology η on E finer than the original ξ that makes $(E, \eta)'$ a given linear space containing $(E, \xi)'$ as an $\aleph_0$-codimensional subspace. This is the more widely appropriate Tweddle-Catalán [32] variation of the original Tweddle-Yeomans definition. The still-open question as to whether every barrelled space that admits a CE must admit a barrelled CE (BCE) remains the same with either definition. If E has a given one of the properties in the table's left column and the CE η enjoys the weaker property to the immediate right, then η also enjoys the given stronger property. For example, if (E, ξ) is C-barrelled and its CE (E, η) is c_0-barrelled, then (E, η) is also C-barrelled. But (E, η) cannot generally advance in this way from either the lowest or second lowest level to the next higher one. For example, every infinite-dimensional Banach space is barrelled (on the highest level), even Baire-like, and admits a CE η that is primitive (thus inductive by the next-to-last sentence) but with η not dlc; i.e., η does not rise to the next higher level. Notwithstanding, if (E, ξ) enjoys a table property at not the lowest nor the middle level, or if (E, ξ) is either quasi-Baire or Baire-like or has either property (C) or (S), then the same stronger property is possessed by any CE (E, η) that is dlc. Again, a good view of the table helps us to discover and assess this maximal extension of Tweddle-Catalán's related result [32, Proposition 6]. Their result says that if (E, ξ) is barrelled, $\aleph_0$-barrelled or ℓ^∞-barrelled, then (E, η) has the same given property if it has the weaker property (S).

We solved positively the metrizable BCE problem, as old as the general BCE problem, by first showing that the least infinite-dimensionality for metrizable barrelled spaces is the bounding cardinal $\mathfrak{b}$ [24] and then assuming in [25] the existence of scales, equivalently, that $\mathfrak{b} = \mathfrak{d}$ (the *d*ominating cardinal), consistent with the usual

ZFC axiom system. In [19] we achieve the same result, namely that every infinite-dimensional metrizable barrelled space admits a BCE, but without making any extra-ZFC assumptions, although the proof still heavily involves $\mathfrak{b}$. In [30], assuming that $\aleph_1 < \mathfrak{b}$ ($\leq \mathfrak{d}$), which is ZFC-consistent, we exhibit an $\aleph_0$-barrelled space with a CE that preserves inductivity but with none that preserves dlc, thus giving negative answers, in general, to all eight CE questions of preserving properties, including (C) and (S), strictly between the lowest and highest levels. Since the eleven conditions condense to only two under metrizability, the lone metrizable CE problem remaining is the one we just solved [30] by proving that every infinite-dimensional metrizable inductive space admits an inductive CE, again without any extra-ZFC assumptions, although, again, $\mathfrak{b}$ plays a major role in the proof.

Let us mention that, under ZFC axioms only, it is unknown what is the largest cardinal $\mathfrak{K}$ such that each subspace of codimension $< \mathfrak{K}$ in a barrelled space is barrelled. Under metrizability, Saxon-Sánchez Ruiz [24, Theorem 2] showed that $\mathfrak{K} = \mathfrak{b}$ (cf. [4, 2.4.4]). For Fréchet spaces $\mathfrak{K}$ is the continuum $\mathfrak{c}$ [21] (cf. [39]). Very recently the authors [28] have found a metrizable inductive space χ of dimension $\aleph_1$, the least infinite dimension possible, since no $\aleph_0$-dimensional metrizable space is inductive. This implies, via [5], that $\mathfrak{K} = \aleph_1$ if inductivity is the property to be preserved under metrizability. The example is quite simple: we take χ to be the subspace of the Fréchet space ω of all scalar sequences spanned by any given $\aleph_1$ sequences of the form $(x^n)_n$. By the fundamental theorem of algebra, a nonzero continuous linear form can vanish on only finitely many of the $(x^n)_n$, which proves that χ is a dense subspace of ω and cannot be covered by countably many proper closed hyperplanes. Therefore χ cannot be covered by an increasing sequence of proper closed subspaces; i.e., χ is non-S_σ; and every non-S_σ space is inductive.

The authors' metrizable barrelled space $\psi_\mathfrak{b}$ [23] of least infinite dimension $\mathfrak{b}$, also a dense subspace of ω, was used [24] to sharpen the classical BCE results [1, 11, 13, 33, 34], and these improvements extend to weak barrelledness conditions [30]. The basic pattern, consistent with the above, employs $\psi_\mathfrak{b}$ and $\mathfrak{b}$ for the conditions above the lowest level, χ and $\aleph_1$ at the lowest level. And, precisely because dense hyperplanes of C- and c_0-barrelled spaces do not always preserve the property, this level of the table, again, exhibits exceptional behavior, resisting proof in two of the following three theorems [30].

Theorem 6.1 *Suppose (E, ξ) and its subspace F have a given one of the properties in* *the table, including properties (C) and (S) and the quasi-Baire and Baire-like cases, but excluding the C- and c_0-barrelled cases, and suppose there is a CE for E that preserves the given property. If the given property is at least as strong as dlc (above the lowest level) and* codim $(F) < \mathfrak{b}$, *then there is a CE for* (F, ξ) *that also preserves the property. The same conclusion holds if the given property is inductivity or primitivity (the lowest level) and* codim $(F) < \aleph_1$.

This theorem actually represents eleven, one of which is [1, Theorem 6] (the barrelled case) with $\mathfrak{b}$ optimally replacing $\aleph_1$, the authors' [24, Theorem 4]. In all eleven cases the codimensional inequality is the best possible, since the metrizable $\psi_{\mathfrak{b}}$ and χ have property-preserving CEs while their subspace $\{0\}$ does not.

The next theorem again leaves the C- and c_0-barrelled cases unresolved and raises the additional question as to whether we could replace $\mathfrak{b}$ with $\aleph_1$ in the lowest level cases. Its roots are traced through [33, 32, 24].

Theorem 6.2 *If E has a given one of the eleven properties of the previous theorem and contains a bounded set of dimension at least* $\mathfrak{b}$, *then there is a CE that preserves the property.*

The next theorem actually represents thirteen, with no indecision about properties (C) and (S) or the use of $\mathfrak{b}$ and $\aleph_1$. The barrelled case is [24, Theorem 5], a sharpening of [13, Theorem 5], the earliest BCE result of its kind. The $\aleph_0$- and ℓ^∞-barrelled and property (S) cases sharpen results of [32].

Theorem 6.3 *If E is either Baire-like or quasi-Baire, has property (C) or (S), or has a given one of the nine properties listed in the table, and contains a dense subspace F of codimension at least* $\mathfrak{b}$ *such that all the subspaces between F and E have the same given property, then there is a CE that preserves the property. If the given property is primitivity or inductivity, one may replace* $\mathfrak{b}$ *with* $\aleph_1$.

Only two basic arguments are needed to prove all thirteen results. In the first instance, where the given property is at least as strong as dlc and $\dim(E/F) \geq \mathfrak{b}$, we find a subspace G between F and E with a $\mathfrak{b}$-dimensional algebraic complement H, identify H algebraically with $\psi_{\mathfrak{b}}$, and let M be the ($\aleph_0$-dimensional) subspace of E^* whose members vanish on G and act on H as the dual of $\psi_{\mathfrak{b}}$. By density, M is transverse to E', and we proceed to prove that the CE η corresponding to the new dual $E' + M$ is dlc, and thus, by the second paragraph of this section, preserves any of the properties except possibly those of the middle level, which are not necessarily inherited by dense hyperplanes. But if L is a finite-dimensional subspace of M, then $L^{\perp}$ is between G and E, hence between F and E, and thus by hypothesis has the given property, even though it be of the middle level. From this and a general result of [30] it follows that η will preserve even the middle level properties.

In the second instance, where the given property is at the lowest level and dim $(E/F) \geq \aleph_1$, we argue exactly as in the first, only replacing $\mathfrak{b}$ with $\aleph_1$, and $\psi_{\mathfrak{b}}$ with χ, and proceed to show that the CE so obtained is primitive, and therefore would also preserve inductivity, the property to the left of primitivity.

The efficiency of proving scores of theorems with just a few arguments should be apparent. Indeed, a fourteenth property, that of being non-S_{σ}, is important to the study of the lowest level, and is fully discussed in [28, 29, 30]. The three-space problem permits a complete proof of the barrelled case more efficient than the one found in [24, Theorem 5], or even in the less sharp classical version [13, Theorem 5].

Theorem 6.4 *If E has a dense barrelled subspace F of codimension at least $\mathfrak{b}$, then E has a BCE.*

Proof. We find G, H, M and η as in the first instance above. In the $\langle E, E' + M\rangle$ duality, $M^{\perp} = G$, so that η induces the original topology on G, which is thus a closed barrelled subspace of (E, η). Since $G^{\perp} = M$, we may jointly identify E/G and its dual with H and M restricted to H, respectively. But then minimality identifies E/G with the barrelled space $\psi_{\mathfrak{b}}$, since they both have the same dual. It follows from the barrelled three-space problem [14] that (E, η) is barrelled. ■

Since the three-space problem has negative solutions in the dlc and primitive cases, it does not prove the full-blown Theorem 6.3. In fact, all four duality invariant properties deny the three space problem, for there is a non-primitive space E with a closed subspace F such that both F and E/F have property (C). There is a space E with closed barrelled subspace F such that E/F is inductive but E is not even primitive. However, we can prove that if F is inductive and E/F is dlc, then E must be primitive. If we keep the conclusion on E and change the hypothesis on F and/or on E/F in the context of our table, the result will be either invalid or inferior to the one stated, as the above-cited examples show.

Perhaps as important as any ten of our results is a structural synthesis which prompts such results and measures their relative merits.

References

[1] J. Bonet and P. Pérez Carreras, Remarks on the stability of barrelled-type topologies, Bull. Soc. Roy. Sci. Liège 52(1983), 313-318.

[2] M. De Wilde and C. Houet, On increasing sequences of absolutely convex sets in locally convex spaces, Math. Ann. 192(1971), 257-261.

[3] J. C. Ferrando and M. López Pellicer, Una propiedad de M. Valdivia, espacios de tipo (L) y condiciones débiles de tonelación, Rev. Re. Ac. Ci. Ex. Fís. y Nat. 83(1991), 25-34.

[4] J. C. Ferrando, M. López Pellicer and L. M. Sánchez Ruiz, Metrizable Barrelled Spaces, Pitman RNMS 332, Longman 1995.

[5] J. C. Ferrando and L. M. Sánchez Ruiz, On sequential barrelledness, Arch. Math. 57(1991), 597-605.

[6] T. Husain, Two new classes of locally convex spaces, Math. Ann. 166(1966), 289-299.

[7] H. Jarchow, Locally Convex Spaces, Stuttgart, Teubner 1981.

[8] N. J. Kalton, Some forms of the closed graph theorem, Proc. Camb. Phil. Soc. 70(1971), 401-408.

[9] G. Köthe, Topological Vector Spaces I, Springer-Verlag 1969.

[10] M. Levin and S. Saxon, A note on the inheritance of properties of locally convex spaces by subspaces of countable codimension, Proc. Amer. Math. Soc. 29(1971), 97-102.

[11] P. Pérez Carreras and J. Bonet, Barrelled Locally Convex Spaces, Math. Studies 131, North Holland 1987.

[12] W. J. Robertson, S. A. Saxon and A. P. Robertson, Barrelled spaces and dense vector subspaces, Bull. Austral. Math. Soc. 37(1988), 383-388.

[13] W. J. Robertson, I. Tweddle and F. E. Yeomans, On the stability of barrelled topologies, III, Bull. Austral. Math. Soc. 22(1980), 99-112.

[14] W. Roelcke and S. Dierolf, On the three-space problem for topological vector spaces, Collec. Math. 32(1981), 3-25.

[15] W. Ruess, A Grothendieck representation for the completion of cones of continuous seminorms, Math. Ann. 208(1974), 71-90.

[16] W. Ruess, Generalized inductive limit topologies and barrelledness properties, Pac. J. Math. 63, no. 2(1976), 499-516.

[17] W. Ruess, Closed graph theorems for generalized inductive limit topologies, Math. Proc. Camb. Phil. Soc. 82(1977), 67-83.

[18] S. Saxon, Nuclear and product spaces, Baire-like spaces, and the strongest locally convex topology, Math. Ann. 197(1972), 87-106.

[19] S. A. Saxon, Metrizable barrelled countable enlargements, preprint.

[20] S. A. Saxon, The codensity character of topological vector spaces, in Topological vector spaces, algebras and related areas, (A. Lau and I. Tweddle, Editors) Pitman RNMS 316, Longman 1994, pp. 24-36.

[21] S. Saxon and M. Levin, Every countable-codimensional subspace of a barrelled space is barrelled, Proc. Amer. Math. Soc. 29(1971), 91-96.

[22] S. A. Saxon and W. J. Robertson, Dense barrelled subspaces of uncountable codimension, Proc. Amer. Math. Soc. 107(1989), 1021-1029.

[23] S. A. Saxon and L. M. Sánchez Ruiz, Optimal cardinals for metrizable barrelled spaces, J. London Math. Soc. (2) 51(1995), 137-147.

[24] S. A. Saxon and L. M. Sánchez Ruiz, Barrelled countable enlargements and the bounding cardinal, J. London Math. Soc. (2) 53(1996), 158-166.

[25] S. A. Saxon and L. M. Sánchez Ruiz, Barrelled countable enlargements and the dominating cardinal, J. Math. Anal. App. 203(1996), 677-685.

[26] S. A. Saxon and L. M. Sánchez Ruiz, Dual local completeness, Proc. Amer. Math. Soc. 125(1997), 1063-1070.

[27] S. A. Saxon and L. M. Sánchez Ruiz, Mackey weak barrelledness, preprint.

[28] S. A. Saxon and L. M. Sánchez Ruiz, Metrizable weak barrelledness and dimension, preprint.

[29] S. A. Saxon and L. M. Sánchez Ruiz, Reinventing weak barrelledness, in preparation.

[30] S. A. Saxon and L. M. Sánchez Ruiz, Countably enlarging weak barrelledness, in preparation.

[31] S. A. Saxon and I. Tweddle, The fit and flat components of barrelled spaces, Bull. Austral. Math. Soc. 51(1995), 521-528.

[32] I. Tweddle and F. X. Catalán, Countable enlargements and weak barrelledness conditions, in Topological vector spaces, algebras and related areas, (A. Lau and I. Tweddle, Editors) Pitman RNMS 316, Longman, 1994, pp. 56-65.

[33] I. Tweddle and F. E. Yeomans, On the stability of barrelled topologies II, Glasgow Math. J. 21(1980), 91–95.

[34] I. Tweddle, S. A. Saxon and L. M. Sánchez Ruiz, Barrelled countable enlargements, in Topological vector spaces, algebras and related areas, (A. Lau and I. Tweddle, Editors) Pitman RNMS 316, Longman 1994, pp. 3-15.

[35] M. Valdivia, Absolutely convex sets in barrelled spaces, Ann. Inst. Fourier 21(1971), 3-13.

[36] M. Valdivia, Mackey convergence and the closed graph theorem, Arch. Math. 25(1974), 649-656.

[37] M. Valdivia, On quasi-completeness and sequential completeness in locally convex spaces, J. Reine Angew. Math. 276(1975), 190-199.

[38] M. Valdivia, Topics in Locally Convex Spaces, Math. Studies 67, North Holland 1982.

[39] M. Valdivia, A property of Fréchet spaces, in Functional Analysis, Holomorphy and Approximation Theory II, Proc. Semin. Rio de Janeiro 1981(G. I. Zapata, Editor) Math. Studies 86, North Holland 1984, pp. 469-477.

[40] J. H. Webb, Sequential convergence in locally convex spaces, Proc. Camb. Phil. Soc. 64(1968), 341-364.

[41] J. H. Webb, Countable-codimensional subspaces of locally convex spaces, Proc. Edinb. Math. Soc. 18(1973), 167-172.

[42] A. Wilansky, Modern Methods in Topological Vector Spaces, McGraw-Hill 1978.

EUITI-Departamento de Matemática Aplicada
Universidad Politécnica de Valencia
E-46071 Valencia, Spain

Department of Mathematics
University of Florida, PO Box 118005
Gainesville, FL 32611-8005, USA

L M SÁNCHEZ RUIZ

Topological vector spaces without local convexity conditions

1 Introduction

The theory of topological vector spaces (t.v.s.) in the absence of local convexity has been developed with less depth than with local convexity, which had the advantage of beeing able to support a strong use of duality theory. A consequence of this situation is that we are able to find several texts that cover the wide spectrum of locally convex spaces (l.c.s.) theory, such as the ones written by Horváth [7], Jarchow [11], Köthe [12], Pérez Carreras/Bonet [15], Robertson/Robertson [17] and Valdivia [37], just to mention some of them. However, without local convexity conditions, very few authors have attempted to summarize in text book form the existing knowledge. In this line the effort made by Adasch/Ernst/Keim [4] and Husain/Khaleelulla [9] deserves our recognition. In our understanding the pioneer attempt of building up a general theory of t.v.s. could have followed the steps of the more developed l.c.s. theory in such a way that under local convexity conditions the new classes of t.v.s. defined were to coincide with the local convex ones. As we will see this was not the case and instead of looking at the l.c.s. existing classes, authors looked at the kind of principles of functional analysis they satisfied and tried to get rid of local convexity. In this paper we will show how some classes of t.v.s. may be considered in such a way that, under the assumption of local convexity, they actually coincide with the existing locally convex classes and in fact they are characterized by verifying the same functional analysis principles.

All spaces considered will be assumed to be Haussdorff and t.v.s. are defined over the real or complex field $\mathbb{K}$. In order to be able to develop a theory of t.v.s. without local convexity conditions, the role played by absorbing absolutely convex subsets in l.c.s. is usually played in general t.v.s. by *strings* (called defining sequences by Iyahen [10]), i.e. sequences $\{U_n\}_n$ of absorbing balanced subsets, called knots, such that $U_{n+1} + U_{n+1} \subseteq U_n, n \in \mathbb{N}$, [4]. When each knot of a string is ⟨a neighbourhood of 0⟩[closed], the string is said to be ⟨topological⟩[closed], and bornivorous if each knot is so. We will follow the terminology of [4] and when dealing with classes of spaces in the category of l.c.s. we will write $\mathcal{C}$- in front of the respective class. So, for instance, $\mathcal{C}$-barrelled and $\mathcal{C}$-bornological space will stand for barrelled and bornological l.c.s. respectively.

2 Closed graph theorems

Let us recall that a linear mapping A between two t.v.s. E and F has *closed graph* if the graph of A, $\{(x, Ax) : x \in E\}$, is closed in the product space $E \times F$. Every continuous mapping has closed graph and closed graph theorems search for conditions on the spaces E and F such that every linear mapping between E and F with closed graph is continuous. One of the earliest results in this line was obtained by Banach when showing that a closed graph theorem holds between complete metrizable t.v.s., which are called (F)-spaces, therefore Fréchet spaces being locally convex (F)-spaces. It is well known that $\mathcal{C}$-barrelled spaces is the maximal class of l.c.s. for which the closed graph theorem holds when Fréchet or Banach spaces are considered in the range, [13]. A t.v.s. E is said to be *$\mathcal{L}$-barrelled* [4], originally called ultrabarrelled by Robertson [18] (cf. [9]), when each closed string of E is topological. The class of spaces so defined is satisfactory in order to get a behaviour parallel to $\mathcal{C}$-barrelled spaces in closed graph theorems when local convexity is not assumed [18, 10] (cf. [4, 9]).

Theorem 2.1 *For a given t.v.s. $E(\tau)$ the following assertions are equivalent:*
(i) *$E(\tau)$ is $\mathcal{L}$-barrelled.*
(ii) *$\tau = \tau^b$, where τ^b denotes the strong topology which is generated by the closed strings of $E(\tau)$.*
(iii) *Every Hausdorff linear topology τ' on E which has a base of 0-neighbourhoods formed by τ-closed sets, is coarser than τ.*
(iv) *Every linear mapping from $E(\tau)$ into any (F)-space with closed graph is continuous.*

Clearly every $\mathcal{L}$-barrelled l.c.s. is $\mathcal{C}$-barrelled, but there are $\mathcal{C}$-barrelled spaces which are not $\mathcal{L}$-barrelled, [18, pp. 256-7], therefore $\mathcal{L}$-barrelled space not being a generalization of $\mathcal{C}$-barrelled spaces. However [4, 19(1)] gives that if a $\mathcal{C}$-barrelled space E has a fundamental sequence $\{B_n\}_n$ of bounded sets (i.e. with each B_n balanced and bounded, $B_n + B_n \subseteq B_{n+1}, n \in \mathbb{N}$, and each bounded subset of E contained in some

B_n) formed by Banach disks [15] then E is $\mathcal{L}$-barrelled. A generalization of $\mathcal{C}$-barrelled spaces to a non-locally convex setting was developed by introducing *subbarrelled spaces* [21, 24] as those t.v.s. in which each closed absolutely convex string is topological. The class of spaces so defined coincides with $\mathcal{C}$-barrelled spaces when locally convex and contains $\mathcal{L}$-barrelled spaces but is stricly bigger than the union of the aforemetioned classes of spaces since if E is an uncountable-dimensional vector space and $\langle \tau^c \rangle$ $[\tau^f]$ denotes the finest $\langle$locally convex$\rangle$ [linear] topology on E, then the topological product $E(\tau^c) \times E(\tau^f)$ is a subbarrelled space which is neither $\mathcal{C}$-barrelled nor $\mathcal{L}$-barrelled [24, Example]. Hence the following result, [24, Theorem 1], generalizes [4, 19(1)].

Theorem 2.2 *Let $E(\tau)$ be a t.v.s. with a fundamental sequence of bounded sets formed by Banach disks. Then $E(\tau)$ is subbarrelled if and only if $E(\tau)$ is $\mathcal{L}$-barrelled.*

Let us note that in the above it is not enough to have just a fundamental sequence of bounded sets since the spaces $\ell^p, 0 < p < 1$, [18], equipped with the topology induced by ℓ^1 is a normed $\mathcal{C}$-barrelled space which is not $\mathcal{L}$-barrelled. Subbarrelled spaces are characterized for satisfying the same closed graph theorems as $\mathcal{C}$-barrelled spaces, [24, Theorem 2],

Theorem 2.3 *For a given t.v.s. E the following assertions are equivalent:*
(i) *E is subbarrelled.*
(ii) *Every linear mapping from E into any Fréchet space with closed graph is continuous.*
(iii) *Every linear mapping from E into any Banach space with closed graph is continuous.*

Remark 2.1 *Each barrel B in a t.v.s. generates the closed absolutely convex string $\{2^{-n}B\}_n$. Thus a t.v.s. is subbarrelled if and only if every barrel in E is a 0-neighbourhood.*

The amazing thing of the above remark and theorem is that they show that in order to get the classical graph theorem with $\mathcal{C}$-barrelled spaces in the domain E, the local convexity of E is not needed. This may lead us into thinking that perhaps it would have been more natural to have originally named barrelled spaces to the subbarrelled spaces without any local convexity assumption and then call either $\mathcal{C}$-barrelled or just l.c.s. barrelled when local convexity is assumed. In favour of this idea is Theorem 2.3

and the fact that the definition would be exactly the same. We have not dared to do this and preferred to follow the terminology of [4] in order to avoid the confussion this might cause and the fact that historically one of the main reasons for the study of $\mathcal{C}$-barrelled spaces was the Uniform Boundedness Principle when any l.c.s. is considered in the range, being yet unknown if this property characterizes subbarrelled spaces. In fact if $E(\tau)$ is a subbarrelled space, then the associated locally convex topology $\tau^{\circ\circ}$ generated by all the absolutely convex neighbourhoods of 0 is $\mathcal{C}$-barrelled [22, p. 144] which extends [18, Proposition 19], and the following Banach-Steinhaus type theorem holds, [22, Theorem 1],

Theorem 2.4 *For a given t.v.s.* $E(\tau)$ *the following assertions are equivalent:*
(i) $E(\tau^{\circ\circ})$ *is* $\mathcal{C}$*-barrelled.*
(ii) *Every pointwise bounded set of continuous linear mappings from* $E(\tau)$ *into any l.c.s. is continuous.*
(iii) *Every pointwise bounded set of continuous linear mappings from* $E(\tau)$ *into any Fréchet space is continuous.*
(iv) *Every pointwise bounded set of continuous linear mappings from* $E(\tau)$ *into any Banach space is continuous.*

When $E(\tau)$ is locally convex, the above theorem becomes the classical Banach-Steinhaus Theorem since in that case condition (i) just asserts that $E(\tau)$ is $\mathcal{C}$-barrelled. If in Theorem 2.4 we replace (i) by (i') $E(\tau)$ *is* $\mathcal{L}$*-barrelled*, then conditions (ii) and (iii) are to be changed into (ii') *Every pointwise bounded set of continuous linear mappings from* $E(\tau)$ *into any t.v.s. is continuous* and (iii') *Every pointwise bounded set of continuous linear mappings from* $E(\tau)$ *into any (F)-space is continuous*, respectively, condition (iv) disappearing (cf. [4, 7(3)]). From these, (i') $\Rightarrow$ (ii') was already given in the pioneer work of Robertson, [18, Theorem 5].

Let us finish this section pointing out that the maximal classes of topological vector spaces characterized by verifying the closed graph theorem in the range when subbarrelled spaces are considered in the domain with its corresponding open mapping theorem has been studied in [26].

3 Open mapping theorems

Let us recall that if $\mathcal{U}$ and $\mathcal{V}$ are the sets of all the neighbourhoods of the spaces E and F, then a linear mapping f of E into F is nearly continuous if $\overline{f^{-1}(V)} \in \mathcal{U}$ for each $V \in \mathcal{V}$; nearly open if $\overline{f(U)} \in \mathcal{V}$ for each $U \in \mathcal{U}$; and weakly singular if the closure of the kernel of f coincides with $\bigcap_{V \in \mathcal{V}} \overline{f^{-1}(V)}$. Pták's theory characterizes *B-complete* spaces as those l.c.s. E for which every continuous nearly open linear mapping of E into any l.c.s. F is open (cf. [8, 32]), where it holds that a l.c.s. E is *B-complete* (or a Pták space) if each subspace L of its dual E', such that $L \cap U^\circ$ is $\sigma(E', E)$-closed for all $U \in \mathcal{U}$, is $\sigma(E', E)$-closed.

On the other hand, let us point out that subbarrelled spaces have also been considered by Rodrigues [19, 20] under the name of semi-barrelled spaces. Rodrigues' focus was to get an extension of Pták's homomorphism theorem to a non-necessarily locally convex setting, obtaining that a nearly semi-open mapping between what he called a *semi-B-complete* space and an arbitrary t.v.s. is semi-open. Ferrer/Sánchez Ruiz/López Pellicer [6] have studied thoroughly this extension and, as consequence of the results obtained, have got an improvement of Pták's homomorphism theorem. Rodrigues' semi-B-complete spaces give Pták's theorem when l.c.s. are considered in the range but they may fail to be complete spaces when locally convex [19, p. 325]. This does not happen with the non-locally convex development of Adasch [1, 2] and Sánchez Ruiz [27], the former under a purely non-locally convex setting and the latter mixing locally convex and non-locally convex spaces. These two authors considered a t.v.s. E to be ⟨*B-complete in* $\mathcal{L}$⟩ [*subB-complete*] if each weakly singular nearly open linear mapping of E onto any ⟨t.v.s.⟩ [l.c.s.] is open, and a t.v.s. F to be ⟨B_r*-complete in* $\mathcal{L}$⟩ [*sub*B_r*-complete*] if each nearly continuous linear mapping with closed graph of each ⟨t.v.s.⟩ [l.c.s.] E into F is continuous. The following results contain characterizations of these spaces, ([2, 2(4)] and [27, Theorem 1] for Theorem 3.1, [2, 4(2)] and [27, Theorem 2] for Theorem 3.2).

Theorem 3.1 *For a given t.v.s. F the following assertions are equivalent:*
(i) *F is* ⟨B_r*-complete in* $\mathcal{L}$⟩ [*sub*B_r*-complete*].
(ii) *Each ono-to-one nearly continuous linear mapping with closed graph of each* ⟨*t.v.s.*⟩ [*l.c.s.*] *E onto F is continuous.*
(iii) *Each ono-to-one nearly continuous linear mapping with closed graph of each* ⟨*t.v.s.*⟩ [*l.c.s.*] *E onto F is a topological isomorphism.*

Theorem 3.2 *For a given t.v.s. E the following assertions are equivalent:*
(i) *E is ⟨B-complete in $\mathcal{L}$⟩ [subB-complete].*
(ii) *Each separated quotient E/H is ⟨B_r-complete in $\mathcal{L}$⟩ [subB_r-complete].*

4 Locally bounded linear mappings

Iyahen [10] introduced $\mathcal{L}$-bornological spaces (originally called ultrabornological, nowadays this term standing for an inductive limit of Banach spaces) in order to get results without local convexity conditions analogous to the ones that characterize $\mathcal{C}$-bornological spaces. A t.v.s. E is *$\mathcal{L}$-bornological* when each locally bounded linear mapping of E into any t.v.s. F, i.e. that maps bounded subsets of E into bounded subsets of F, is continuous. Clearly every metrizable space is $\mathcal{L}$-bornological. For them we have (cf. [4, 11(2)]),

Theorem 4.1 *For a given t.v.s. E the following assertions are equivalent:*
(i) *E is $\mathcal{L}$-bornological.*
(ii) *Every locally bounded linear mapping from E into any (F)-space is continuous.*
(iii) *Each bornivorous string in E is topological.*

Every $\mathcal{L}$-bornological l.c.s. is $\mathcal{C}$-bornological, but there are $\mathcal{C}$-bornological spaces which are not $\mathcal{L}$-bornological, for instance any uncountable-dimensional vector space E endowed with the $\mu(E, E^*)$-topology, [10, p. 299]. Hence, the relationship between these two classes of spaces is the same as with $\mathcal{L}$-barrelled and $\mathcal{C}$-barrelled spaces in the sense that $\mathcal{L}$-bornological spaces are not a real generalization of $\mathcal{C}$-bornological spaces. For the converse to hold [4, 19(2)] gives as a sufficient condition that if the $\mathcal{C}$-bornological space E has a fundamental sequence of bounded subsets, then E is $\mathcal{L}$-bornological. In a parallel way as with subbarrelled spaces, a generalization of $\mathcal{C}$-bornological spaces to a non-locally convex setting may be done by considering *subbornogical spaces* [21, 25] as those t.v.s. E such that each locally bounded linear mapping of E into any l.c.s. is continuous. For them we have the following characterizations, [25, Proposition 1].

Theorem 4.2 *For a given t.v.s. E the following assertions are equivalent:*
(i) *E is subbornological.*
(ii) *Every locally bounded linear mapping from E into any Fréchet space is continuous.*

(iii) *Every locally bounded linear mapping from E into any Banach space is continuous.*
(iv) *Each bornivorous absolutely convex string of E is topological.*
(v) *τ^{oo} is the finest locally convex topology τ^* for which each τ-bounded subset of E is τ^*-bounded.*

Remark 4.1 *A t.v.s. is subbornological if and only if every bornivorous absolutely convex subset of E is a 0-neighbourhood.*

Note that in Theorem 4.2 and Remark 4.1 local convexity is not required for the space E. Clearly subbornological spaces coincide with $\mathcal{C}$-bornological spaces when locally convex and contains $\mathcal{L}$-bornological spaces as subclass of spaces but there exist subbornological spaces which are neither $\mathcal{C}$-bornological nor $\mathcal{L}$-bornological [25, Example 1]. However, [25, Theorem 1] gives the following,

Theorem 4.3 *Let E be a t.v.s. with a fundamental sequence of bounded absolutely convex subsets. Then E is subbornological if and only if E is $\mathcal{L}$-bornological.*

With the same aim as when introducing subbarrelled and subbornological spaces, *subquasibarrelled* spaces [28] may me defined as those t.v.s. E in which every absolutely convex closed bornivorous string is topological or, equivalently, if every bornivorous barrel in E is a 0-neighbourhood. The relationship with its cousins the $\mathcal{C}$-quasibarrelled and $\mathcal{L}$-quasibarrelled spaces [4] (originally called quasiultrabarrelled by Iyahen [10]), is exactly the same as between subbornological spaces and the $\mathcal{C}$-bornological and $\mathcal{L}$-bornological spaces. Subquasibarrelled spaces are characterized by satisfying the same closed graph theorems, but without any local convexity assumption on them, as quasibarrelled spaces, [28, Theorem 2],

Theorem 4.4 *For a given t.v.s. E the following assertions are equivalent:*
(i) *E is subquasibarrelled.*
(ii) *Every locally bounded linear mapping from E into any Fréchet space with closed graph is continuous.*
(iii) *Every locally bounded linear mapping from E into any Banach space with closed graph is continuous.*

A space E is *boundedly summing* (cf. [4, S. 14]) if for each bounded subset B of E there is a sequence of non-null scalars $\{\lambda_n\}_n$ such that $\bigcup_{m=1}^{\infty} \sum_{n=1}^{m} \lambda_n B$ is bounded. The following result gives a sufficient condition for subquasibarrelledness to imply subbarrelledness, [28, Proposition 5], analogous to the one existing between $\mathcal{L}$-quasibarrelledness and $\mathcal{L}$-barrelledness (cf. [4, 14(6)]).

Theorem 4.5 *If E is a boundedly summing sequentially complete subquasibarrelled space, then E is subbarrelled.*

Since every l.c.s. is boundedly summing, Theorem 4.5 generalizes the well known result that asserts that every sequentially complete $\mathcal{C}$-quasibarrelled space is $\mathcal{C}$-barrelled. Let us note that Theorem 4.5 is not valid in general if the condition of being boundedly summing is dropped from the hypothesis since the space Φ of all the finite sequences of scalars endowed with the finest linear topology for which the set formed by the sequences $\{e_n\}_n$, where e_n is the sequence with all 0's but one 1 as n^{th} term, is bounded, [33], is complete and $\mathcal{L}$-bornological, and therefore subquasibarrelled, but not subbarrelled, [28, p. 880].

5 Inheritance properties

The property of being subbarrelled, subbornological and subquasibarrelled is inherited by finite-dimesional subspaces, $\mathcal{L}$-inductive limits, topological direct sums and separated quotients, [24, 25, 27]. A topological product of sub[quasi]barrelled spaces is always sub[quasi]barrelled [24, 27] and, by the Mackey-Ulam theorem, the topological product of a number of subbornological spaces, smaller than the smallest strongly inaccesible cardinal, is always subbornological, [25, Proposition 4].

The existance of non-$\mathcal{C}$-bornological countable-codimensional subspaces of $\mathcal{C}$-quasibarrelled spaces, [35], implies that the property of being subbornological or subquasibarrelled is not inherited by countable-codimensional subspaces.

Every finite-codimensional subspace of an ⟨$\mathcal{L}$-bornological⟩ [$\mathcal{L}$-quasibarrelled] space is ⟨$\mathcal{L}$-bornological⟩ [$\mathcal{L}$-quasibarrelled] [3, 29]. Analogously as it happens with $\mathcal{C}$-bornological and $\mathcal{C}$-quasibarrelled spaces, this result is not true for countable-codimensional subspaces since Mas/Sánchez Ruiz, [14, Example 1], have shown that there exists an $\mathcal{L}$-bornological space with a countable-codimensional subspace which is not $\mathcal{L}$-quasibarrelled. However it is possible to obtain a version of [34, Theorem 3] in the absence of local convexity by means of the following result, [14, Corollary 4], which gives a sufficient condition that assures the inheritance of the property of being $\mathcal{L}$-bornological by countable-codimensional subspaces.

Theorem 5.1 *Let E be an $\mathcal{L}$-inductive limit of a family of (F)-spaces. If G is a countable-codimensional subspace of E, then G is $\mathcal{L}$-bornological.*

Among other existing examples in the literature let us mention that Valdivia's immersion theorem [36] may be carried over to a non-locally convex setting and provide a non-$\mathcal{C}$-bornological $\mathcal{L}$-barrelled space, and consequently non-subbornological $\mathcal{L}$-barrelled, space, [23, Theorem 1],

Theorem 5.2 *If E is an infinite-dimensional separable Banach space, then the l.c.s. $E(\sigma(E, E'))$ is a dense subspace of a non-$\mathcal{C}$-bornological $\mathcal{L}$-barrelled space.*

After seeing how many results of l.c.s. theorey may be obtained in the absence of local convexity, we end this work by pointing out a question that has not found yet an answer without local convexity conditions. It is the parallel result to the Saxon-Levin-Valdivia Theorem which states that every countable-codimensional subspace of a $\mathcal{C}$-barrelled space is $\mathcal{C}$-barrelled, [30, 34] (see [31, Theorem 3.2]).
Open problem. *Is $\langle\mathcal{L}$-barrelled$\rangle$ [subbarrelled] each countable-codimensional subspace of an $\langle\mathcal{L}$-barrelled$\rangle$ [subbarrelled] space?*

References

[1] N. Adasch, Der Graphensatz in Topologischen Vectorräumen, Math. Z. **119** (1971), 131-142.

[2] N. Adasch, Vollständigkeit und der Graphensatz, J. reine angew. Math. **249** (1971), 217-220.

[3] N. Adasch and B. Ernst, Teilräume gewisser topologischer vektorräume, Collect. Math. **24** (1973), 27-39.

[4] N. Adasch, B. Ernst and D. Keim, **Topological Vector Spaces**, Lecture Notes in Math. **639**. Springer Verlag 1978.

[5] J. C. Ferrando, M. López Pellicer and L. M. Sánchez Ruiz, **Metrizable Barrelled Spaces**, Pitman RNMS **332**, Longman 1995.

[6] J. R. Ferrer, M. López Pellicer and L. M. Sánchez Ruiz, Pták Homomorphism Theorem Revisited, preprint.

[7] J. Horváth, **Locally Convex Spaces**, Lecture Notes in Math. **331**. Springer-Verlag 1973.

[8] T. Husain, **The open mapping and closed graph theorems in topological vector spaces**, Clarendon Press, Oxford 1965.

[9] T. Husain and S. M. Khaleelulla, **Barrelledness in Topological and Ordered Vector Spaces**, Lecture Notes in Math. **692**. Springer Verlag 1978.

[10] S. O. Iyahen, On certain classes of linear topological spaces, Proc. London Math. Soc. (3) **18** (1968), 285–307.

[11] H. Jarchow, **Locally Convex Spaces**, Stuttgart, Teubner 1981.

[12] G. Köthe, **Topological Vector Spaces I**, Springer-Verlag 1969.

[13] M. Mahowald, Barrelled spaces and the closed graph theorem, J. London Math. Soc. **36** (1961), 108–110.

[14] J. Mas and L. M. Sánchez Ruiz, On final topologies, An. Sti. Univ. Al. I. Cuza Iasi Matematica **36** (1990), 87-92.

[15] P. Pérez Carreras and J. Bonet, **Barrelled Locally Convex Spaces**, Math. Studies **131**, North Holland 1987.

[16] V. Pták, Completeness and the open mapping theorem, Bull. Soc. Math. France **86** (1958), 41-74.

[17] A. P. Robertson and W. Robertson, **Topological Vector Spaces**, Cambridge University Press 1973.

[18] W. Robertson, Completions of topological vector spaces, Proc. London Math. Soc. (3) **8** (1958), 242–257.

[19] B. Rodrigues, On Pták Homomorphism Theorem, J. Austral. Math. Soc. (Series A) **47** (1989), 322–333.

[20] B. Rodrigues, Some new classes of topological vector spaces with closed graph theorems, Comment. Math. Univ. Carolinae **32** (1991), 287–296.

[21] L. M. Sánchez Ruiz, **Condiciones de tonelación en espacios vectoriales topológicos**. Tesis Doctoral, Universidad de Valencia 1988.

[22] L. M. Sánchez Ruiz, On the Banach-Steinhaus theorem between topological vector spaces and locally convex spaces, Math. Japonica **36** (1991), 143–145.

[23] L. M. Sánchez Ruiz, Some examples on $\mathcal{L}$-barrelled spaces, Bull. Soc. Math. Belg. **43** (1991), 211–215.

[24] L. M. Sánchez Ruiz, On the closed graph theorem between topological vector spaces and Fréchet spaces, Math. Japonica **36** (1991), 271–275.

[25] L. M. Sánchez Ruiz, On locally bounded linear mappings between topological vector spaces and Fréchet spaces, Math. Japonica **36** (1991), 277–282.

[26] L. M. Sánchez Ruiz, Two maximal classes of topological vector spaces, Math. Japonica **36** (1991), 643–646.

[27] L. M. Sánchez Ruiz, On completeness and the closed graph theorem, Math. Japonica **36** (1991), 891–894.

[28] L. M. Sánchez Ruiz, The closed graph theorem for locally bounded linear mappings between topological vector spaces and Fréchet spaces, Math. Japonica **36** (1992), 877–881.

[29] L. M. Sánchez Ruiz and M. López Pellicer, On a result of Adasch and Ernst, Collect. Math. **40** (1989), 13-17.

[30] S. Saxòn and M. Levin, Every countable-codimensional subspace of a barrelled space is barrelled, Proc. Amer. Math. Soc. **29** (1971), 91-96.

[31] S. A. Saxon and L. M. Sánchez Ruiz, Dual local completeness, Proc. Amer. Math. Soc. **125** (1997), 1063-1070.

[32] H. H. Schaefer, **Topological Vector Spaces**. Springer-Verlag 1986.

[33] P. Turpin, **Convexité dans les espaces vectoriels topologiques généraux**. Thése, Orsay 1974, Published in Dissert. Math. **131**, Warszawa 1976.

[34] M. Valdivia, Absolutely convex sets in barrelled spaces, Ann. Inst. Fourier **21** (1971), 3-13.

[35] M. Valdivia, Some examples on Quasi-barrelled spaces, Ann. Inst. Fourier **22** (1972), 21-26.

[36] M. Valdivia, Un teorema de inmersión en espacios tonelados que no son bornológicos, Collect. Math. **24** (1973), 3-7.

[37] M. Valdivia, **Topics in Locally Convex Spaces**, Math. Studies 67, North Holland 1982.

Departamento de Matemática Aplicada
Universidad Politécnica de Valencia
E-46071 Valencia, Spain
E-mail address: LMSR@mat.upv.es

Partially supported by IVEI 003/033.

A EL KINANI AND M OUDADESS

Unified inequality and commutativity in Banach algebras with or without involution

Introduction. C. Le Page considered ([11]) in complex and unitary Banach algebras the following condition (C): $\|xy\| \leq \alpha\|xy\|$, (x, y in A, with a given $\alpha > 0$), and showed that it implies commutativity. Several authors worked on this condition in different settings (cf. references). The principal ingredient in the proofs is Liouville's theorem for holomorphic functions. Only few results have been obtained in involutive Banach algebras. This is probably due to the fact that involution does not keep holomorphy. We came ([3]) with a generalised criterion of Le Page from which follow most of commutativity results issued by condition (C). Here we use Liouville's theorem for harmonic functions to obtain results on involutive Banach algebras satisfying inequalities of type (C). Finally we come with a general criterion of Le Page type which applies to both involutive and non involutive Banach algebras, unitary or not. Besides new results, all known ones are corollaries of a single theorem. In the sequel A will always be a complex associative Banach algebra with a norm $\| \ \|$; and $C(A)$ the center of A. The numerical radius and the spectral radius of an element x will be denoted by $\nu(x)$ and $\rho(x)$ respectively. Recall that a linear involution is any antilinear mapping from A to A.

Principal theorem. Here is the principal result:

Theorem 1: Let $(A, \| \ \|)$ be a complex Banach algebra, M a t.v.s. which is a topological left A-module, X a t.v.s., $h : A \times M \to X$ a continuous bilinear mapping, $k : A \times M \to X$ a continuous bi-antilinear mapping, (E, p) a semi normed space, S and T two linear continuous maps from X to E, and $q : M \to R^+$ any function. If for every $a \in A$ and $m \in M$,

$$p\Big[S(h(a,m)) + T(k(a,m))\Big] \leq q(am)\,, \tag{1}$$

then

1) **(i)** $p\Big[S(h(b,am)) - S(h(ba,m))\Big] = 0$, for every $a, b \in A$ and $m \in M$.

(ii) $p\Big[T(k(b,am)) - T(k(ba,m))\Big] = 0$, for every $a, b \in A$ and $m \in M$.

2) If A has a left approximate identity $(e_i)_i$, then

(i) $p\Big[S(h(a,m)) - \lim_i S(h(e_i,am))\Big] = 0$, for every $a \in A$ and $m \in M$.

(ii) $p\Big[T(k(a,m)) - \lim_i T(k(e_i,am))\Big] = 0$, for every $a \in A$ and $m \in M$.

3) If **(1)** is verified for every $a \in A^1 = A \oplus C$ and $m \in M$, then

(i) $p\Big[S(h(a,m)) - S(h(e,am))\Big] = 0$, for every $a \in A$ and $m \in M$,

(ii) $p\Big[T(k(a,m)) - T(k(e,am))\Big] = 0$, for every $a \in A$ and $m \in M$,

where e is the unit of A^1.

Proof: If A is not unitary, consider its unitization $A^1 = A \oplus C$; then M is a left A^1-module.

1) For a, b in A and $m \in M$, let f be the map defined, on C and with values in E, by

$$f(\lambda) = S\left(h\left(be^{-\lambda a}, e^{\lambda a}m\right)\right) + T\left(k\left(be^{-\lambda a}, e^{\lambda a}m\right)\right).$$

One checks that, for any φ in the topological dual of (E,p), $\varphi \circ f$ is harmonic. It is also bounded for $|\varphi \circ f(\lambda)| \leq \|\varphi\| q(bm)$. By Liouville's theorem, it is constant. Differentiating relatively to the real part and the imaginary part of λ; we have

$$\varphi\Big[S(h(b,am)) - S(h(ba,m)) + T(k(b,am)) - T(k(ba,m))\Big] = 0,$$

$$\varphi\Big[S(h(b,am)) - S(h(ba,m)) + T(k(ba,m)) - T(k(b,am))\Big] = 0.$$

Whence **(i)** and **(ii)**, by the Hahn-Banach theorem.

2) The conclusion follows from the equalities

$$p\Big[S(h(e_i,am)) - S(h(e_ia,m))\Big] = 0$$

and

$$p\Big[T(k(e_i,am)) - T(k(e_ia,m))\Big] = 0$$

since S, h, T and k are continuous.

3) For $a \in A$ and $m \in M$, consider the map g defined by

$$g(\lambda) = S\left(h\left(e^{-\lambda a}, e^{\lambda a}m\right)\right) + T\left(k\left(e^{-\lambda a}, e^{\lambda a}m\right)\right)$$

and proceed as for **1)**.

Remark 2: a) If M is a right A-module, we obtain analogous conclusions under the condition

$$p\Big[S(h(a,m)) + T(k(a,m))\Big] \leq q(ma)\,. \tag{2}$$

b) The assertions **(i)** of **1)** is the analog, in the general case, of the fact that $A^2 \subset C(A)$, when $\|xy\| \leq \alpha\|yx\|$ ([14]), while **(ii)** of **1)** is the analog of $x^*(yz)^* = (xy)^*z^*$, under the condition $\|x^*y^*\| \leq \alpha\|xy\|$, where $*$ is a continuous linear involution ([5]).

c) Situations where $h = 0$ or $k = 0$ have been studied in [5] and [4] respectively.

To apply Theorem 1, we particularise hypotheses.

Corollary 3: Let $X = B$ be an algebra with two Banach algebra products Δ and $\Diamond$, $h(a,m) = \varphi_1(a)\Delta\psi_1(m)$, $k(a,m) = \varphi_2(a)\Diamond\psi_2(m)$, with φ_1, ψ_1 (resp. φ_2, ψ_2) continuous linear (resp. antilinear) mappings and p is a norm. If for every $a \in A$ and $m \in M$,

$$p\Big[S(\varphi_1(a)\Delta\psi_1(m)) + T(\varphi_2(a)\Diamond\psi_2(m))\Big] \leq q(am) \tag{3}$$

then

1) **(i)** $S[\varphi_1(b)\Delta\psi_1(am)] = S[\varphi_1(ba)\Delta\psi_1(m)]$, for every $a, b \in A$ and $m \in M$.

(ii) $T[\varphi_2(b)\Diamond\psi_2(am)] = T[\varphi_2(ba)\Diamond\psi_2(m)]$, for every $a, b \in A$ and $m \in M$.

2) If A has a left approximate identity $(e_i)_i$, then

(i) $\lim_i S[\varphi_1(e_i)\Delta\psi_1(am)] = S[\varphi_1(a)\Delta\psi_1(m)]$, for every $a \in A$ and $m \in M$.

(ii) $\lim_i T[\varphi_2(e_i)\Diamond\psi_2(am)] = T[\varphi_2(a)\Diamond\psi_2(m)]$, for every $a \in A$ and $m \in M$.

3) If **(3)** is verified for every $a \in A^1 = A \oplus C$ and $m \in M$, then

(i) $S[\varphi_1(e)\Delta\psi_1(am)] = S[\varphi_1(a)\Delta\psi_1(m)]$, for every $a \in A$ and $m \in M$,

(ii) $T[\varphi_2(e)\Diamond\psi_2(am)] = T[\varphi_2(a)\Diamond\psi_2(m)]$, for every $a \in A$ and $m \in M$,

where e is the unit of A^1.

We now give some consequences.

If $M = A$, $E = X = B$, $S = Id$ and $T = Id$, the conclusion **(1)** of Corollary 2 becomes

(i) $\varphi_1(x)\Delta\psi_1(yz) = \varphi_1(xy)\Delta\psi_1(z)$, for every $x, y, z \in A$.

(ii) $\varphi_2(x)\Diamond\psi_2(yz) = \varphi_2(xy)\Diamond\psi_2(z)$, for every $x, y, z \in A$.

a) If $(e_i)_i$ is a left approximate identity of A such that $(\varphi_1(e_i))_i$ and $(\psi_1(e_i))_i$ are approximate identities of B for Δ, then, by **(i)**, $\varphi_1 = \psi_1$ and $\varphi_1(xy) = \varphi_1(x)\Delta\varphi_1(y)$ for every $x, y \in A$. Whence, if also $A = B$ and $\varphi_1 = Id$, then $x\Delta y = xy$ for every $x, y \in A$.

b) If $(f_i)_i$ is a left approximate identity of A such that $(\varphi_2(f_i))_i$ and $(\psi_2(f_i))_i$ are also approximate identities of B for $\Diamond$, then, by **(ii)**, $\varphi_2 = \psi_2$ and $\varphi_2(xy) = \varphi_2(x)\Diamond\varphi_2(y)$ for every $x, y \in A$. Hence, if $A = B$, φ_2 is one to one and $\varphi_2(x\Diamond y) = \varphi_2(x)\Diamond\varphi_2(y)$, then $x\Diamond y = xy$ for every $x, y \in A$.

The following consequence is worthwile to mention.

Corollary 4: Let $(A, \|\ \|)$ be a Banach algebra endowed with algebra products Δ and $\Diamond$ which are continuous with respect to $\|\ \|$. Let also $\psi : A \to A$ be a continuous linear map and two continous linear involutions $*$ and $\#$ such that $*$ is one to one and verifies $(xy)^* = x^*\Diamond y^*$. Suppose moreover that $(e_i)_i$ and $(f_i)_i$ are two approximate identities in $(A, .)$ such that $(e_i)_i$ and $(\psi(e_i))_i$ are approximate identities in A for Δ and $(f_i^*)_i$ and $(f_i^\#)_i$ are approximate identities in A for $\Diamond$. If the following inequality holds

$$p\left[x\Delta\psi(y) + x^*\Diamond y^\#\right] \leq q(xy) \tag{4}$$

where $q : A \to R^+$ is any function, then

1) ψ is the identity map and the two linear involutions $*$ and $\#$ coincide.

2) $x\Delta y = xy = x\Diamond y$, $\forall x, y \in A$.

3) A is commutative.

4) If moreover $(x^*)^* = x$ for every $x \in A$, then $*$ is a continuous algebra involution.

Remark 5: The existence of an approximate identity in A is an essential hypothesis in the previous corollary. Indeed, let e_1 and e_2 be two symbols verifying the relations $e_1^2 = e_2^2 = 0$ and $e_ie_je_i = 0$ $(i, j = 1, 2)$. Consider the complex algebra A generated by e_1 and e_2 subject to these relations. The algebra A is nothing else than the 4-dimensional vector space for which a base is $\{e_1, e_2, e_1e_2, e_2e_1\}$. For $x = \alpha_1e_1 + \alpha_2e_2 + \alpha_3e_1e_2 + \alpha_4e_2e_1$, put $\|x\| = \sum_{i=1}^4 |\alpha_i|$ and $x^* = \bar{\alpha}_1e_1 + \bar{\alpha}_2e_2$. Then A is a Banach algebra endowed with a continous linear involution such that $\|xy+y^*x^*\| \leq 2\|xy\|$. But A is not commutative ([8]) nor the involution is an antimorphism.

A particular case of (**4**) is

$$\|xy + y^*x^*\| \leq 2\|xy\|$$

Considered in [6], theorem III.1 of which is now a consequence of corollary 2.

Remark 6: Let A be an involutive Banach algebra and p a seminorm on A such that

$$p\left(\frac{x+x^*}{2}\right) \leq \rho(x)\,, \quad \forall x \in A$$

We, of course, have

$$p\left(\frac{uv+v^*u^*}{2}\right) \leq \rho(vu)\,, \quad \forall u \in A^1\,, \forall v \in A$$

Then, by **3)** of theorem 1, $p(xy - yx) = 0$, for every x and y in A.
As an application, notice that, if

$$\rho\left(\frac{x+x^*}{2}\right) \leq \rho(x)\,, \quad \forall x \in A\,,$$

then

$$P\left(\frac{x+x^*}{2}\right) \leq \rho(x)\,, \quad \forall x \in A\,,$$

where P is Palmer's seminorm. Inequality (5) has been considered by B. Aupetit ([2]) who showed that $A/\mathrm{Ker}P$ is commutative.
In case of a given norm p, we have the following results

Corollary 7: Each of the following conditions implies the commutativity of A.
1) $\nu\left(\frac{x+x^*}{2}\right) \leq \alpha\rho(x)\,, \forall x \in A$.
2) $\|\frac{x+x^*}{2}\| \leq \alpha\rho(x)\,, \forall x \in A$.

References

[1] **J.W. Baker, J.S. Pym:** *Remark on continuous bilinear mappings. Proc. Edinburgh. Math. Soc. (2) 17, (1970-1971), 245 - 248.*

[2] **B. Aupetit:** *Propriétés spectrales des algèbres de Banach, Lecture Notes in Math., 735 (1979).*

[3] **A. El kinani, M. Oudadess:** *Un Critère généralisé de Le Page et commutativité. C.R. Math. Rep. Acad. Sc. Canada, Vol. XVIII, no. 2,3, April-June (1996), 71 - 74.*

[4] **A. El kinani, M. Oudadess:** *Bilinear mappings and commutativity in Banach algebras, Rapport de recherche, E.N.S. Takaddoum, Rabat, Maroc, (1996).*

[5] **A. El kinani, M. Oudadess:** *Bi-antilinear mappings and commutativity in involutive Banach algebras. Rapport de recherche, E.N.S. Takaddoum, Rabat, Maroc, (1996).*

[6] **A. El kinani, M. Oudadess:** *Criteria of Le Page type in involutive Banach algebras. Rapport de recherche, E.N.S. Takaddoum, Rabat, Maroc, (1996).*

[7] **F.F. Bonsall and J. Duncan:** *Numerical ranges of operators on normed spaces and of elements of normed algebras. London Math. Soc. Lecture Note Series, no. 2 (1971).*

[8] **O.H. Cheikh, M. Oudadess:** *On a commutativity question in Banach algebras. Arab. Gulf J. Sc. Res. Math. Phys. Sc. A6 (2), (1988), 173 - 179.*

[9] **R.A. Hirschfeld, W. Zelazko:** *On spectral norm Banach algebras. Bull. Acad. Polonaise. Sc. Math. Astr. Phys. Vol XVI, no. 3, (1968), 195 - 199.*

[10] **E. Illoussamen, M. Oudadess:** *Sur des critères de commutativité dans les algèbres de Banach. C.R. Math. Rep. Acad. Sc. Canada, Vol XIV, no. 5, (1992), 183 - 188.*

[11] **C. Le Page:** *Sur quelques conditions impliquant la commutativité dans les algèbres de Banach. C.R.A.S. Paris Ser. A-B; 265, (1967), A 235 - A 237.*

[12] **Gh. Mocanu:** *A remark on a commutativity criterion in Banach algebras. Stud. Cerc. Mat. 21, (1969), 947 - 952.*

[13] **G. Niestegge:** *A note on criterion of Le Page and Hirschfeld-Zelazko for the commutativity of Banach algebras. Studia Math. T. LXXIX, (1984), 87 - 90.*

[14] **M. Oudadess:** *Commutativité de certaines algèbres de Banach. Bol. Soc. Math. Mexicana, Vol. 28, no. 1, (1983), 9 - 14.*

[15] **K. Srinivasacharyulu:** *Remark on Banach algebras. Bull. Soc. Roy. Sci. Liège, (1973), 107 - 108.*

Ecole Normale Supérieure de Takaddoum.
B.P.: 5118
10105 Rabat (MAROC)

M BERKANI AND M SARIH

Extension de certaines inegálités dans les algébres de Banach

1 Introduction

Un élément idempotent dans une algébre de Banach est un élément x tel que $x^2 = x$. Une suite $(e_n)_n$ d'éléments d'une algébre de Banach A est dite unité approché borné squentielle (u.a.b.s) s'il existe une constante K telle que
$\|e_n\| \leq K \quad \forall n \in \mathbb{N}$ et telle que

$$\lim_{n\to\infty} e_n x = \lim_{n\to\infty} x e_n = x \quad \forall x \in A.$$

Dans ce papier, on montre que dans une algébre de Banach sans idempotent non nul
on a:

$$\inf_{\|x\|\geq\alpha} \|x^2 - x\| \geq \beta, \qquad \text{où } \alpha \in [0, \frac{1}{2}] \;\; et \;\; \beta = \alpha - \alpha^2,$$

et on donne deux exemples d'algébres de Banach où

$$\min_{\|x\|=\alpha} \|x^2 - x\| = \beta,$$

et un exemple d'algébre de Banach où

$$\inf_{\|x\|\geq\alpha} \|x^2 - x\| > \beta.$$

On donne aussi une condition nécessaire et suffisante pour que dans une algébre de Banach on ait

$$\min_{\|x\|=\alpha} \|x^2 - x\| = \beta,$$

et deux conditions suffisantes pour que dans une algébre de Banach on ait:

$$\inf_{\|x\|=\alpha} \|x^2 - x\| = \beta.$$

On montre aussi que si A est une algébre de Banach ayant une (u.a.b.s) $(e_n)_n$ tel qu'il existe $\lambda \in]1/2, \frac{1+\sqrt{2}}{2}[$ et vérifiant $\|\lambda^2 e_n^2 - \lambda e_n\| < \beta$, alors A posséde une (u.a.b.s) formée d'idempotents.
Enfin, on démontre que tout homomorphisme continu d'une algébre de Banach A dans une algébre de Banach commutative radicale est nul si A posséde une (u.a.b.s) $(e_n)_n$ vérifiant:

$$\|(\lambda e_n)^2 - \lambda e_n\| \leq \beta,$$

pour n assez grand et pour une constante $\lambda > \alpha$.

Rappelons que J.Esterle a montré dans [5], que dans une algébre de Banach sans idempotent non nul, on a:

$$\inf_{\|x\|\geq \frac{1}{2}} \| x^2 - x \| \geq \frac{1}{4},$$

et M.Berkani a caractérisé les algébres de Banach telle que

$$\min_{\|x\|=\frac{1}{2}} \|x^2 - x\| = \frac{1}{4},$$

et divers résultats liés à cette inégalité ont été obtenus par Berkani dans [1],[2] et Berkani-Esterle [3]. Nous utiliserons souvent le fait que la fonction g (resp h) définie par:

$$g : u \longrightarrow \sum_{n=1}^{\infty} \frac{(2n-1)}{(n-1)!2^{n-2}} \frac{u^n}{n!},$$

$$(resp\ \ h : v \longrightarrow \sum_{n=1}^{\infty} (-\frac{1}{2})(-\frac{1}{2}-1)\ldots(-\frac{1}{2}-n+1)\frac{v^n}{n!}),$$

est bien définie et continue sur $\overline{D}$ (resp D) et qu'elle y coincide avec la fonction:

$$u \longrightarrow -1+\sqrt{1-u} \quad (resp\ \ v \longrightarrow -1+\frac{1}{\sqrt{1+v}}),$$

où $D = \{z \in \mathbf{C} \ / \mid z \mid < 1\}$, $\overline{D} = \{z \in \mathbf{C} \ / \mid z \mid \leq 1\}$ et $\sqrt{\ }$ désigne la détermination principale de la racine carrée. Notons que , si $u \in D$ on a:

$$|g(u)| \leq 1 \ \ et \ \ |1+h(v)| = |\frac{1}{\sqrt{1-v}}| \leq \frac{1}{\sqrt{1-|v|}}.$$

Si x est un élément d'une algébre de Banach A tel que $\|x\| \leq 1$, (resp $\|x\| < 1$) on peut donc définir l'élément $g(x)$ (resp $[\ e + h(x)\ ]$) dans A, (resp dans $A^{\#}$) où $A^{\#}$ désigne l'algébre obtenue par adjonction d'une unité à A et e son unité. Remarquons qu'alors:

$$\|g(x)\| \leq 1 - \sqrt{1-\|x\|} \ \ et \ \ \|e + h(x)\| \leq \frac{1}{\sqrt{1-\|x\|}},$$

et on a aussi l'identité importante:

$$[\frac{-g(x)}{2}]^2 - [\frac{-g(x)}{2}] = \frac{x}{4}. \qquad (*)$$

2 Inégalités et exemples

2.1 Inégalités

Thèoréme 1 *Soit A une algébre de Banach sans idempotent non nul on a:*

$$\inf_{\|x\|\geq \alpha} \|x^2 - x\| \geq \beta, \qquad o\grave{u}\ \ \alpha \in [0, \frac{1}{2}] \ \ et \ \ \beta = \alpha - \alpha^2.$$

Démonstration:
Supposons qu'il existe un élément x de A tel que: $||x|| \geq \alpha$ *et* $||x^2 - x|| < \beta$.

Alors $||4x^2 - 4x|| < 1$, on peut donc définir dans $A^\#$ l'élément:

$$p(x) = \frac{e}{2} + (x - \frac{e}{2})(e + 4x^2 - 4x)^{-\frac{1}{2}}.$$

On a $p(x) \in A$, et

$$[p(x)]^2 = \frac{e}{4} + (x - \frac{e}{2})(e + 4x^2 - 4x)^{-\frac{1}{2}} + (x - \frac{e}{2})^2 \, (e + 4x^2 - 4x)^{-1}.$$

Or

$$(e - 2x)^2 = e - 4x + 4x^2, \; alors \; [p(x)]^2 = \frac{e}{2} + (x - \frac{e}{2})(e + 4x^2 - 4x)^{-\frac{1}{2}}.$$

Donc $[p(x)]^2 = p(x)$ et $p(x)$ est un idempotent. Montrons que $p(x) \neq 0$.
Supposons par l'absurde que $p(x) = 0$. Alors $\frac{e}{2} = -(x - \frac{e}{2})(e + 4x^2 - 4x)^{-\frac{1}{2}}$,
donc:

$$x = [\frac{e - [(e + 4x^2 - 4x)^{\frac{1}{2}}]}{2}], et||x|| = ||\frac{[e - [(e + 4x^2 - 4x)^{\frac{1}{2}}]]}{2}||.$$

Or

$$||\frac{[e - [(e + 4x^2 - 4x)^{\frac{1}{2}}]]}{2}|| \leq \frac{1 - \sqrt{1 - ||4x^2 - 4x||}}{2} < \frac{1 - \sqrt{1 - 4\beta}}{2} = \alpha.$$

Donc $||x|| < \alpha$, ce qui est absurde car on a supposé que $||x|| \geq \alpha$. D'où $p(x) \neq 0$ et ceci est une contradiction avec l'hypothése du thèoréme.

Remarque1 On peut démontrer ce théoréme par deux autres méthodes
a) en utilisant les mêmes techniques que celles utilisées par J. Esterle dans [5].
b) en utilisant le fait que la fonction $t \longrightarrow t - t^2$ est strictement croissante sur $[0, \frac{1}{2}]$ et l'inégalité déjà démontrée par J.Esterle dans [5] à savoir : $\inf_{||x|| \geq \frac{1}{2}} || \, x^2 - x \, || \geq \frac{1}{4}$.

Proposition 1 *Soient A une algébre de Banach sans idempotent non nul et $\alpha \in [0, \frac{1}{2}[$. Pour tout élément x dans A tel que $||x|| > \alpha$ on a $||x^2 - x|| > \beta$.*

Preuve Soit $x \in A$ tel que $||x|| > \alpha$. Si $||x|| \geq \frac{1}{2}$, on a: $||x^2 - x|| \geq \frac{1}{4} > \alpha - \alpha^2$.
Si $||x|| < \frac{1}{2}$, on a $||x^2 - x|| \geq ||x|| - ||x||^2 > \alpha - \alpha^2$,
car la fonction $t \longrightarrow t - t^2$ est strictement croissante sur $[0, \frac{1}{2}]$.

Remarque2 Soit α un élément de $[0, \frac{1}{2}]$ et $f \in L^1(\mathbb{R})$ tel que $||f|| \geq \alpha$, alors $||f^2 - f|| > \beta$ où $f^2 = f * f$ et $*$ désigne le produit de convolution. En effet:
Si $\alpha = \frac{1}{2}$ on a le résultat voir [1, Prop.6.3]. Si $\alpha \in [0, \frac{1}{2}[$ la proposition précédente donne le résultat.

2.2 Exemples d'algèbres de Banach où $\min\limits_{||x||=\alpha} ||x^2 - x|| = \beta$

Exemple1 Soit $D = \{z \in \mathbb{C} \,/\, |z| \leq 1\}$ et $A(D)$ l'algébre des fonctions continues sur $\overline{D}$ et holomorphe sur $\overset{o}{D}$, considérons la sous-algébre fermée B de $A(D)$ définie par: $B = \{f \in A(D)/f(0) = 0\}$,
alors B est une algébre de Banach sans idempotent non nul et B contient la fonction g définie par: $g : z \longrightarrow -1 + \sqrt{1+z}$. Posons $h(z) = -\alpha g(z)$. Comme la norme de la fonction $k : z \longrightarrow z$ vaut 1, nous avons déja $||h|| \leq \alpha$, or $|h(-1)| = \alpha$, donc $||h|| = \alpha$ et par suite,

$$||h^2 - h|| \geq \beta. \qquad (1)$$

Posons $g = h^2 - h$, on a $g = \alpha^2(-1 + \sqrt{1+z})^2 - \alpha(-1 + \sqrt{1+z})$, ou encore

$$g = (2\alpha)^2[(\frac{-1+\sqrt{1+z}}{2})^2 - \frac{-1+\sqrt{1+z}}{2}] + [(2\alpha)^2 - 2\alpha]\frac{-1+\sqrt{1+z}}{2}.$$

L'identité $(*)$ donne

$$g = 4\alpha^2\frac{z}{4} + [(2\alpha)^2 - 2\alpha]\frac{-1+\sqrt{1+z}}{2}$$

Alors:

$$||g|| \leq \alpha^2 + (\alpha - 2\alpha^2) = \beta. \qquad (2)$$

(1) et (2) montrent que $||h^2 - h|| = \beta$. Donc $\min\limits_{||f||=\alpha, f\in B} ||f^2 - f|| = \beta$.

Exemple2
$L^1(\mathbb{R}) = \{ f : \mathbb{R} \longrightarrow \mathbb{C}$ mesurable $/ \int_{-\infty}^{+\infty} |f(x)|\, dx < \infty \}$
Considérons la fonction

$$h : \mathbb{R} \longrightarrow \mathbb{R}$$
$$x \longrightarrow h(x) = \frac{\alpha}{\pi(1+x^2)},$$

h est une fonction continue et

$$\int_{-\infty}^{+\infty} |h(x)|\, dx = \alpha.$$

Donc $h \in L^1(\mathbb{R})$, d'autre part $h * h = \dfrac{4\alpha^2}{2\pi(4+x^2)}$,
où $*$ désigne le produit de convolution voir [6, p.177,prop.9–8]. On a

$$(h^2 - h)(x) = \frac{4\alpha^2}{2\pi(4+x^2)} - \frac{2\alpha}{\pi(1+x)} \qquad \text{avec} \qquad h^2 = h * h,$$

et par suite

$$(h^2 - h)(x) = 4\alpha^2[\frac{1}{2\pi(4+x^2)} - \frac{1}{2\pi(1+x^2)}] + (4\alpha^2 - 2\alpha)\frac{1}{2\pi(1+x^2)}.$$

Alors

$$||h^2 - h|| \leq \frac{4\alpha^2}{4} + (\alpha - 2\alpha^2) = \beta. \tag{1}$$

Or

$$\forall f \in L^1(\mathbb{R}) \ ||f|| = \alpha \Rightarrow ||f^2 - f|| \geq ||f|| - ||f||^2 = \beta. \tag{2}$$

(1) et (2) montrent que $||h^2 - h|| = \beta$. Donc

$$\min_{||f||=\alpha, f\in L^1(\mathbb{R})} ||f^2 - f|| = \beta.$$

2.3 Exemple d'algébre de Banach où $\min_{||x|| \geq \alpha} ||x^2 - x|| > \beta$

Soient $D = \{z \in C \ / \ | z | \leq 1\}$ et h l'élément de $A(D)$ définie par: $h(z) = kz$ où k est un réel tel que $0 < k < 1$.
Soit $B_1 = \{g \in A(D)/g(0) = g'(0) = 0\}, B_1$ est une sous-algébre fermée de $A(D)$ et on remarque que l'application

$$\Phi : \quad \begin{array}{l} B \longrightarrow B_1 \ , \\ f \longrightarrow \Phi(f) = hf \end{array}$$

où B est la sous-algébre fermée de $A(D)$ introduite dans l'exemple 1 du paragraphe précédent, est un isomorphisme d'algébres de B dans B_1. On définit alors une norme sur B_1 par: $||g||_1 = ||\Phi^{-1}(g)||, \ g \in B_1$.
B_1 est donc un espace de Banach pour la norme $|| \ ||_1$, c'est aussi une algébre de Banach car, si g_1 et g_2 sont deux éléments de B_1 , on a : $g_i = hf_i$ avec $f_i \in B(i = 1, 2)$ et

$$||g_1 g_2||_1 = ||h^2 f_1 f_2||_1 \leq k||f_1|| ||f_2|| \leq ||f_1|| ||f_2|| \leq ||g_1||_1 ||g_2||_1.$$

Soit l un élément de B_1 tel que: $||l||_1 \geq \alpha$ et soit $f = \Phi^{-1}(l)$ alors

$$\alpha \leq ||f|| = \max_{z \in D} |f(z)|.$$

Or D est compact, donc il existe $z_1 \in D$ tel que $|f(z_1)| \geq \alpha$. Comme $|f(0)| = 0$, il existe $z_0 \in D$ tel que: $|f(z_0)| = \alpha$, car D est connexe et $z \longrightarrow |f(z)|$ est continue. Donc $\{|f(z)|, z \in D\}$ est connexe dans IR, c'est donc un intervalle. Alors

$$||l^2 - l||_1 = ||f^2 h - f|| = \sup_{z \in D} |kf^2(z) - f(z)|,$$

Donc

$$\begin{aligned} ||l^2 - l|| &\geq |kf^2(z_0) - f(z_0)| \\ &\geq |f(z_0)| - k|f(z_0)|^2|z_0| \\ &\geq \alpha - k\alpha^2|z_0| \\ &\geq \alpha - k\alpha^2. \end{aligned}$$

Donc

$$\inf_{||l||_1 \geq \alpha} ||l^2 - l||_1 \geq \alpha - k\alpha^2 > \alpha - \alpha^2 = \beta.$$

3 Condition pour qu'on ait $\min_{||x||=\alpha} ||x^2 - x|| = \beta$

Thèoréme 2 *Soit A une algébre de Banach. Pour qu'il existe un élément $b \in A$ tel que $||b|| = \alpha$ et $||b^2 - b|| = \beta$ il faut et il suffit qu'il existe un élément a de A tel que $||a|| = \rho(a) = 1$, où ρ désigne le rayon spectral.*

Preuve Sans perte de généralité on peut se limiter au cas où A est commutative. Supposons qu'il existe un élément a de A tel que $||a|| = \rho(a) = 1$.
On sait que le spectre de a , $\sigma(a) = \{\chi(a)/\chi \in \hat{A}^{\#}\}$, où $\hat{A}^{\#}$ désigne l'ensemble des caractéres de $A^{\#}$. Comme $\rho(a) = max_{\chi \in A^{\#}} |\chi(a)|$ et $\sigma(a)$ est compact, alors il existe un caractére χ_0 de $A^{\#}$ tel que $\rho(a) = |\chi_0(a)|$. Posons $\chi_0(a) = e^{i\theta}, \theta$ un réel et $h = 4\beta e^{-i\theta} a$. Alors

$$||h|| = \chi_0(h) = \rho(h) = 4\beta.$$

Considérons dans $A^{\#}$ l'élément défini par:

$$b = \frac{e - \sqrt{e - h}}{2}.$$

b est bien défini, car $||h|| \leq 1$ et on a $b \in A$. Par ailleurs,

$$||b|| \leq \frac{1 - \sqrt{1 - ||h||}}{2} = \frac{1 - \sqrt{1 - 4\beta}}{2} = \alpha. \tag{1}$$

Or

$$\chi_0(b) = \chi_0(\frac{e - \sqrt{e-h}}{2}) = \frac{1 - \sqrt{1 - \chi_0(h)}}{2}.$$

Donc

$$\chi_0(b) = \frac{1 - \sqrt{1 - 4\beta}}{2} = \alpha. \qquad (2).$$

(1) et (2) montrent que $||b|| = \alpha$. D'autre part l'identité (*) donne $b^2 - b = h$, donc $||b^2 - b|| = \beta$.

Réciproquement

Supposons qu'il existe $b \in B$ tel que $||b|| = \alpha$ et $||b^2 - b|| = \beta$. Posons $C = \frac{b}{2\alpha}$ et $f = C^2 - C$, on a $||C|| = \frac{1}{2}$ donc $||C^2 - C|| \geq \frac{1}{4}$ car $||f|| \geq ||C|| - ||C||^2 = \frac{1}{4}$. Or

$$f = \frac{b^2 - b}{(2\alpha)^2} + (\frac{1}{(2\alpha)^2} - \frac{1}{2\alpha})b,$$

Donc

$$||f|| \leq \frac{\beta}{4\alpha} + (\frac{1}{2\alpha} - \frac{1}{4\alpha^2})\alpha = \frac{1}{4}.$$

D'où $||C^2 - C|| = \frac{1}{4}$.

Posons alors $L = -\frac{g(4f)}{2}$. Par la relation (*) on a $L^2 - L = C^2 - C$ donc $(L - C)(e - (L + C)) = 0$ où e désigne l'unité de $A^\#$.

On va distinguer deux cas, selon que $L = C$ où $L \neq C$.

1er cas : $L = C$

On va montrer que $\rho(4f) = 1$. Pour cela, on remarque que

$$||C|| = \frac{1}{2}||g(4f)|| \leq \frac{1}{2}\sum_{n=1}^{\infty} \frac{(2n-2)}{(n-1)!2^{n-1}} \frac{||4f||^n}{n!} \leq \frac{1}{2}(1 - \sqrt{1 - ||4f||}) = \frac{1}{2}.$$

D'autre part $||L|| = ||C|| = 1/2$, de sorte que

$$\sum_{n=1}^{\infty} \frac{(2n-2)}{(n-1)!2^{n-1}} \frac{||4f||^n}{n!} = 1.$$

En comparant avec

$$\sum_{n=1}^{\infty} \frac{(2n-2)}{(n-1)!2^{n-1}} \frac{1}{n!} = 1,$$

(Valeur en 1 de la fonction $u \longrightarrow -g(-u)$) on voit qu'on a $||(4f)^n|| = 1$ pour tout n , donc

$$\rho(4f) = \|4f\| = 1.$$

$2^{\text{éme}}$ **cas :** $L \neq C$

On va montrer que $\rho(L+C) = \|L+C\| = 1$. Comme $(L-C)(L+C) = L-C$ alors

$$(L-C)(e-(L+C)) = 0.$$

Donc $e-(L+C)$ n'est pas inversible dans $A^{\#}$. Soit χ un caractére de $A^{\#}$ tel que $\chi(e-(L+C)) = 0$ alors $\chi(L+C) = 1$. Comme

$$\|L+C\| \leq \|L\| + \|C\|, \quad \|L\| = \frac{1}{2} \quad et \quad \|C\| \leq \frac{1}{2},$$

on conclut alors que $1 \leq \rho(L+C) \leq \|L+C\| \leq 1$. Ce qui démontre le théoréme.

Corollaire 1 *Soit A une algébre de Banach commutative radicale. Alors pour tout élément x de A tel que $\|x\| \geq \alpha$ on a $\|x^2 - x\| > \beta$.*

Preuve Dire que A est une algébre de Banach commutative radicale c'est équivalent à dire que tout élément est de rayon spectral égal à zéro.
Soit x un élément de A tel que $\|x\| \geq \alpha$.
Si $\|x\| = \alpha$ alors $\|x^2 - x\| > \beta$, car sinon on aurait d'aprés le théoréme précédent un élément de rayon spectral égal à 1 et ceci est impossible dans une algébre de Banach commutative radicale.
Si $\|x\| > \alpha$, supposons que $\|x^2 - x\| = \beta$. Posons $f = x^2 - x$ et $k = \dfrac{e - \sqrt{e-4f}}{2}$, on a : $k^2 - k = x^2 - x$ c'est -à dire : $(k-x)(e-(k+x)) = 0$.
De plus $\|f\| = \beta \implies \|k\| \leq \dfrac{1-\sqrt{1-4\beta}}{2} = \alpha$, or $\|x\| > \alpha$ donc $k - x \neq 0$.
D'où $e-(k+x)$ n'est pas inversible dans $A^{\#}$. Ce qui est absurde, car A est radicale.

4 Condition pour qu'on ait $\inf_{\|x\|=\alpha} \|x^2 - x\| = \beta$

Thèoréme 3 *Soit B une algébre de Banach, s'il existe dans B une suite $(e_n)_n$ bornée en norme par 1 et un élément x non nul tel que: $\lim_{n\to\infty} e_n x = x$ ou $\lim_{n\to\infty} x e_n = x$ alors:*

$$\inf_{\|x\|=\alpha} \|x^2 - x\| = \beta.$$

Preuve Supposons que $\lim_{n\to\infty} e_n x = x$ (l'autre cas se traite de la même maniére).
Pour tout $n \geq 1$, on définit un élément v_n de B par $v_n = e - \sqrt{e + e_n}$. D'aprés [1, th 5-1] on a:

$$\lim_{n\to\infty} v_n x = x \quad et \quad \lim_{n\to\infty} \|v_n\| = \lim_{n\to\infty} \|e_n\| = 1$$

Posons, $U_n = \frac{\alpha v_n}{\|v_n\|}$ (on peut quitter à tronquer la suite, supposer $\|v_n\| \neq 0 \quad \forall n \in \mathbb{N}$) et posons $g = {U_n}^2 - U_n$, on a:

$$g = \frac{\alpha^2 {v_n}^2}{\|v_n\|^2} - \frac{\alpha v_n}{\|v_n\|}$$

$$= \frac{4\alpha^2}{\|v_n\|^2}(\frac{{v_n}^2}{4} - \frac{v_n}{2}) + (\frac{2\alpha^2}{\|v_n\|} - \alpha)\frac{v_n}{\|v_n\|} \, .$$

L'identité (*) montre que

$$g = -\frac{4\alpha^2}{\|v_n\|^2}(\frac{-e_n}{4}) + (\frac{2\alpha^2}{\|v_n\|} - \alpha)\frac{v_n}{\|v_n\|}.$$

Donc

$$\|g\| \leq \frac{\alpha^2}{\|v_n\|^2}\|e_n\| + |(\frac{2\alpha^2}{\|v_n\|} - \alpha)|.$$

Or si $\|y\| = \alpha$ on a $\|y^2 - y\| \geq \|y\| - \|y\|^2 \geq \alpha - \alpha^2$. Comme $\|U_n\| = \alpha$ on a

$$\beta \leq \|{U_n}^2 - U_n\| \leq \frac{\alpha^2}{\|v_n\|^2}\|e_n\| + |(\frac{2\alpha^2}{\|v_n\|} - \alpha)|. \qquad (1)$$

En faisant tendre n vers l'infini dans (1) on voit que:

$\lim_{n\to\infty} \|{U_n}^2 - U_n\| = \beta$. Donc $\inf_{\|y\|=\alpha} \|y^2 - y\| = \beta$.

Corollaire 2 *Soit B une algébre de Banach. Si B admet une (u.a.b.s) bornée par 1 alors :*

$$\inf_{\|y\|=\alpha} \|y^2 - y\| = \beta.$$

Preuve D'aprés la définition d'une (u.a.b.s) $(e_n)_n$ on a:
$\lim_{n\to\infty} e_n x = x$, $\forall x \in B$. Le thèoréme précédent démontre alors le corollaire.

Thèoréme 4 *Soit B une algébre de Banach, s'il existe une suite $(a_n)_n$ dans B telle que $\|a_n\| = 1$ et $\lim_{n\to\infty} \rho(a_n) = 1$, alors*

$$\inf_{\|y\|=\alpha} \|y^2 - y\| = \beta.$$

Preuve Pour tout $n \geq 1$, il existe un caractére χ_n de la sous-algébre commutative B_n de B engendrée par a_n tel que $\rho(a_n) = |\chi_n(a_n)|$.
Posons $\chi_n(a_n) = \rho(a_n)e^{i\theta_n}, \theta_n \in \mathbb{R}$ et $f_n = a_n e^{-i\theta_n}$, alors :

$$\rho(f_n) = \rho(a_n) = \chi_n(f_n) \quad et \quad \lim_{n\to\infty} \chi_n(f_n) = 1.$$

Comme $f_n = a_n e^{-i\theta_n}$ on a $||f_n|| = ||a_n|| = 1$. Posons alors $v_n = \dfrac{e - \sqrt{e - f_n}}{2}$.
Donc $\chi_n(v_n) = \dfrac{1 - \sqrt{1 - \chi_n(f_n)}}{2}$, on a alors: $\lim\limits_{n\to\infty} \chi_n(v_n) = 1$. Comme

$$||v_n|| \leq 1, \forall n \in \mathbb{N} \quad ona \quad \lim_{n\to\infty} \rho(v_n) = 1.$$

Posons, $U_n = \dfrac{\alpha v_n}{||v_n||}$ (on peut quitter à tronquer la suite, supposer $\rho(v_n) \neq 0 \quad \forall n \in \mathbb{N}$)
et posons $g_n = {U_n}^2 - U_n$, on a :

$$\begin{aligned} g_n &= \frac{\alpha^2 {v_n}^2}{||v_n||^2} - \frac{\alpha v_n}{||v_n||} \\ &= \frac{4\alpha^2}{||v_n||^2}(\frac{{v_n}^2}{4} - \frac{v_n}{2}) + (\frac{2\alpha^2}{||v_n||} - \alpha)\frac{v_n}{||v_n||}. \end{aligned}$$

L'identité (*) montre que

$$g_n = \frac{4\alpha^2}{||v_n||^2}\frac{f_n}{4} + (\frac{2\alpha^2}{||v_n||} - \alpha)\frac{v_n}{||v_n||}.$$

Donc

$$||g_n|| \geq \frac{\alpha^2}{||v_n||^2}||f_n|| + |(\frac{2\alpha^2}{||v_n||} - \alpha)|.$$

Or $\forall y \in A \quad ||y|| = \alpha$ on a $||y^2 - y|| \geq \beta$.

Comme $||U_n|| = \alpha$ on a $||{U_n}^2 - U_n|| \geq \beta$. D'oú

$$\beta \leq ||{U_n}^2 - U_n|| \leq \frac{\alpha^2}{||v_n||^2}||f_n|| + |\frac{2\alpha^2}{||v_n||} - \alpha|. \tag{1}$$

En faisant tendre n vers l'infini dans (1) on voit que:

$\lim\limits_{n\to\infty} ||{U_n}^2 - U_n|| = \beta$. Donc

$$\inf_{||y||=\alpha} ||y^2 - y|| = \beta.$$

5 Applications

5.1 Algébre de Banach ayant une (u.a.b.s) formée d'idempotents

Thèoréme 5 *Soit A un algébre de Banach. S'il existe dans A une (u.a.b.s) $(e_n)_{n\geq 1}$ et une constante λ tels que $\lambda \in]1/2, \frac{1+\sqrt{2}}{2}[$ et $||\lambda e_n - \lambda^2 e_n^2|| < \beta, \forall n \in \mathbb{N}^*$ avec $\beta \in]0, 1/4[$, alors A admet une (u.a.b.s) formée d'idempotents.*

Preuve

Comme $\lambda \in]1/2, \frac{1+\sqrt{2}}{2}[$ on a $|\lambda - \lambda^2| < 1/4$ et comme $(e_n)_n$ est une (u.a.b.s) alors il existe une constante K telle que $\|e_n\| \leq K$. Posons $f_n = \lambda e_n$, on a alors $\|f_n^2 - f_n\| < \beta$, donc on peut définir:

$$p(f_n) = \frac{e}{2} - (\frac{e}{2} - f_n)(e - 4f_n + 4f_n^2)^{\frac{-1}{2}}.$$

On a $p(f_n) \in A$, $p(f_n)$ est un idempotent pour tout n et

$$\begin{aligned}\|p(f_n)\| &\leq 1/2 + (1/2 + \lambda K)\|(e - 4f_n + 4f_n^2)^{\frac{-1}{2}}\| \\ &\leq 1/2 + (1/2 + \lambda K)(1 - 4\|f_n^2 - f_n\|)^{\frac{-1}{2}} \\ &\leq 1/2 + (1/2 + \lambda K)(1 - 4\beta)^{\frac{-1}{2}} = M.\end{aligned}$$

D'où la suite $[p(f_n)]_n$ est bornée. Or on a:

$$\begin{aligned}p(f_n).u - u &= (f_n - e/2)[e - 4(f_n - f_n^2)]^{\frac{-1}{2}}u - (\lambda - 1/2)[1 - 4(\lambda - \lambda^2)]^{\frac{-1}{2}}u \\ &= [e - 4(f_n - f_n^2)]^{\frac{-1}{2}}[(f_n - e/2)u - (\lambda - 1/2)u] \\ &\quad +(\lambda - 1/2)[(e - 4f_n + 4f_n^2)^{\frac{-1}{2}}u - (1 - 4(\lambda - \lambda^2))^{\frac{-1}{2}}u].\end{aligned}$$

Donc

$$\begin{aligned}\|p(f_n).u - u\| &\leq \|[e - 4(f_n - f_n^2)]^{\frac{-1}{2}}\|\|f_n u - \lambda u\| \\ &\quad +(\lambda - 1/2)[\|(e - 4f_n + 4f_n^2)]^{\frac{-1}{2}}u - (1 - 4(\lambda - \lambda^2)]^{\frac{-1}{2}})u\|] \\ &\leq \frac{\lambda}{\sqrt{1 - 4\beta}}\|e_n u - u\| + (\lambda - 1/2)[\|(e - 4f_n + 4f_n^2)]^{\frac{-1}{2}}u \\ &\quad -(1 - 4(\lambda - \lambda^2)]^{\frac{-1}{2}})u\|].\end{aligned}$$

On pose

$$[e - 4(f_n - f_n^2)]^{\frac{-1}{2}} = \sum_{p \geq 1} a_p(4f_n - 4f_n^2)^p = \sum_{p \geq 1} a_p(4\lambda e_n - 4\lambda^2 e_n^2)^p,$$

avec $a_p \geq 0$ et

$$[e - 4(\lambda_n - \lambda_n^2)]^{\frac{-1}{2}} = \sum_{p \geq 1} a_p(4\lambda - 4\lambda^2)^p.$$

Soit $\epsilon \geq 0$, il existe N_1, N_2 et N_3 tels que:

$$\|e_n u - u\| \leq \frac{\epsilon}{4\lambda}\sqrt{1 - 4\beta} \quad \forall n \geq N_1,$$

$$\sum_{p \geq N_2} a_p(4\beta)^p \leq \frac{\epsilon}{4(\lambda - 1/2)}$$

$$et \sum_{p \geq N_3} a_p|4\lambda - 4\lambda^2|^p < \frac{\epsilon}{4(\lambda - 1/2)}.$$

Or
$$\sum_{p\geq N_2} a_p\|4\lambda e_n - 4\lambda^2 e_n^2\|^p \leq \sum_{p\geq N_2} a_p(4\beta)^p \leq \frac{\epsilon}{4(\lambda-1/2)}.$$
Donc pour $N=\max\{N_2,N_3\}$ on a:
$$\|p(f_n).u-u\| \leq \frac{\lambda}{\sqrt{1-4\beta}}\|e_n u-u\| + (\lambda-1/2)[\sum_{p<N} a_p\|(4\lambda e_n - 4\lambda^2 e_n^2)^p u - (4\lambda-4\lambda^2)^p u\| + \sum_{p\geq N} a_p(4\beta)^p + \sum_{p\geq N} a_p|4\lambda-4\lambda^2|^p].$$

Or $(e_n)_n$ est une (u.a.b.s) alors $\forall p\in \mathbb{N}, (e_n^p)_n$ est une (u.a.b.s) et par la suite on aura:
$$\lim_{n\to\infty}\|(4\lambda e_n - 4\lambda^2 e_n^2)^p u - (4\lambda-4\lambda^2)^p u\| = 0 \quad \forall p.$$
Donc pour chaque p il va exister un entier M'_p tel que $\forall n\geq M'_p$ on a :
$$\|(4\lambda e_n - 4\lambda^2 e_n^2)^p u - (4\lambda-4\lambda^2)^p u\| \leq \frac{\epsilon}{4N(\lambda-1/2)}.$$
Posons $N' = \max_{p<N}\{M'_p\}$ et $N_0=\max\{N',N_1\}$ alors
$$\|p(f_n).u-u\| \leq \frac{\epsilon}{4}+\frac{\epsilon}{4}+\frac{\epsilon}{4}+\frac{\epsilon}{4}=\epsilon \quad \forall n\geq N_0.$$
D'où
$$\lim_{n\to\infty}\|p(f_n)u-u)\|=0.$$
Donc $[p(f_n)]_n$ est une (u.a.b.s) formée d'idempotents.

Remarque 3 Pour $\lambda=1$ on retrouve le thèoréme [3, Theorem 3.4].

Corollaire 3 *Soit A une algébre de Banach et soit $\hat{A}$ l'ensemble des caractéres de A. S'il existe dans A une (u.a.b.s) $(e_n)_{n\geq 1}$ et une constante λ tels que $\lambda\in]1/2,\frac{1+\sqrt{2}}{2}[$ et $\|\lambda^2 e_n^2 - \lambda e_n\| < \beta$ avec $\beta\in]0,1/4[$ alors il existe une suite de sous-ensembles (U_n), non vides , deux á deux disjoints ,ouverts et compacts , tels que $\cup_n U_n = \hat{A}$.*

Preuve

D'aprés le théoréme précédent on aura l'existence d'une (u.a.b.s) $(f_n)_n$ formée d'idempotents , alors
$$\liminf_{n\to\infty}\|f_n^2-f_n\| = 0 < \frac{1}{4}.$$
et d'aprés [3, Corollary 3.5] on a le résultat.

5.2 Homomorphismes d'une algébre de Banach a valeurs dans une algébre de Banach radicale

Soit A une algébre de Banach ayant une (u.a.b.s) formée d'idempotents , alors il est facile de voir que tout homomorphisme continu de A dans une algébre de Banach radicale B est nul. Dans ce paragraphe nous montrons que sous certaines conditions, ce résultat est valable dans les algébres de Banach ayant une (u.a.b.s) quelconque.

Thèoréme 6 *Soit A une algébre de Banach ayant une (u.a.b.s) $(e_n)_n$ telle qu'il existe $\lambda > \alpha$ et $\|(\lambda e_n)^2 - \lambda e_n\| \leq \beta$, pour n assez grand; Alors tout homomorphisme continu de A dans une algébre de Banach radicale est nul.*

Preuve

Soient B une algébre de Banach radicale et ϕ un homomorphisme continu de A dans B. Supposons que $\phi(A) \neq \{0\}$, et soit $S = \{x \in A / \|x\| \leq 1\}$ et $\phi(S)$ l'image de S par ϕ.

Comme ϕ est continue et S est borné alors $\phi(S)$ est borné, d'autre part $\phi(S)$ est stable par produit, alors en vertu de [4, lemme.7,p.21] , il existe une norme p sur B équivalente à sa norme initiale et tel que $p[\phi(s)] \leq 1$ pour tout élément s de S. Donc ϕ est une contraction de A dans B. Soit a un élément de A tel que $\phi(a) \neq 0$. Comme $(e_n)_n$ est une (u.a.b.s) de A on a

$$\lim_{n \to \infty} a e_n = a.$$

Comme ϕ est continu on a

$$\lim_{n \to \infty} \phi(a)\phi(e_n) = \phi(a),$$

par conséquent on a

$$\lim_{n \to \infty} inf\, p[\phi(e_n)] \geq 1.$$

Donc $p[\lambda\phi(e_n)] > \alpha$ pour n assez grand. Comme ϕ est une contraction on a

$$p[\lambda^2\phi(e_n{}^2) - \lambda\phi(e_n)] \leq \beta \quad \textit{pour n assez grand.}$$

D'aprés le corollaire1 du paragraphe3, B ne peut pas être dans ce cas radicale, ce qui est absurde. D'oú $\phi(A) = \{0\}$.

Exemple

Soit $C[0,1]$ l'algébre des fonctions continues sur $[0,1]$. Alors l'algébre

$$C_0[0,1] = \{f \in C[0,1] / f(0) = 0\}$$

admet une (u.a.b.s) $(e_n)_n$ vérifiant les hypothéses du théoréme, en effet:
Posons

$$e_n(t) = 1 \quad si \ \frac{1}{n} \leq t \leq 1,$$
$$e_n(t) = nt \quad si \ \ 0 \leq t \leq \frac{1}{n},$$

Pour tout $n \geq 1$, e_n est un élément de $C_0[0,1]$ et la suite $(e_n)_n$ converge uniformément vers 1 sur tout compact de]0,1], donc $(e_n)_n$ est (u.a.b.s) de $C_0[0,1]$. De plus on a

$$\|e_n^2 - e_n\| = 1/4 \quad \forall n \in \mathbb{N}^*.$$

Donc $C_0[0,1]$ vérifie les conditions du théoréme.

Corollaire 4 *Soit A une algébre de Banach vérifiant les hypothéses du thèoréme précédent. Si I un idéal fermé de A tel que l'algébre A/I est radicale alors $A = I$.*

Preuve Soit $\pi : A \longrightarrow A/I$ la surjection canonique de A sur A/I, π est un homomorphisme continu surjectif de A dans A/I. D'aprés le théoréme précédent on a $\pi(A) = A/I = 0$; donc $A = I$.

Remarque 4 Malheuresement, d'aprés la remaque 2 du paragrphe $2-1$ ce thèoréme ne s'applique pas à $L^1(\mathbb{R})$.

References

[1] M.Berkani: Inégalités dans les algébres de Banach.
Bull. Soc. Math. Belg, 42(1990),1,ser.B,105-116.

[2] M.Berkani: Inégalités et propriétés spectrales dans les algébres de Banach.
Thése de 3^{éme} cycle. Bordeaux. 1983.

[3] M.Berkani. J.Esterle: Banach algebras with left sequential approximate identities close to their square.
Operator theory: advances and application ,Birkhäuser Verlag , Basel,vol.24 , 1987 , p.29-40.

[4] F.F. Bonsal. J.Duncan: Numerical Range of Operators on Normed space and of elements of normed algebras.
LMS lectures Notes series 2 CAMBRIDGE University Press, 1971.

[5] J.Esterle: Quasimultipliers, Representation of H^∞ and the closed ideal problem for commutative Banach algebras.
Lecture notes in mathematics 975, Springer verlag. 65-162.

[6] W.Rudin: Analyse réelle et complexe.
Masson. Paris.1987.

Adresse:
Département de Mathématiques
Faculté des Sciences
Université Mohammed I
Oujda - Maroc.

P K JAIN, D P SINHA AND K K ARORA

On Grothendieck spaces with PRI

1. INTRODUCTION

A Banach space is called a *Grothendieck space* if weak and weak* convergence of sequences in the dual space coincide. Every separable Banach space is a Grothendieck space if and only if it is reflexive. However, if Q is an extremely disconnected compact Hausdorff space, then C(Q) is a non separable Grothendieck space which is not reflexive.

A long sequence of bounded linear projections on a Banach space E which satisfies certain conditions to be specified later is called a projectional resolution of the identity (in short, PRI) for E. In this paper, we prove that every PRI of a Grothendieck space is of a special type, called type I. Further, we show that every Grothendieck space belonging to a class $\mathcal{P}$ (to be defined in Section 2) is reflexive. However, not every Grothendieck space admits a PRI. Indeed, the Grothendieck space l^∞ provides such an example.

In Section 2, we give the necessary notations and definitions which include that of a PRI, a shrinking PRI and the class $\mathcal{P}$. In Section 3 we define a PRI of type I. Every shrinking PRI is of type I. However, we give an example to show that a PRI of type I may not be shrinking. Every PRI of a Grothendieck space is shown to be of type I. Finally, we prove that a Grothendieck space belonging to the class $\mathcal{P}$ is reflexive.

2. PRELIMINARIES

By E and E^* we shall denote a Banach space and its dual, respectively, and B_{E^*} denotes the dual unit ball of E^*. The cardinality of the smallest dense subset of E is denoted by dens E. The symbols α, β and μ are used to denote ordinal numbers in general and ω denotes the first infinite ordinal with cardinality $\aleph_0$. The brackets [ ] shall denote the closed linear span of the set they enclose.

Let μ be the first ordinal of cardinality dens E. A long sequence $\{p_\alpha\}_{\omega \leq \alpha \leq \mu}$ of linear projections on E is said to be a **projectional resolution of the identity** (in short, PRI) of E, if it satisfies the following

(2.1) $\|p_\alpha\| = 1,$ $(\omega \leq \alpha \leq \mu)$

(2.2) $p_\alpha \cdot p_\beta = p_\beta \cdot p_\alpha = p_\beta,$ $(\omega \leq \beta \leq \alpha \leq \mu)$

(2.3) $\text{dens } p_\alpha(E) \leq \text{card } \alpha,$ $(\omega \leq \alpha \leq \mu)$

(2.4) $p_\alpha(E) = \overline{\bigcup \{p_{\beta+1}(E) : \omega \leq \beta < \alpha\}}$ $(\omega \leq \alpha \leq \mu)$

(2.5) p_μ is the identity operator on E.

If a Banach space E has a PRI $\{p_\alpha\}_{\omega \leq \alpha \leq \mu}$, then for each $x \in E$ the map $\alpha \longrightarrow p_\alpha(x)$ from $[\omega, \mu]$ into E is continuous and if $\beta \leq \mu$ is a limit ordinal, then $p_\beta(x) = \lim_{\alpha < \beta} p_\alpha(x)$ $(x \in E)$ (see [6, p.646]). Every separable Banach space admits a PRI. In fact, Amir and Lindenstrauss [1] first proved the existence of a PRI in every weakly compactly generated (in short, WCG) Banach space.

A PRI $\{p_\alpha\}_{\omega \leq \alpha \leq \mu}$ of a Banach space E is said to be shrinking if the long sequence $\{p^*_\alpha\}_{\omega \leq \alpha \leq \mu}$ of projections on E^* forms a PRI of the dual space E^*. John and Zizler [5] were the first to establish the existence of a shrinking PRI in a WCG space with WCG dual. Several more results regarding the existence of PRI and shrinking PRI can be found in [2] and the references therein. For a PRI $\{p_\alpha\}_{\omega \leq \alpha \leq \mu}$ of E we write

$$V = \overline{\bigcup \left\{ p^*_{\alpha+1}(E^*) \ : \ \omega \leq \alpha < \mu \right\}} \ .$$

A Banach space E is said to belong to the class $\mathcal{P}$, if it has a PRI $\{p_\alpha\}_{\omega \leq \alpha \leq \mu}$ such that $p_\alpha(E)$ (respectively, $(p_{\alpha+1} - p_\alpha)(E)$) belongs to $\mathcal{P}$ for each $\omega \leq \alpha < \mu$. This class of Banach spaces was defined in [2, p.286], where it has also been shown that every Banach space in the class $\mathcal{P}$ admits an equivalent locally uniformly rotund norm. Recently, the authors [4] have proved that for every Banach space of the class $\mathcal{P}$, the dual unit ball contain a weak* sequentially precompact norming subset.

A system $(x_i, f_i)_{i \in I}$, where $\{x_i\}_{i \in I} \subset E$, $\{f_i\}_{i \in I} \subset E^*$ and I is an index set, is said to be an extended biorthogonal system if $f_i(x_j) = \delta_{ij}$ for all $i, j \in I$, where δ_{ij} is the Kronecker delta. An extended biorthogonal system $(x_i, f_i)_{i \in I}$ is said to be complete in E if $[x_i : i \in I] = E$, and total on E if $f_i(x) = 0$ for all $i \in I$ imply $x = 0$. A family $\{x_i\}_{i \in I}$ in E is said to be an extended Markuševič basis (in short, M-basis) if there is a corresponding family $\{f_i\}_{i \in I}$ in E^* such that $(x_i, f_i)_{i \in I}$ is an extended biorthogonal system which is both complete and total. In this case the family $\{f_i\}_{i \in I}$ is uniquely determined and is called the associated family of functionals (in short, a.f.f) to the extended M-basis $\{x_i\}_{i \in I}$. Further, if $\{f_i\}_{i \in I}$ forms an M-basis of the dual E^*, then $\{x_i\}_{i \in I}$ is called an extended shrinking M-basis. Every M-basis of a reflexive Banach space is shrinking.

3. MAIN RESULTS

To begin with we shall define a special type of PRI. We shall call a PRI $\{p_\alpha\}_{\omega \le \alpha \le \mu}$ of E to be of type I if $V = E^*$.

If $\{p_\alpha\}_{\omega \le \alpha \le \mu}$ is a shrinking PRI of E, then $\{p_\alpha^*\}_{\omega \le \alpha \le \mu}$ is a PRI of E^* whence, by condition (2.4) in the definition of a PRI, it follows that $\{p_\alpha\}_{\omega \le \alpha \le \mu}$ is of type I. However, a PRI of type I may not be a shrinking PRI. An example to this effect can be constructed as follows.

Let B be a non separable reflexive Banach space. Now, we consider the Banach space $E = l^1 \oplus_2 B$; i.e., if P and Q are bounded linear projections of E onto l^1 along B and onto B along l^1, respectively, then the norm on E is given by

$$\|x\| = \left(\|Px\|^2 + \|Qx\|^2 \right)^{1/2} .$$

Define a long sequence $\{q_\alpha\}_{\omega \le \alpha \le \mu}$ of linear operators on E by

$$q_\alpha = P + p_\alpha \cdot Q \qquad (\omega \le \alpha \le \mu) .$$

It is easy to see that $\{q_\alpha\}_{\omega \le \alpha \le \mu}$ is a PRI of E. Since $\{p_\alpha\}_{\omega \le \alpha \le \mu}$ is a shrinking PRI of B, $\{p^*_\alpha\}_{\omega \le \alpha \le \mu}$ is a PRI of B^*. It now follows that

$$\overline{\bigcup \left\{q^*_{\alpha+1}(E^*) \ : \ \omega \le \alpha < \mu\right\}} = (l^1)^* \oplus_2 B^* = E^* .$$

Thus, $\{q_\alpha\}_{\omega \le \alpha \le \mu}$ is a PRI of E of type I. However, this PRI is not shrinking since

$$\text{dens } q^*_\omega (E^*) = \text{dens } l^\infty > \text{card } \omega .$$

One may observe that every PRI of a reflexive Banach space is of type I. Indeed, for a Banach space with a PRI $\{p_\alpha\}_{\omega \le \alpha \le \mu}$, V is a norming subspace of E^* [4, Lemma 2] and hence coincides with E^*. In what follows, we prove a more general result.

Theorem 1. *Every PRI of a Grothendieck space is of type I.*

Proof. Let $\{p_\alpha\}_{\omega \le \alpha \le \mu}$ be a PRI of a Grothendieck space E, where μ is the first ordinal with cardinality dens E. Then, for any $f \in E^*$, we have

$$\lim_{\alpha < \mu} (p_\alpha^* . f)(x) = f(x) \qquad (x \in E)$$

We shall now split the discussion in two cases. Fix $f \in E^*$.

Case 1. If there exists a sequence of ordinals $\{\alpha_n\}$ with $\omega < \alpha_n < \alpha_{n+1} < \mu$ $(n = 1,2,\ldots)$ which is cofinal in μ; i.e., for any $\omega < \beta < \mu$, there is an integer n such that $\alpha_n \le \beta$. In this case

$$\lim_{n \to \infty} (p_{\alpha_n}^* . f)\ (x) = f(x) \qquad (x \in E) .$$

Since weak and weak* convergence of sequences in E^* coincide, we have

$$\lim_{n \to \infty} \phi(p_{\alpha_n}^* . f) = \phi(f) \qquad (\phi \in E^{**}) .$$

Thus $\{p_{\alpha_n}^* . f\}$ is weakly convergent to f. Consequently, if $\varepsilon > 0$ is given, and since $\{p_{\alpha_n}^*\}$ is a bounded sequence in norm, there is a finite linear combination (see [3], p.422)

$$f_\varepsilon = \sum_{\alpha \in \Lambda} t_\alpha (p_\alpha^* . f) ,$$

where Λ is a finite subset of $\{\alpha_n : n = 1,2,\ldots\}$ and t_α $(\alpha \in \Lambda)$ are scalars such that

$$\left\| f - f_\varepsilon \right\| < \frac{\varepsilon}{1 + \sup_{1 \le n < \infty} \left\| p^*_{\alpha_n} \right\|} = \frac{\varepsilon}{2} .$$

Let $\alpha_0 = \max \{\alpha : \alpha \in \Lambda\}$. Then, by the definition of PRI, $p^*_\alpha(f_\varepsilon) = f_\varepsilon$ $(\alpha \ge \alpha_0)$. Thus

$$\begin{aligned} \left\| f - p^*_\alpha . f \right\| &\le \left\| f - f_\varepsilon \right\| + \left\| p^*_\alpha . f_\varepsilon - p^*_\alpha . f \right\| \\ &\le \left\| f - f_\varepsilon \right\| \left(1 + \sup_{\omega \le \alpha < \mu} \left\| p^*_\alpha \right\| \right) \\ &< \varepsilon \qquad (\alpha \ge \alpha_0) . \end{aligned}$$

This proves that in the norm topology of E^*, we have

$$\lim_{\alpha < \mu} p^*_\alpha . f = f .$$

Case 2. If there does not exist any sequence of ordinals $\{\alpha_n\}$ with $\omega < \alpha_n < \alpha_{n+1} < \mu$ $(n = 1, 2, \ldots)$ which is cofinal in μ. In this case, for each n, there is an $\omega < \beta_n < \mu$ with $\beta_n < \beta_{n+1}$ such that

$$\left\| p^*_{\beta_n} . f - f \right\| < 1/n \qquad (n = 1, 2, \ldots) .$$

Since $\{\beta_n\}$ is not cofinal in μ, there exists an ordinal $\omega < \beta_0 < \mu$ such that $\beta_n \le \beta_0$ $(n = 1, 2, \ldots)$. It follows that $p^*_\alpha . f = f$ $(\alpha \ge \beta_0)$, whence

$$\lim_{\alpha < \mu} p^*_\alpha . f = f$$

in the norm topology of E^*.

Thus in both the cases we find that $\lim_{\alpha < \mu} p_\alpha^*.f = f$ $(f \in E^*)$ in the norm topology of E. Hence, $V = E^*$, i.e., $\{p_\alpha\}_{\omega \le \alpha \le \mu}$ is of type I.

Finally, we prove

Theorem 2. *A Grothendieck space in the class $\mathcal{P}$ is reflexive.*

Proof. Let E be a Grothendieck space in the class $\mathcal{P}$. We shall prove the result by transfinite induction on dens E. If E is separable, then it is reflexive. Let E be non separable. We set up an induction hypothesis as follows : We assume that every Grothendieck space F in the class $\mathcal{P}$ such that dens F < dens E is reflexive. Let μ be the first ordinal with cardinality dens E. Then, E has a PRI $\{p_\alpha\}_{\omega \le \alpha \le \mu}$ such that for each $\omega \le \alpha < \mu$, both $p_\alpha(E)$ and $(p_{\alpha+1}-p_\alpha)(E)$ are also in the class $\mathcal{P}$.

Firstly, we note that if $\{g_n\}$ is a sequence in E^*, $g \in E^*$ and q is any of the projections p_α or $p_{\alpha+1} - p_\alpha$ such that $\{q^*(g_n)\}$ is $\sigma(q^*(E^*), q(E))$-convergent to $q^*(g)$, then $\{q^*(g_n)\}$ is also $\sigma(E^*, E)$-convergent to $q^*(g)$. It follows that $\{q^*(g_n)\}$ is $\sigma(q^*(E^*), q^{**}(E^{**}))$-convergent to $q^*(g)$, whence q(E) is also a Grothendieck space.

Let $E_\omega = p_\omega(E)$ and $E_{\alpha+1} = (p_{\alpha+1} - p_\alpha)(E)$ for each $\omega \le \alpha < \mu$. Now, by the definition of a PRI, we have

$$\text{dens } E_\omega \le \text{card } \omega < \text{card } \mu = \text{dens } E ,$$

and

$$\text{dens } E_{\alpha+1} \le \text{card } (\alpha+1) < \text{card } \mu = \text{dens } E \quad (\omega \le \alpha < \mu).$$

Thus, by induction hypothesis, E_ω and $E_{\alpha+1}$ for $\omega \le \alpha < \mu$ are reflexive.

Let $\{x_i^{(\omega)}\}_{i \in I_\omega}$ be a shrinking M-basis of E_ω with a.f.f. $\{f_i^{(\omega)}\}_{i \in I_\omega}$ in E_ω^* and, for each $\omega \le \alpha < \mu$, $\{x_i^{(\alpha+1)}\}_{i \in I_{\alpha+1}}$ be an extended shrinking M-basis of $E_{\alpha+1}$ with the a.f.f. $\{f_i^{(\alpha+1)}\}_{i \in I_{\alpha+1}}$ in $E_{\alpha+1}^*$, where card $I_\omega = \aleph_0$ and card $I_{\alpha+1}$ = dens $E_{\alpha+1}$ for each $\omega \le \alpha < \mu$. It follows that $\left\{\{x_i^{(\omega)}\}_{i \in I_\omega} \cup \left(\cup_{\omega \le \alpha < \mu} \{x_i^{(\alpha+1)}\}_{i \in I_{\alpha+1}} \right) \right\}$ is an extended M-basis of E with the a.f.f. $\left\{\{g_i^{(\omega)}\}_{i \in I_\omega} \cup \left(\cup_{\omega \le \alpha < \mu} \{g_i^{(\alpha+1)}\}_{i \in I_{\alpha+1}} \right) \right\}$ in E^*, where $g_i^{\omega} = f_i^{\omega}.p_\omega$ $(i \in I_\omega)$ and $g_i^{\alpha+1} = f_i^{\alpha+1}.(p_{\alpha+1} - p_\alpha)$ $(i \in I_{\alpha+1}, \ \omega \le \alpha < \mu)$.

Note that since $\{x_i^{(\omega)}\}_{i \in I_\omega}$ and $\{x_i^{(\alpha+1)}\}_{i \in I_{\alpha+1}}$ for $\omega \le \alpha < \mu$ are shrinking M-bases, we have

$$[g_i^{\omega} : i \in I_\omega] = E_\omega^*, \quad [g_i^{\alpha+1} : i \in I_{\alpha+1}] = E_{\alpha+1}^* \quad (\omega \le \alpha < \mu),$$

so that

$$\begin{aligned} V &= [\, E_\omega^* \cup (\cup_{\omega \le \alpha < \mu} E_{\alpha+1}^*) \,] \\ &= [\, \{g_i^{\omega}\}_{i \in I_\omega} \cup (\cup_{\omega \le \alpha < \mu} \{g_i^{\alpha+1}\}_{i \in I_{\alpha+1}}) \,] . \end{aligned}$$

However, by Theorem 1, $\{p_\alpha\}_{\omega \le \alpha \le \mu}$ is a PRI of type I, whence $V = E^*$. Thus, $\left\{\{x_i^{(\omega)}\}_{i \in I_\omega} \cup (\cup_{\omega \le \alpha < \mu} \{x_i^{(\alpha+1)}\}_{i \in I_{\alpha+1}})\right\}$ is an extended shrinking M-basis of E with the a.f.f. $\left\{\{g_i^{(\omega)}\}_{i \in I_\omega} \cup (\cup_{\omega \le \alpha < \mu} \{g_i^{(\alpha+1)}\}_{i \in I_{\alpha+1}}) \right\}$. Let us denote this extended shrinking M-basis by $\{x_i\}_{i\in I}$ and its a.f.f. by $\{g_i\}_{i\in I}$.

Finally, let $\{h_n\}$ be a sequence in the dual unit ball B_{E^*}. Since $(x_i, f_i)_{i\in I}$ is an extended biorthogonal system and each h_n is the limit of a sequence in the linear span of $\{f_i\}_{i\in I}$, it follows that, for each n, $I_n = \{i\in I : h_n(x_i) \neq 0\}$ is countable. Let $J = \bigcup_{n=1}^{\infty} I_n = \{m_k\}_{k=1}^{\infty}$.

Now, there is a subsequence $\{h_{1,n}\}$ of $\{h_n\}$ such that $\lim_{n\to\infty} h_{1,n}(x_{m_1})$ exists. Again, there is a subsequence $\{h_{2,n}\}$ of $\{h_{1,n}\}$ such that $\lim_{n\to\infty} h_{2,n}(x_{m_2})$ exists. Continuing in this way, for each k, there is a subsequence $\{h_{k,n}\}$ of $\{h_{k-1,n}\}$ such that $\lim_{n\to\infty} h_{k,n}(x_{m_i})$ exist, for $1 \le i \le k$.

Let $f_n = h_{n,n}$ $(n = 1,2,\ldots)$. Then $\{f_n\}$ is a subsequence of $\{h_n\}$ such that $\lim_{n\to\infty} f_n(x_i)$ exists, for each $i \in J$. Since $f_n(x_i) = 0$ for each k and $i \in I\backslash J$, it follows that $\lim_{n\to\infty} f_n(x_i)$ exists for each $i \in I$.

Now $\{x_i\}_{i\in I}$ being complete in E and since $\{f_n\}$ is a bounded sequence, it follows that $\{f_n\}$ is weak* convergent to some $f \in E^*$. But E being a Grothendieck space, $\{f_n\}$ converges to f in the weak topology and so $f \in B_{E^*}$. Thus,

B_{E^*} is weakly sequentially compact, whence, by Eberlin-Shmulian theorem, E is reflexive.

Corollary. *A Grothendieck space is reflexive if and only if it is in the class $\mathcal{P}$.*

REFERENCES

1. **D. Amir and J. Lindenstrauss,** The structure of weakly compact sets in Banach spaces, Ann. of Math. 88 (1968), 35-46.

2. **R. Deville, G. Godefroy and V. Zizler,** Smoothness and renormings in Banach spaces, Longman Scientific and Technical, Harlow, 1993.

3. **N. Dunford and J. T. Schwartz,** Linear operators, Part I, General Theory, Interscience Publ., New York 1958.

4. **P. K. Jain, K. K. Arora and D. P. Sinha,** Weak* sequential compactness of dual unit balls of Banach spaces with PRI, Jour. Indian Math. Soc. 63 (1997) - in print.

5. **K. John and V. Zizler,** Projections in dual weakly compactly generated Banach spaces, Studia Math. 49 (1973), 41-50.

6. **I. Singer,** Bases in Banach spaces II, Springer-Verlag, Berlin - Heidelberg - New York, 1981.

Department of Mathematics,
University of Delhi,
Delhi-110007, INDIA

Department of Mathematics,
Dyal Singh College (University of Delhi),
Lodi Road, New Delhi-110003, INDIA

Department of Mathematics,
Rajdhani College (University of Delhi)
Ring Road, Raja Garden,
New Delhi-110015, INDIA

P MANCHANDA

Hensel's lemma for normed rings

The concept of residue class ring for a non-archimedean normed ring is introduced and an analogue of Hensel's lemma for a complete non-archimedean normed ring is established.

1. Introduction:

Hensel introduced an irreducibility criterion for polynomials over the field of p-adic numbers, which goes by the name of 'Hensel's lemma'. Later validity of Hensel's lemma was proved for complete non-archimedean absolute valued field and this is used for prolongation of non-archimedean absolute value of a complete field to any algebraic extension. An analogous of Hensel's lemma is not available in the literature for non-archimedean normed rings. It is due to non-availability of suitably defined residue class ring of a non-archimedean normed ring. Our aim in this paper is to define a suitable residue class ring and to prove Hensel's lemma for a non-archimedean complete normed ring. All rings considered in this paper are commutative with identity element 1.

2. Basic Definitions:

Let R be a ring and let T be a **ring topology** on R, that is, T is a topology on R making $(x,y) \to x-y$ and $(x,y) \to xy$ continuous from $R \times R$ to R. A subset A of R is **bounded** for T if given any neighbourhood U of zero, there exists a neighbourhood V of zero such that that $AV \subseteq U$. T is **locally bounded topology** on R if there exists a fundamental system of neighbourhoods of zero for T consisting of bounded sets. An element c of R is said to be a topological nilpotent for T if lim $c^n = 0$.

A **norm** N on a ring R is a function from R to the non-negative reals satisfying $N(x) = 0$ if and only if $x = 0, N(xy) \leq N(x)N(y)$ and $N(x-y) \leq N(x) + N(y)$ for all x, y in R. A norm on R is called **homogeneous** or **(spectral)** if $N(x^n) = [N(x)]^n$ for all x in R and for all positive integers n. A norm N on R is called an **absolute value** if $N(xy) = N(x)\ N(y)$ for all x, y in R. A norm (absolute value) on R is said to be **non-archimediean** if $N(x - y) \leq \max\{N(x), N(y)\}$. By a **normed ring (homogeneous normed ring, non-archimedean normed ring,** $\cdots$) we mean a pair (R, N) where R is a ring and N is a norm (homogeneous norm, non-archimedean norm, $\cdots$) defined on it.

If N is a norm on R, for each $\mathcal{E} > 0$ define $B_{\mathcal{E}} = \{x \in R : N(x) < \mathcal{E}\}$. Then $(B_{\mathcal{E}} : \mathcal{E} > 0)$ is a fundamental system of neighbourhoods of zero for a Hausdorff locally bounded ring topology T_N on R. Two norms on R are **equivalent** if they define the same topology. One can check that if N is a non-trivial norm on a field K (that is T_N is non-discrete), then a subset A of K is bounded for the topology T_N if and only if A is bounded in norm (that is, there exists a real number $\delta > 0$ such that $N(x) \leq \delta$ for all x in A).

3. Complete Rings and Completions:

Let (R, N) be a normed ring, a sequence $\{a_n\}$ of elements of R is said to be **convergent** if there is an element a in R such that given $\mathcal{E} > 0$ there is an integer $n = n_0(\mathcal{E})$ such that $N(a_n - a) < \mathcal{E}$ whenever $N \geq n_0$. In this case we say that sequence $\{a_n\}$ converges to a and we write lim $a_n = a$. A sequence $\{a_n\}$ of elements of R is called a **null sequence** if lim $a_n = 0$.

A sequence $\{a_n\}$ of elements of R is called a **Cauchy sequence** if for any $\mathcal{E} > 0$ there is an integer $n_0 = n_0(\mathcal{E})$ such that $N(a_m - a_n) \leq \mathcal{E}$ whenever $m, n \geq n_0$. A normed ring (R, N) is said to be complte if every Cauchy sequence of elements of R is a convergent sequence. A finite set $(N_1, N_2 \cdots, N_m)$ of norms on a ring R is said to be **independent** if to each of m elements $a_1, a_2 \cdots, a_m$ from R and to each real number $\mathcal{E} > 0$, there exists an element a in R such that $N_i(a - a_i) < \mathcal{E}$ for $i = 1, 2, \cdots$. From the Approximation Theorem (Theorem 5, Chapter 5[3]) it follows that a finite set of inequivalent nontrivial absolute values on a field is independent.

A normed ring (S, M) is said to be an extension of a normed ring (R, N) [in symbol $(R, N) < (S, M)$] if S is an over ring of R and $N(a) = M(a)$ for all a in R. If (R, N) and (S, M) are normed rings, then an isomorphism σ from R into S is said to be **analytic** if $N(a) = M(\sigma(a))$ for all a in R. A normed ring (S, M) is said to be a **completion** of normed ring (R, N) if

(S, M) is an exension of (R, N), (S, M) is complete and every element of S is the limit of some Cauchy sequence of elements of R.

The following theorems were proved by Kurt Mahler [2].

Theorem 1. Let (R, N) be a normed ring. Then

1. There exists a completion of (R, N)
2. If (S, M) is a completion of (R, N) and (S', M') is a complete normed ring which is an extension of (R, N) then there is an analytic R-isomorphism from S onto some subring of S' and in particular.
3. Any two completions of (R, N) are analytically R-isomorphic.
4. If the norm N on R is non-archimedean or homogeneous then so is the norm on the completion of R.

Theorem 2. If $(N_1, N_2 \cdots, N_m)$ is a finite set of independent norms on a ring R, then N defined by $N(x) = \max\{N_1(x), N_2(x), \cdots, N_m(x)\}$ for all x in R is a norm on R and

$$S = S_1 \oplus S_2 \oplus \cdots \oplus S_m$$

where S is completion of (R, N) and S_i is completion of (R, N_i) for $i = 1, 2, \cdots, m$.

4. Value sets and residue class rings:

Let (R, N) be a normed ring. These set $N(R) = \{N(a) : a \in R\}$ is called the value set of the normed ring (R, N). If normed ring (S, M) is an extension of (R, N), then $N(R) \subseteq M(S)$.

The proof of the following theorem is similar to the proof of classical case of absolute value (Theorem 5, Chapter 3[3]).

Theorem 3 If (R, N) is a non-archimedean normed ring and (S, M) is the completion of (R, N) then $N(R) = M(S)$.

Theorem 4 Let (R, N) be a non-archimedean normed ring. Define

$$P = \{a \in R : N(a^n) < 1 \text{ for some positive integer } n\}$$

and

$$D = \{a \in R : aP \subseteq P\}$$

Then D is a subring of R and P is an ideal of D.

Proof. Clearly $0 \in P$. Now we prove that P is closed under addition and subraction. Suppose a, $b \in P$. There exist positive integers n_1 and n_2 such that $N(a^{n_1}) < 1$ and $N(b^{n_2}) < 1$. Taking $n = n_1 n_2$, we get that $N(a^n) < 1$ and $N(b^n) < 1$.

Let $\delta = \max\{1, N(a) \cdots N(a^{n-1}), N(b), \cdots N(b^{n-1})\}$. Every positive integer m can be written uniquely as

$$m = nq + r, 0 \leq r \leq n-1,$$

Therefore

$$N(a^m) = N(a^{nq+r}) \leq N(a^n)^q N(a^r) \leq \delta.$$

Similarly

$$N(b^m) \leq \delta$$

Next we choose a positive ineger p such that

$$N(a^{np}) \leq N(a^n)^p < 1/\delta^2 \text{ and } N(b^{np}) \leq N(b^n)^p < 1/\delta^2$$

Taking $n_0 = np$, one can check that for every positive integer

$$t \geq n_0, N(a^t) < 1/\delta \text{ and } N(b^t) < 1/\delta$$

Now from non-archimedean property of separate norm N, we have

$$N(a \pm b)^{2n_0} = N(\sum_{i+j=2n_0} \pm a^i b^j) \leq \max_{i+j=2n_0} \left\{N(a^i)N(b^j)\right\} < \delta 1/\delta = 1$$

Therefore $a \pm b \in P$ and hence P is an additive subgroup of R.

To prove that D is a subring of R we note that $1 \in D$ and whenever $a, b \in P$

$$(a \pm b)P = aP \pm bP \subseteq P + P \subseteq P$$

and

$$(ab)P = a(bP) \subseteq aP \subseteq P,$$

which implies that $a \pm b \in D$ and $ab \in D$. Therefore D is a subring of R.

By definition of D, P is an ideal of D. This completes the proof of the theorem.

Definition 1. This ring D as defined in theorem 4 is called the ring of integers of the non-archimedean normed ring (R, N). The ideal P of D is precisely the set of topological nilpotent elements of (R, N). The quotient ring D/P is called the **residue class ring** of the non-archimedean normed ring (R, N) and is denoted by $\overline{R}_N$ or simply $\overline{R}$ if there is no confusion of norm N on R.

We now determine residue class rings of some particular normed rings.

Example 1. Let N be trivial norm on a ring R. Then

$$P = \{a \in R : N(a^n) < 1 \text{ for some positive integer } n\} = (0)$$

and

$$D = \{a \in R : aP \subseteq P\} = R$$

Therefore $\overline{R} = R/(O) = R$.

Example 2. If R is a ring and N_1 and N_2 are two equivalent norms on R then $\overline{R}_{N_1}$ is same as $\overline{R}_{N_2}$ because $N_1(a^n) < 1$ for some positive integer n if and only if $N_2(a^m) < 1$ for some positive integer m.

Example 3. Let K be a field with a non-archimedean absolute value N. Then D, the ring of integers is the usual valuated ring associated with N and P is the maximal ideal of D consisting of non-units of D. Therefore, D/P is the usual residue class field of K associated with N.

Example 4. Let N be a non-archimedean, homogeneous norm on a ring R which contains an invertible element d satisfying $0 < N(d) < 1$ (in particular if R is field and N is a nontrivial, non-archimedean, homogeneous norm on R then there always exists a non-zero element d in R which is of course invertible satisfying $0 < N(d) < 1$).

Then

$$P = \{a \in R : N(a^n) < 1 \text{ for some positive integer n }\}$$

$$= \{a \in R : N(a) < 1\}, \text{ because } N(a)^n = N(a^n) < 1 \text{ if and only if} N(a) < 1.$$

$$D = \{a \in R : aP \subseteq P\}$$

$$= \{a \in R : N(a) \leq 1)$$

Because $d \in P$ and for $a \in D, a^n \in D$ for every positive integer n, $a^n d \in P$. Now

$$N(a)^n = N(a^n) = N(a^n dd^{-1}) \leq N(a^n d)N(d^{-1}) < N(d^{-1}),$$

Therefore $N(a) < [N(d^{-1})]^{1/n}$ for every positive integer n. Taking limit $n \to \infty$, we get $N(a) \leq 1$.

Example 5. If for a prime p, N_p denotes the p-adic absolute value on Q-the field of rational numbers, then $\overline{Q}_{N_p} = Z/(p)$. More generally if $p_1, p_2 \cdots, p_m$ are distinct primes and

$$N(x) = \max\{N_{p_1}(x), N_{p_2}(x) \cdots, N_{p_m}(x)\} \text{ for all } x \in Q,$$

then $\overline{Q}_N = Z/(p_1 \cdots p_m)$.

Theorem 5. Let (R, N) be a non-archimedean normed ring and let P and D be defined as in Theorem 4. Then P is open as well as closed and Disclosed in R under the norm topology T_N.

Proof First we prove that P is an open subset of R. For this it is sufficient to prove that for $a \in P$ there exists a real number $\mathcal{E} > 0$ such that if $b \in R$ with $N(b-a) < \mathcal{E}$, then $b \in P$. Take an element $a \in P$. There exists a positive integer n such that $N(a^n) < 1$. Now

$$N(b^n) = N(\sum_{i+j=n} (b-a)^i a^j) \leq \max_{i+j=n} \left\{N(b-a)^i (a^j)\right\}$$

$$= \max_{i=1 \text{ to } n-1} \left\{N(a^n), \mathcal{E}^i N(a^{n-i})\right\} < 1 \text{ for suitable small } \mathcal{E} > 0.$$

Therefore P is an open subset of R.

Since P is a subgroup of R and $R - P = \underset{\substack{a \notin P \\ a \in R}}{U}(a+P)$

P is a closed subset of R.

We now prove that D is a closed subset of R. Let a be a limit point of D in R. There exists a sequence $\{a_n\}$ of elements of D such that $a_n \to a$. Let p be an arbitrary element of P. Now each $a_n p$ is an element of P and P is closed, therefore $a_p \in P$ which implies that $a \in D$. Therefore D is a closed subset of R.

Theorem 6. If (S, M) be the completion of a non-archimedean normed ring (R, N) then $\overline{R}_N \cong \overline{S}_M$.

Proof. Let

$$P = \{a \in R : N(a^n) < 1 \text{ for some positive integer n}\}$$

$$P' = \{a \in S : M(a^n) < 1 \text{ for some positive integer n}\}$$

$$D = \{a \in R : aP \subseteq P\} \text{ and}$$

$$D' = \left\{a \in S : aP' \subseteq P'\right\}$$

Therefore $\overline{R}_N = D/P$ and $\overline{S}_M = D'/P'$. It is clear that $P \subseteq P'$ and $R \cap P' = P$. We now prove that $D \subseteq D'$. Take an element a $\in D$. Let α be an arbitrary element of P'. Since S is completion of R there exists a cauchy sequence $\{a_n\}$ in R such that $\alpha = \lim a_n$. There exists a positive integer m such that $M(\alpha^m) < 1$. Since $\alpha^m = \lim a_n^m$, we can assume without loss of generality that $N(a_n^m) < 1$ for all $n \geq 1$, that is $a_n \in P$ for all $n \geq 1$. Since $a \in D$ and DP $\subseteq P, aa_n \in P \subseteq P'$. By Theorem 5, P' is a closed subset of S and $\{aa_n\}$ is a sequence from P' which converges to $a\alpha$ in S, we get that $a\,\alpha \in P'$. Therefore $D \subseteq D'$.

Consider the map $\psi : D/P \to D'/P'$ defined by $\psi(a+p) = a + p'$. One can easily check that ψ is well defined and is an isomorphism from $\overline{R}_N$ to $\overline{S}_M$. Finally we prove ψ is subjective. Take an element α in D'. There exists a Cauchy sequence $\{a_n\}$ in R such that $\alpha = \lim a_n$. By assumption, there exists a positive integer n_0 such that $M(a_n - \alpha) < 1$ for all $n \geq n_0$. Choose one such n and fix it. For any $x \in P, a_n x = (a_n - \alpha)x + \alpha x$. By definition of $D', \alpha\, x \in P'$. Since $a_n - \alpha \in P', (a_n - \alpha)x \in P'$. Therefore $a_n x \in P' \cap R = P$. This proves $a_n \in D$. Therefore $\alpha + P' = a_n + P'$ and hence ψ is subjective.

5. Hensel's lemma:

In this section our aim is to prove an analogous of Hensel's lemma for complete non-archimedean normed rings. For this first we collect some definitions and well known results to be used subsequently.

Definition 2. Let R be a topological ring, we say that the topology on R is **linear topology** if there exists a fundamental system of neighbourhoods of zero consisting of ideals.

Let $R[[x]]$ be the ring of formal power series in one indeterminate x over the ring R. We say that a formal power series $f = \sum_{n=0}^{\infty} a_n x^n$ of $R[[x]]$ is restricted (with respect to topology on R) if $\lim a_n = 0$. The collection of all restricted power series forms a subring of $R[[x]]$ which we denote by $R\{x\}$.

Definition 3. Let R be a ring. Two elements a and b of R are called **strongly relatively prime** if there exist elements p and q of R such that $pa + qb = 1$.

Theorem 7. (4.3 Theorem 1[1]). Let R be a complete Hausdorff linearly topologized commutative ring. Let A be a closed ideal of R whose elements are topological nilpotents. Let $B = R/A$ be the quotient topological ring and $\phi : R \to B$ be the canonical mapping. Let f be an element of $R\{x\}$, $\overline{p}$ be a monic polynomial in $B[x]$ and $\overline{q}$ be an element of $B\{x\}$ such that $\phi(f) = \overline{p}$ $\overline{q}$ and $\overline{p}$, $\overline{q}$ are strongly relatively prime in $B\{x\}$, then there exists a unique ordered pair (p, q) consisting of a monic polynomial $p \in R[x]$ and $q \in R\{x\}$ such that $f = pq, \phi(p) = \overline{p}, \phi(q) = \overline{q}$. Moreover p and q are strongly relatively prime in $R\{x\}$ and if f is a polynomial so is q.

Theorem 8. (Hensel's lemma). Let (R, N) be a complete non-archimedean normed ring, D/P be its residue class ring and D be bounded as a subset of itself with respect to induced topology on D. Let $f(x)$ be a polynomial in $D[x]$. Suppose $P(x)$ (the image of $f(x)$ in the residue class ring D/P), factors into two strongly relatively prime polynomials $\overline{p}(x)\overline{q}(x)$ one of which say $\overline{p}(x)$ is monic and has degree $i > 0$. Then there exist $p(x)$ and $q(x)$ in $D[x]$ satisfying the following
(1) $f(x) = p(x)q(x)$
(2) $p(x)$ is monic
(3) the image of $p(x)$ in the residue class ring is $\overline{p}(x)$ and the image of $q(x)$ in the residue class ring is $\overline{q}(x)$.

Proof. We know that

$$P = \{a \in R : N(a^n) < 1 \text{ for some positive integer n}\}$$

$$D = \{a \in R : aP \subseteq P\}$$

and $\overline{R} = D/P$.

Since R is complete with respect to N and D is closed subring of R,D is a complete topological ring with respect to the topology induced by the norm N on D,

Put $U_n = \{a \in D : N(a) < 1/n\}$.

One can easily check that U_n is a subring of D and moreover $\{U_n\}, n = 1, 2, 3 \cdots$ form a fundamental system of neighbourhoods of zero in D.

Let A_n be the ideal generated by U_n in D. We claim that $\{A_n\}, n = 1, 2, 3 \cdots$ form a fundamental system of neighbourhoods of zero in D. Take a neighbourhood U of zero in D. There exists a positive integer n such that

$U_n \subseteq U$. Since D is bounded there exists an integer m such that $DU_m \subseteq U_n$. Clearly the subgroup generated by $D\ U_m$ in D is the ideal A_m. Therefore $A_m \subseteq U_n$.

Now D is a complete Hausdorff linearly topologized ring and the set P of nilpotents of D is a closed ideal of D. Therefore using the above theorem we get the result.

REFERENCES

1. Bourbaki, N. Commutative Algebra, Addison-Weseley, 1972.
2. Mahler, K. Über Pseudo bewertungen I, Acta Math. 66(1935), 79–119.
3. McCarthy, *P.J.* Algebraic Extensions of fields, Chelsea, New York, 1976.

Department of Mathematics
Guru Nanak Dev University
Amritsar, India 143005.

A GUPTA AND B S KOMAL

Composite integral operators on $L^2(\mu)$

1. Introduction

In recent years considerable attention has been given to weighted composition operators on $L^p(\mu)$ spaces. So far as we know, composition operator found its first appearance in a paper of Schroeder [10] in 1871, in which it was asked to find a function f and a scalar α such that

$$(f \circ g)(z) = \alpha f(z)$$

for every z in the appropriate domain, whenever g is given. Koenigs [6] gave a solution of this problem in 1884. The composition operators in the beginning were known as substitution operators or translation operators or pointwise induced operators. They make contact with several branches of mathematical sciences and played an important role in the study of classical mechanics, stochastic processes, ergodic theory etc. Especially to worth mention here is the work of Koopman [7] on classical mechanics. Banach [1] used these operators to study the isometries of Banach spaces of continuous functions. The adjoint of the unilateral shift, the bilateral shift, and the weighted shifts on $l^2(Z)$ are well studied examples of weighted composition operators. Whitley [13] established the Lyubic's conjecture and generalised it to Volterra composition operator on $L^p[0,1]$. The Volterra composition operator is a composition of Volterra integral operator V and a composition operator C_Φ defined as

$$(V_\Phi f)(x) = (C_\Phi V f)(x) = (Vf) \circ \Phi(x) = \int_0^{\Phi(x)} f(t)\, dt\,,$$

for every $f \in L^p[0,1]$, where $\Phi : [0,1] \to [0,1]$ is a measurable function.

In this paper we are interested to explore some properties of a more general operator called a composite integral operator. For more details about integral operators and composition operators, we refer to Halmos and Sunder [5], Cowen and Barbara [4], Shapiro [11] and Singh and Manhas [12].

Let (X, S, μ) be a σ-finite measure space and let $\Phi : X \to X$ be a nonsingular measurable transformation ($\mu(E) = 0 \Rightarrow \mu\Phi^{-1}(E) = 0$). Then a composition transformation, for $1 \le p < \infty$, $C_\Phi : L^p(\mu) \to L^p(\mu)$ is defined by

$$C_\Phi f = f \circ \Phi\,,$$

for every $f \in L^p(\mu)$. In case C_Φ is continuous, we call it a composition operator induced by Φ. It is easy to see that C_Φ is a bounded operator if and only if

$$f_0 = \frac{d\mu\Phi^{-1}}{d\mu}\,,$$

the Radon-Nikodym derivative of the measure $\mu\Phi^{-1}$ with respect to the measure μ, is essentially bounded. A measurable function $K : X \times X \to C$ is called a kernel. A bounded linear operator

$$T_K : L^p(\mu) \to L^p(\mu)$$

defined by

$$(T_K f)(x) = \int K(x,y) f(y)\, d\mu(y)$$

is called a bounded integral operator. Given a kernel K and a non-singular measurable function $\Phi : X \to X$, the composite integral operator T_{K_Φ} induced by (K, Φ) is a bounded linear operator

$$T_{K_\Phi} : L^p(\mu) \to L^p(\mu)$$

defined by

$$(T_K f)(x) = \int K(x,y) f(\Phi(y))\, d\mu(y)\,.$$

For each $f \in L^p(X, S, \mu)$, $1 \le p < \infty$, there exists a unique $\Phi^{-1}(S)$-measurable function $E(f)$ such that $\int gf\, d\mu = \int gE(f)\, d\mu$ for every $\Phi^{-1}(S)$-measurable function g for which left integral exists. The function $E(f)$ is called conditional expectation of f with respect to the subalgebra $\Phi^{-1}(S)$. The operator E has the property that $E(hf) = hE(f)$ for all $\Phi^{-1}(S)$-measurable functions h for which hf is integrable. Set

$$K_\Phi(x,y) = (E(K(x,.)) \circ \Phi^{-1} f_0)(y) = E(K(x, \Phi^{-1}(y)) f_0(y)) = E(K_x(\Phi^{-1}(y)) f_0(y))\,,$$

where $E(f) \circ \Phi^{-1} = g$ if and only if $E(f) = g \circ \Phi$. For additional properties of the expectation operator, see Lambert [8].

It is natural to ask that if A is an integral operator, does it follow that A^* is an integral operator or is A^* an integral operator without A being an integral operator ? Many related questions were mentioned at 1977 conference at Long Beach. In this paper we prove that both a composition operator and its adjoint cannot be integral operators. A sufficient condition for the boundedness of an integral operator is obtained. It is often very difficult to compute the adjoint of a composition operator on $L^2(\mu)$ which for many years was unknown to us. Recently, Campbell and Jamison [2] obtained the adjoint of a weighted composition operator on $L^p(\mu)$. In this present paper, we obtain the adjoint of a composition integral operator. Hermitian and normal composite integral operators are characterized.

The symbol $B(L^2(\mu))$ denotes the Banach algebra of all bounded linear operators from $L^2(\mu)$ to $L^2(\mu)$.

2. Bounded composite integral operators

In this section we first give a sufficient condition for a composite integral operator to be bounded and then we prove that neither a composition operator nor its adjoint is an integral operator on $L^2(\mu)$ of a non-atomic σ-finite measure space.

Theorem 2.1: Let $K_\Phi \in L^2(\mu \times \mu)$. Then T_{K_Φ} is a bounded integral operator.

Proof: For every $f \in L^2(\mu)$,

$$\begin{aligned} |(T_{K_\Phi} f)(x)| &= \left|\int_X K(x,y) f(\Phi(y))\, d\mu(y)\right| = \left|\int_X (E(K(x,y)) \circ \Phi^{-1} f_0) f(y)\, d\mu(y)\right| \\ &\leq \int_X (E(|K(x,y)|) \circ \Phi^{-1}(y) f_0(y)) |f(y)|\, d\mu(y) \\ &= \int_X |K_\Phi(x,y)||f(y)|\, d\mu(y) \\ &\leq \left(\int_X |K_\Phi(x,y)|^2\, d\mu(y)\right)^{\frac{1}{2}} \left(\int_X |f(y)|^2\, d\mu(y)\right)^{\frac{1}{2}}. \end{aligned}$$

Therefore,

$$\int |(T_{K_\Phi}) f(x)|^2\, d\mu(x) \leq \int\int |K_\Phi(x,y)|^2\, d\mu(x)\, d\mu(y) \|f\|^2 = \|K_\Phi\|_2^2 \, \|f\|^2 < \infty .$$

Hence T_{K_Φ} is a bounded integral operator.

We know that if (X, S, μ) is an atomic measure space and if A is an integral operator on $L^2(\mu)$, then A^* is also an integral operator. But this result is no longer true, if the underlying measure space is non-atomic. In the following theorem, the measure space under consideration is non-atomic and σ-finite.

Theorem 2.2: Let $C_\Phi \in B(L^2(\mu))$. Then neither C_Φ nor its adjoint is an integral operator.

Proof: Suppose, if possible that C_Φ is an integral operator. Then there is a measurable function $K : X \times X \to C$ such that

$$(C_\Phi f)(x) = (T_K f)(x) = \int_X K(x,y) f(y)\, dy\,.$$

Let E and F be any two measurable sets in X. Then

$$\langle \chi_E, C_\Phi \chi_F \rangle = \mu(E \cap \Phi^{-1}(F)).$$

Therefore

$$\begin{aligned} \mu(E \cap \Phi^{-1}(F)) &= \int_X \chi_E(x)\chi_F(\Phi(x))\, dx = \int_X \chi_E(x) \int_X K(x,y)\chi_F(y)\, dy\, dx \\ &= \int_{E\times F} K(x,y)\, d(\mu \times \mu)\,, \end{aligned} \tag{1}$$

(by Fubini theorem). Since measure space is non-atomic σ-finite measure space, we can choose a sequence $\{F_n\}$ of measurable sets of finite measures such that

$$\mu(F_n) < \infty\,, \quad \bigcup_{n=1}^{\infty} F_n = X\,, \quad \text{and} \quad \mu(X \setminus \Phi^{-1}(F_n)) > 0\,.$$

Let E_n be any measurable set of a measure in $X \setminus \Phi^{-1}(F)$ with $\mu(E_n) > 0$. Then from (1)

$$\int_{E\times F} K(x,y)\, d(\mu \times \mu) = 0\,.$$

Since countably many such rectangles covers $X \times X$, we can conclude that $K = 0$ a.e., which is a contradiction. Hence C_Φ is not an integral operator.

Next, we prove that adjoint of a composition operator is also not an integral operator. For $E, F \in S$

$$\langle C_\Phi^* \chi_E, \chi_F \rangle = \langle \chi_E, C_\Phi \chi_F \rangle = \mu(E \cap \Phi^{-1}(F))$$

and

$$\begin{aligned} \langle C_\Phi \chi_E, \chi_F \rangle &= \int C_\Phi \chi_E(x) \chi_F(x)\, d\mu(x) = \int\!\!\int K(x,y)\chi_E(y)\, d\mu(y)\, \chi_F(x)\, d\mu(x) \\ &= \int_{E\times F} K(x,y)\, d(\mu \times \mu) = \mu(E \times F)\,. \end{aligned}$$

Again, by using the first part of the proof, we can see that C_Φ^* is not an integral operator. This completes the proof of the theorem.

Corollary 2.3: The identity operator on $L^2(X)$ is not an integral operator.

3. Hermitian and normal composition integral operators.

The Hermitian and normal integral operators are studied in this section. The adjoint of a composite integral operator is obtained.

For $K_\Phi \in L^2(\mu \times \mu)$, define $K_\Phi^*(x,y) = \overline{K_\Phi(y,x)}$. Let $A : L^2(\mu) \to L^2(\mu)$ be defined by

$$(Af)(x) = \int_X K_\Phi^*(x,y) f(y)\, d\mu(y).$$

Theorem 3.1: Let $T_{K_\Phi} \in B(L^2(\mu))$. Then $A = T_{K_\Phi}^*$.

Proof: Consider

$$\begin{aligned}
\langle f, T_{K_\Phi} f\rangle &= \int_X f(x)\overline{(T_{K_\Phi} g)(x)}\, d\mu(x) = \int_X f(x)\overline{\int_X K_\Phi(x,y) g(y)\, d\mu(y)}\, d\mu(x) \\
&= \int_X\int_X f(x)\overline{K_\Phi(x,y) g(y)}\, d\mu(x)\, d\mu(y) \qquad \text{(By Fubini theorem)} \\
&= \int_X\int_X f(x) K_\Phi^*(y,x)\, d\mu(x)\, \overline{g(y)}\, d\mu(y) = \int_X (Af)(y)\overline{g(y)}\, d\mu(y) \\
&= \langle Af, g\rangle
\end{aligned}$$

for every $f, g \in L^2(\mu)$. Hence $A = T_{K_\Phi}$. This completes the proof of the theorem.

Theorem 3.2: Let $T_{K_\Phi} \in B(L^2(\mu))$. Then T_{K_Φ} is Hermitian if and only if

$$K_\Phi(x,y) = \overline{K_\Phi(y,x)}.$$

Proof: Suppose first that the condition is true. For $f, g \in L^2(\mu)$,

$$\begin{aligned}
\langle f, T_{K_\Phi} g\rangle &= \int f(x)\overline{(T_{K_\Phi} g)(x)}\, d\mu(x) = \int f(x) \int \overline{K_\Phi(x,y) g(y)}\, d\mu(y)\, d\mu(x) \\
&= \int\int f(x) K_\Phi(y,x)\overline{g(y)}\, d\mu(x)\, d\mu(y) \\
&= \int\int K_\Phi(y,x) f(x)\, d\mu(x)\overline{g(y)}\, d\mu(y) \\
&= \int (T_{K_\Phi} f(y)\overline{g(y)}\, d\mu(y) = \langle T_{K_\Phi} f, g\rangle. \qquad (2)
\end{aligned}$$

Hence T_{K_Φ} is an Hermitian operator.

Conversely, if T_{K_Φ} is Hermitian, then for any measurable rectangle $E \times F$ of finite measure,

$$\langle T_{K_\Phi}\chi_E, \chi_F\rangle = \int_E\int_F (E(K_x)\circ\Phi^{-1})(y) f_0(y)\, d\mu(y)\, d\mu(x), \qquad (3)$$

$$\langle \chi_E, T_{K_\Phi}\chi_F\rangle = \int_E \int_F (E(K_y)\circ\Phi^{-1})(x) f_0(x)\, d\mu(x)\, d\mu(y)\,. \tag{4}$$

From (2) and (3) we can conclude that

$$(E(K_x)\circ\Phi^{-1})(y) f_0(y) = (E(K_y)\circ\Phi^{-1})(x) f_0(x)\,.$$

Thus $K_\Phi(x,y) = \overline{K_\Phi(y,x)}$. This completes the proof of the theorem.

Example 3.3: Let $X = [0,1]$ and μ be the Lebesgue measure. Define $\Phi : [0,1] \to [0,1]$ by $\Phi(x) = 1 - x$, $0 \le x \le 1$. Let $K : [0,1]\times[0,1] \to \mathbb{R}$ be defined by

$$K(x,y) = \frac{1}{1+|x-y|}\,.$$

Then T_{K_Φ} is an Hermitian operator, since $K_\Phi(x,y) = \overline{K_\Phi(y,x)}$.

Example 3.4: Let $X = \mathbb{N}$, the set of natural numbers and μ be the counting measure. Define $\Phi : \mathbb{N} \to \mathbb{N}$ as $\Phi(n) = n+1$ for every $n \in \mathbb{N}$, and

$$K(m,n) = \frac{1}{m+n+1}$$

for every $m, n \in \mathbb{N}$. Then

$$\begin{aligned}
\|T_{K_\Phi} f\|^2 &= \sum_{m=1}^{\infty} \left|T_{K_\Phi} f(m)\right|^2 = \sum_{m=1}^{\infty}\left|\sum_{n=1}^{\infty} K(m,n) f(\Phi(n))\right|^2 \\
&= \sum_{m=1}^{\infty}\left|\sum_{n=1}^{\infty} \frac{1}{m+n+1} f(n+1)\right|^2 \\
&= \sum_{m=1}^{\infty}\left|\sum_{n=1}^{\infty} \frac{(n+\frac{1}{2})^{\frac{1}{4}} f(n+1)}{(m+n+1)^{\frac{1}{2}}(n+\frac{1}{2})^{\frac{1}{4}}(m+n+1)^{\frac{1}{2}}}\right|^2 \\
&= \sum_{m=1}^{\infty}\left(\sum_{n=1}^{\infty} \frac{1}{(m+n+1)(n+\frac{1}{2})^{\frac{1}{2}}} \sum_{n=1}^{\infty} \frac{(n+\frac{1}{2})^{\frac{1}{2}}|f(n+1)|^2}{(m+n+1)}\right) \\
&\le \sum_{m=1}^{\infty} \frac{\pi}{(m+\frac{1}{2})^{\frac{1}{2}}} \sum_{n=1}^{\infty} \frac{(n+\frac{1}{2})^{\frac{1}{2}}|f(n+1)|^2}{m+n+1} \\
&= \pi \sum_{n=1}^{\infty}\left(n+\frac{1}{2}\right)^{\frac{1}{2}} |f(n+1)|^2 \sum_{m=1}^{\infty} \frac{1}{(m+\frac{1}{2})^{\frac{1}{2}}} \frac{1}{m+n+1} \\
&\le \pi \sum_{n=1}^{\infty}(n+\frac{1}{2})^{\frac{1}{2}}|f(n+1)|^2 \frac{\pi}{(n+\frac{1}{2})^{\frac{1}{2}}} = \pi^2 \sum_{n=1}^{\infty}|f(n+1)|^2 \le \pi^2\sum_{n=1}^{\infty}|f(n)|^2 \\
&= \pi^2\|f\|^2
\end{aligned}$$

Thus $\|T_{K_\Phi} f\|^2 \leq \pi^2 \|f\|^2$ or $\|T_{K_\Phi}\| \leq \pi$. Thus T_{K_Φ} is bounded. But T_{K_Φ} is not Hermitian, since

$$K_\Phi(2,1) = \sum_{p \in \Phi^{-1}(1)} K(2,p) = 0$$

and

$$K_\Phi(1,2) = \sum_{p \in \Phi^{-1}(2)} K(1,p) = K(1,1) \neq 0 .$$

Theorem 3.5: Let $T_{K_\Phi} \in B(L^2(\mu))$. Then T_{K_Φ} is normal if and only if

$$\int K_\Phi^*(x,y) K_\Phi(y,z)\, dy = \int K_\Phi(x,y) K_\Phi^*(y,z)\, dy .$$

Proof: If the condition is true, then it is easy to prove that T_{K_Φ} is normal. Conversely, suppose T_{K_Φ} is normal. Then for any measurable rectangle $E \times F$ of finite measure, we have

$$\begin{aligned}
\langle T_{K_\Phi}^* T_{K_\Phi} \chi_E, \chi_F \rangle &= \int (T_{K_\Phi}^* (T_{K_\Phi} \chi_E))(x) \chi_F(x)\, d\mu(x) \\
&= \int \int K_\Phi^*(x,y) (T_{K_\Phi} \chi_E)(y)\, d\mu(y) \chi_F(x)\, d\mu(x) \\
&= \int \int K_\Phi^*(x,y) \int K_\Phi(y,z) \chi_E(z)\, d\mu(z)\, d\mu(y) \chi_F(x)\, d\mu(x) \\
&= \int_{E \times F} \int K_\Phi^*(x,y) K_\Phi(y,z)\, dy\, d(\mu \times \mu)
\end{aligned}$$

and similarly

$$\langle T_{K_\Phi} T_{K_\Phi}^* \chi_E, \chi_F \rangle = \int_{E \times F} \int K_\Phi(x,y) K_\Phi^*(y,z)\, dy\, d(\mu \times \mu)$$

Hence it follows that

$$\int K_\Phi{}^*(x,y) K_\Phi(y,z)\, dy = \int K_\Phi(x,y) K_\Phi^*(y,z)\, dy$$

This proves the result.

References

[1] Banach, S., (1932): Theorie des operations linéaires. Monografic Matematyczne Warsaw.

[2] Campbell, J.T. and Jamison, J.E., (1990): On some classes of weighted composition operators, Glasgow Math. J. 32, 87 - 94.

[3] Campbell, J.T. and Horner, W.E., (1994): Localising and seminormal composition operators on L^2, Proc. of Royal Soc. of Edinburgh, 124(A), 301 - 316.

[4] Cowen, C.C. and Maccluer, B.D., (1995): Composition operators on spaces of analytic functions, Studies in Advanced mathematics, CRC Press, New York, London, Tokyo.

[5] Halmos, P.R. and Sunder, V.S., (1978): Bounded integral operators on L^2-spaces, Springer-Verlag, New York.

[6] Koenigs, G., (1884): Recherches sur le integrals de Certcuns equations Fontionalles, Anneles Sci. de l'ECO Normale Superieur, 1, 3 - 41.

[7] Koopman, B.O. and Neumann, J. von, (1932): Dynamical systems of continuous spectra, Proc. Nat. Acad. Sci. U.S.A., 18, 255 - 263.

[8] Lambert, A., (1986): Hyponormal composition operators, Bull. London Math. Soc., 18, 395 - 400.

[9] Nordgren, E.A., (1978): Composition operators on Hilbert spaces, Lecture Notes in Mathematics 693, Springer-Verlag, New York, 37 - 68.

[10] Schroeder, E., (1871): Über Iteratierte Funcktionen, Math. Anal., 3, 296 - 322.

[11] Shapiro, J.H., (1993): Composition operators and classical function theory, Springer-Verlag.

[12] Singh, R.K., and Manhas, J.S., (1993): Composition operators on function spaces, North-Holland, Mathematics Studies no. 178.

[13] Whitley, R., (1987): The spectrum of a Volterra composition operator, Integral equations and operator theory, 10, 146 - 149.

Dept. of Mathematics
University of Jammu
Jammu - 180004
India

Part B

Variational Methods

S KESAVAN

Homogenization and optimal control

1 Introduction

The methods of homogenization permit us to study the global behaviour of bodies of heterogeneous materials with a lot of heterogeneities (like fibres, inclusions, perforations etc.) whose dimensions are small compared to the dimensions of the body. Given the microscopic structure of the body, we attempt to describe its effective or macroscopic behaviour. Let us look at an example.

Example 1.1 Consider a beam made of a homogeneous material with uniform cross section Ω (cf. fig.1.1(a)). To study its torsional rigidity, we need to solve the following boundary value problem.

$$\left.\begin{array}{rcll} -\alpha\Delta u & = & 2 & \text{in } \Omega \\ u & = & 0 & \text{on } \partial\Omega \end{array}\right\} \tag{1.1}$$

where $\alpha > 0$ is a constant depending on the elastic properties of the material. Let us now consider a single fibre of a different material introduced along the length of the beam so that the cross section looks as in fig.1.1(b). Then the problem (1.1) is replaced by

$$\left.\begin{array}{rcll} -\text{div}(A\nabla u) & = & 2 & \text{in } \Omega \\ u & = & 0 & \text{on } \partial\Omega \end{array}\right\} \tag{1.2}$$

where $A(x) = (a_{ij}(x))$ is the matrix $a(x)I$ with

$$a(x) = \left\{\begin{array}{ll} \alpha & \text{for material 1 (unshaded region)} \\ \beta & \text{for material 2 (shaded region).} \end{array}\right. \tag{1.3}$$

Let us now progressively increase the number of fibres while reducing their thickness (cf. fig.1.1(c) and (d)). Again the boundary value problem is given by equations (1.2)-(1.3).

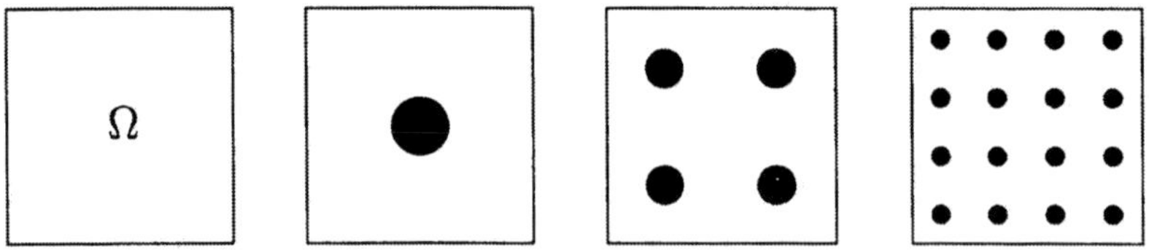

(a) (b) (c) (d)

Figure 1.1

The boundary value problem (1.2)-(1.3) is mathematically a well-posed second order elliptic boundary value problem and has a unique solution. However the coefficients $a_{ij}(x)$ of the matrix A oscillate rapidly when the number of fibres becomes very large. Thus, when we try to solve the problem numerically, we need to use a very fine grid or mesh to get a good approximation of the solution, and this is very expensive. One would like to "average out " the heterogeneities and study an "equivalent" homogeneous, albeit fictitious, material whose behaviour reflects that of the original material, when the number of fibres is very large.□

Such problems are studied by the methods of homogenization.

Let $\Omega \subset \mathbb{R}^n$ be a domain. We say that an $n \times n$ matrix $A = (a_{ij})$ belongs to the class $\mathcal{M}(\alpha_m, \alpha_M, \Omega)$ if, for all $\xi = (\xi_i) \in \mathbb{R}^n$,

$$\alpha_m|\xi|^2 \le a_{ij}\xi_i\xi_j \le \alpha_M|\xi|^2. \tag{1.4}$$

(Here, and throughout the sequel, we use the convention of summation over repeated indices.) We can now formulate the problem described in Example 1.1 in the following manner. Let $f \in H^{-1}(\Omega)$ be given. Let $\varepsilon > 0$ be a parameter which tends to zero. Assume that we have a family of matrices $A_\varepsilon = (a^\varepsilon_{ij}(x))$ in $\mathcal{M}(\alpha_m, \alpha_M, \Omega)$. Let $w_\varepsilon \in H^1_o(\Omega)$ be the (weak) solution of the problem

$$\left.\begin{aligned} \text{-div}(A_\varepsilon \nabla w_\varepsilon) &= f \quad \text{in } \Omega \\ w_\varepsilon &= 0 \quad \text{on } \partial\Omega. \end{aligned}\right\} \tag{1.5}$$

It is easy to see that the family $\{w_\varepsilon\}$ is bounded in $H^1_o(\Omega)$ and so will contain a weakly convergent subsequence. Let $w_\varepsilon \rightharpoonup w_o$ in $H^1_o(\Omega)$. The problem then is to find a matrix A_o such that w_o is the weak solution of

$$\left.\begin{array}{rcl} -\text{div}(A_o \nabla w_o) & = & f \quad \text{in } \Omega \\ w_o & = & 0 \quad \text{on } \partial\Omega. \end{array}\right\} \tag{1.6}$$

The "limit" matrix A_o is called the homogenized matrix and should be independent of f.

Let us look at another example.

Example 1.2 Let $\Omega = (0,1) \subset I\!R$. Let $a_\varepsilon(x)$ be functions such that

$$0 \; < \; \alpha \; \leq \; a_\varepsilon(x) \; \leq \; \beta.$$

Consider the problem

$$\left.\begin{array}{rcll} -(a_\varepsilon {w_\varepsilon}')' & = & f & \text{in } \Omega \\ w_\varepsilon(0) & = & w_\varepsilon(1) & = 0 \end{array}\right\} \tag{1.7}$$

where $f \in L^2(\Omega)$.

Since the functions a_ε are uniformly bounded in $L^\infty(\Omega)$, there exists a subsequence which converges in $L^\infty(\Omega)$ weak-*. It is tempting to conclude that the limit problem will be the analogue of (1.7) with a_ε replaced by that limit. However, this is not the case. It can be proved that (cf. Bensoussan *et al* [2] or Murat [12]) if $w_\varepsilon \rightharpoonup w_o$ in $H_o^1(\Omega)$ weakly, then

$$\left.\begin{array}{rcll} -(a_o w_o')' & = & f & \text{in } \Omega \\ w_o(0) & = & w_o(1) & = 0 \end{array}\right\} \tag{1.8}$$

where

$$\frac{1}{a_\varepsilon} \rightharpoonup \frac{1}{a_o}$$

in $L^\infty(\Omega)$ weak-*. Note that the limit of the reciprocal of a_ε is *not* the reciprocal of the limit of a_ε in general.□

In the sequel, we will briefly present the different techniques used in homogenization. We now describe the optimal control problem which is the object of study in this paper.

Let $\Omega \subset I\!R^n$ be a bounded domain. Let $A_\varepsilon \in \mathcal{M}(\alpha_m, \alpha_M, \Omega)$ and $B_\varepsilon \in \mathcal{M}(\beta_m, \beta_M, \Omega)$. Let $U_{ad} \subset L^2(\Omega)$ be a closed convex set, which will be the set of admissible controls. For $\theta \in U_{ad}$, we define the state u_ε as the weak solution in $H_o^1(\Omega)$ of the problem

$$\left.\begin{array}{rcll} -\operatorname{div}(A_\varepsilon \nabla u_\varepsilon) & = & f+\theta & \text{in } \Omega \\ u_\varepsilon & = & 0 & \text{on } \partial\Omega, \end{array}\right\} \tag{1.9}$$

where $f \in L^2(\Omega)$ is fixed. Then the associated cost functional is defined by

$$J_\varepsilon(\theta) = \frac{1}{2}\int_\Omega B_\varepsilon \nabla u_\varepsilon \nabla u_\varepsilon dx + \frac{N}{2}\int_\Omega {\theta_\varepsilon}^2 dx \tag{1.10}$$

where $N > 0$ is a fixed constant. This is a standard problem in the theory of optimal control and (cf. Lions [11]) there exists a unique optimal control ${\theta_\varepsilon}^* \in U_{ad}$ which satisfies

$$J_\varepsilon({\theta_\varepsilon}^*) = \min_{\theta \in U_{ad}} J_\varepsilon(\theta).$$

It is easy to see that $\{{\theta_\varepsilon}^*\}$ is uniformly bounded in $L^2(\Omega)$. Hence, for a subsequence, ${\theta_\varepsilon}^* \rightharpoonup \theta_o^*$ weakly in $L^2(\Omega)$. We show that θ_o^* can be characterized as the optimal control of a problem analogous to (1.9)- (1.10) and identify the corresponding "limit" matrices.

There is an interesting by-product which comes out of this analysis. Let us go back to the original problem of homogenizing equation (1.5). We know that $\{\int_\Omega |\nabla w_\varepsilon|^2 dx\}$ is uniformly bounded and hence converges for a subsequence. If $w_\varepsilon \rightharpoonup w_o$, we would like to express the limit as a function of w_o. We know that the limit is *not* $\int_\Omega |\nabla w_o|^2 dx$, since we know that w_ε does not converge to w_o strongly in $H_o^1(\Omega)$.

More generally, if $B_\varepsilon \in \mathcal{M}(\beta_m, \beta_M, \Omega)$, we could ask about the limit of

$$\int_\Omega B_\varepsilon \nabla w_\varepsilon \nabla w_\varepsilon dx.$$

This question is also answered by our study. (Note that if $B_\varepsilon = A_\varepsilon$, then from the homogenization of (1.5), we can prove that

$$\int_\Omega A_\varepsilon \nabla w_\varepsilon \nabla w_\varepsilon dx \rightarrow \int_\Omega A_o \nabla w_o \nabla w_o dx$$

where A_o is the homogenized matrix. The result is different if $B_\varepsilon \neq A_\varepsilon$.)

All the problems mentioned hitherto could also be studied over perforated domains;imagine, in figure 1.1, that the shaded regions are holes. Thus we have an additional mathematical difficulty, *viz.*, the $\{w_\varepsilon\}$ are now defined over different domains. A rich theory exists for this class of problems as well.

2 Homogenization Techniques

There are several techniques in homogenization theory. The mathematically important ones amongst them are the methods of multiple scale expansions, two - scale convergence, the energy method, H-convergence, G-convergence and Γ-convergence.

The method of multiple scale expansions and that of two - scale convergence are specially tailored for the case of periodically oscillating coefficients. Let us assume that $Y(=[0,1]^n$, for example) is a reference cell in $I\!R^n$ and that $a(y)$ is a Y-periodic function. Now we partition $I\!R^n$ into cells of size ε by translating the cell εY. This gives a partition of Ω into ε-cells. The function $a(y)$ now gives the function $a(x/\varepsilon)$ on the cell εY and can be translated to each of the other cells. Thus we get a periodically oscillating function, with period ε, on $I\!R^n$. We continue to denote this function by $a(x/\varepsilon)$.

Example 2.1 Let $Y = \Omega = (0,1) \subset I\!R$. Let $a(y)$ be a function on Y such that $a(0) = a(1)$. We now partition Ω into N equal subintervals and set $\varepsilon = 1/N$. Let $x_k = k\varepsilon, 0 \leq \varepsilon \leq N+1$. Then if $x \in [x_k, x_{k+1}]$, we have

$$a(x/\varepsilon) \;=\; a((x-x_k)/\varepsilon)$$

for $0 \leq k \leq N$. If $a_\varepsilon(x) = a(x/\varepsilon)$, then it is easy to verify that

$$a_\varepsilon \rightharpoonup \frac{1}{|Y|}\int_Y a(y)dy$$

in $L^\infty(\Omega)$ weak-*, where $|Y|$ denotes the (Lebesgue) measure of the reference cell Y.□

If $A = (a_{ij})$ has Y-periodic coefficients and if $A_\varepsilon = A(x/\varepsilon)$, then the method of multiple scale expansions consists in expanding the solution w_ε of (1.5) as a formal power series of the form

$$w_\varepsilon = w_o(x, \frac{x}{\varepsilon}) + \varepsilon w_1(x, \frac{x}{\varepsilon}) + \varepsilon^2 w_2(x, \frac{x}{\varepsilon}) + \ldots \tag{2.1}$$

where the functions $w_i(x,y)$ are all Y-periodic in the variable y. It can then be shown that w_o depends only on x; by substituting this form of w_ε in the equation and identifying the different powers of ε, we can deduce the equation satisfied by w_o which will be the homogenized equation.

Let $X_k^\varepsilon(x) = X_k(x/\varepsilon)$ for $1 \leq k \leq n$, where $X_k(y)$ is a Y-periodic function defined on the reference cell Y as the solution of the following problem:

$$\left.\begin{array}{rcl} -\text{div}(A\nabla(X_k - y_k)) & = & 0 \quad \text{in } Y, \\ \int_Y X_k(y)dy & = & 0 \quad X_k \text{ is } Y\text{-periodic.} \end{array}\right\} \tag{2.2}$$

Here, y_k is the k-th coordinate function. The homogenized matrix $A_o = (a_{ij}^o)$ is given by

$$a_{ij}^o = \frac{1}{|Y|}\int_Y a_{kl}(y)\frac{\partial}{\partial y_l}(X_j - y_j)\frac{\partial}{\partial y_k}(X_i - y_i)dy \tag{2.3}$$

While the above method gives correct results and helps us to guess the form of A_o, it nevertheless remains a purely formal one. To justify it, we need to use the energy method of L. Tartar. More recently, Allaire [1] and Nguetseng [13] have developed the method of two - scale convergence which can be used to justify the method of multiple scale expansions in a simple way.

The energy method of L. Tartar (cf. Murat [12]) is more general and applies to larger classes of problems. It is subsumed in the theory of H-convergence. Given a family $A_\varepsilon \in \mathcal{M}(\alpha_m, \alpha_M, \Omega)$, we say that $\{A_\varepsilon\}$ H-converges to A_o if the following holds: given $f_\varepsilon \in H^{-1}(\Omega)$ such that $f_\varepsilon \to f$ strongly in $H^{-1}(\Omega)$, and $w_\varepsilon \rightharpoonup w_o$ weakly in $H_o^1(\Omega)$, where $w_\varepsilon \in H_o^1(\Omega)$ is the weak solution of

$$\left.\begin{array}{rcl} -\,\text{div}(A_\varepsilon\nabla w_\varepsilon) & = & f_\varepsilon \quad \text{in } \Omega \\ w_\varepsilon & = & 0 \quad \text{on } \partial\Omega, \end{array}\right\} \tag{2.4}$$

then

$$A_\varepsilon\nabla w_\varepsilon \rightharpoonup A_o\nabla w_o$$

in $(L^2(\Omega))^n$ weakly and consequently

$$-\,\text{div}(A_o\nabla w_o) = f \quad \text{in } \Omega.$$

The main result is that any family of matrices $A_\varepsilon \in \mathcal{M}(\alpha_m, \alpha_M, \Omega)$ contains a H- convergent subsequence. The coefficients of the homogenized matrix can be described using some standard test functions which are generalizations of the functions X_k defined above in the periodic case. Further

$A_o \in \mathcal{M}(\alpha_m, \alpha_M^2/\alpha_m, \Omega)$. If, in addition, the A_ε are all symmetric, so is A_o and, in this case, $A_o \in \mathcal{M}(\alpha_m, \alpha_M, \Omega)$.

The notions of G- convergence and Γ- convergence deal with the convergence of energy integrals associated with elliptic problems. For a description of these methods, see Dal Maso [6], De Giorgi [7] and Spagnolo [15].

The study of homogenization problems posed over perforated domains, where the perforations increase in number and decrease in size, has also given rise to a vast amount of literature. See, in particular, Cioranescu and Saint Jean Paulin [5] and Cioranescu and Murat [4].

Homogenization has been applied to study the asymptotic behaviour of a host of other problems. For an introduction and other references, the interested reader is referred to Bensoussan *et al* [2] and to Sanchez-Palencia [14]. Several references can also be found in Allaire [1].

3 The Optimal Control Problem

Let us now consider the optimal control problem given by (1.9) - (1.10). A standard technique in the study of such problems is to introduce the adjoint state. Thus, for $f \in L^2(\Omega)$, and for $\theta \in U_{ad}$, we consider the pair $(u_\varepsilon, p_\varepsilon) \in H_o^1(\Omega) \times H_o^1(\Omega)$ as the weak solution of the system

$$\left.\begin{array}{rcll} -\operatorname{div}(A_\varepsilon \nabla u_\varepsilon) & = & f+\theta & \text{in } \Omega \\ \operatorname{div}({}^tA_\varepsilon \nabla p_\varepsilon - B_\varepsilon \nabla u_\varepsilon) & = & 0 & \text{in } \Omega \\ u_\varepsilon = p_\varepsilon & = & 0 & \text{on } \partial\Omega. \end{array}\right\} \tag{3.1}$$

(Here, and throughout the sequel, ${}^tA$ will stand for the transpose of a matrix A.)

The optimal control $\theta_\varepsilon{}^*$ is then characterized as follows: if $(u_\varepsilon{}^*, p_\varepsilon{}^*)$ is the solution of the system (3.1) for $\theta = \theta_\varepsilon{}^*$, then

$$\int_\Omega (p_\varepsilon{}^* + N\theta_\varepsilon{}^*)(\theta - \theta_\varepsilon{}^*)dx \ \geq\ 0 \tag{3.2}$$

for all $\theta \in U_{ad}$. We prove the following result for the homogenization of (3.1).

Theorem 3.1 *Let $A_\varepsilon \in \mathcal{M}(\alpha_m, \alpha_M, \Omega)$ and $B_\varepsilon \in \mathcal{M}(\beta_m, \beta_M, \Omega)$. For $\theta \in L^2(\Omega)$, let $(u_\varepsilon, p_\varepsilon)$ be the solution of the system (3.1). Let $u_\varepsilon \rightharpoonup u_o$ and $p_\varepsilon \rightharpoonup p_o$*

weakly in $H_o^1(\Omega)$. *Let* A_o *be the usual homogenized matrix corresponding to* $\{A_\varepsilon\}$. *Then there exists a matrix* $B^\#$ *such that*

$$ {}^tA_\varepsilon \nabla p_\varepsilon - B_\varepsilon \nabla u_\varepsilon \rightharpoonup {}^tA_o \nabla p_o - B^\# \nabla u_o $$

in $(L^2(\Omega))^n$ *weakly. The pair* (u_o, p_o) *is the solution of the system*

$$ \left. \begin{array}{rcll} -\operatorname{div}(A_o \nabla u_o) & = & f+\theta & in\ \Omega \\ \operatorname{div}({}^tA_o \nabla p - B^\# \nabla u) & = & 0 & in\ \Omega \\ u_o = p_o & = & 0 & on\ \partial\Omega.\Box \end{array} \right\} \qquad (3.3) $$

The matrix $B^\#$ is different from the usual homogenized matrix B_o corresponding to $\{B_\varepsilon\}$. It also reflects the properties of the B_ε. There exists a $\beta > 0$ such that $B^\# \in \mathcal{M}(\beta_m, \beta, \Omega)$ and if the B_ε are all symmetric, so is $B^\#$.

Regarding the optimal control problem, we have the following result.

Theorem 3.2 *Let* $\theta_\varepsilon{}^* \rightharpoonup \theta_o^*$ *weakly in* $L^2(\Omega)$. *Then* θ_o^* *is the optimal control of the problem corresponding to* A_o *and* $B^\#$. *Further,* $\theta_\varepsilon{}^* \to \theta_o^*$ *strongly in* $L^2(\Omega)$ *and*

$$ J_\varepsilon(\theta_\varepsilon{}^*) \to J_o(\theta_o^*) $$

where

$$ J_o(\theta_o^*) = \frac{1}{2}\int_\Omega B^\# \nabla u_o^* \nabla u_o^* dx + \frac{N}{2}\int_\Omega \theta_o^{*2} dx.\Box $$

Thus, in fact, we have strong convergence, in $L^2(\Omega)$, of the optimal control.

The proofs of these results can be found in Kesavan and Saint Jean Paulin [8]. It can now be proved that if w_ε is the solution of (1.5), then

$$ \int_\Omega B_\varepsilon \nabla w_\varepsilon \nabla w_\varepsilon dx \to \int_\Omega B^\# \nabla w_o \nabla w_o dx. $$

We give below the homogenization results in some special cases.

Example 3.1 The one - dimensional case.
Let $0 < \alpha_m \leq a_\varepsilon \leq \alpha_M$ and $0 < \beta_m \leq b_\varepsilon \leq \beta_M$ on an interval $(c,d) \subset I\!R$. Consider the system

$$ \left. \begin{array}{rcll} (a_\varepsilon u_\varepsilon{}')' & = & f+\theta & \text{in } (c,d) \\ (a_\varepsilon p_\varepsilon{}' - b_\varepsilon u_\varepsilon{}')' & = & 0 & \text{in } (c,d) \\ u_\varepsilon(c) = u_\varepsilon(d) & = & p_\varepsilon(c) = p_\varepsilon(d) & = 0. \end{array} \right\} \qquad (3.4) $$

Then, if $u_\varepsilon \rightharpoonup u_o$ and $p_\varepsilon \rightharpoonup p_o$ in $H_o^1(c,d)$ weakly, the pair (u_o, p_o) satisfies

$$\left.\begin{array}{rcll} (a_o u_o')' &=& f+\theta & \text{in } (c,d) \\ (a_o p_o' - b^{\#} u_o')' &=& 0 & \text{in } (c,d) \\ u_o(c) = u_o(d) &=& p_o(c) = p_o(d) & = 0 \end{array}\right\} \quad (3.5)$$

where

$$(a_\varepsilon)^{-1} \rightharpoonup (a_o)^{-1}$$

in $L^\infty(c,d)$ weak-* and

$$b^{\#} = a_o^2/g_o$$

where

$$\frac{1}{g_\varepsilon} = \frac{b_\varepsilon}{{a_\varepsilon}^2} \rightharpoonup \frac{1}{g_o}$$

in $L^\infty(c,d)$ weak-*.□

Example 3.2 The Periodic Case.
Let $A_\varepsilon = A(x/\varepsilon)$ and $B_\varepsilon = B(x/\varepsilon)$ where $A = (a_{ij})$ and $B = (b_{ij})$ are symmetric matrices with Y-periodic coefficients, defined on a reference cell $Y \subset I\!R^n$. For $1 \le k \le n$, define X_k as in (2.2) and Y_k analogously with B replacing A in (2.2). If A_o is the usual homogenized matrix corresponding to A_ε, defined as in (2.3), and B_o the usual homogenized matrix corresponding to B_ε (replace a by b and X_k by Y_k in (2.3)), the matrix $B^{\#}$ is given by

$$(B^{\#})_{ij} = (B_o)_{ij} + \int_\Omega b_{kl}(y)\frac{\partial}{\partial y_l}(X_j - Y_j)\frac{\partial}{\partial y_k}(X_i - Y_i)dy.$$

This result was first proved by Kesavan and Vanninathan [10].□.

Example 3.3 Layered Material.
Let $n = 2$ and let A and B be diagonal matrices of order 2 such that the coefficients a_{ii} and b_{ii} for $i = 1, 2$ depend periodically on x_1 and are independent of x_2. Thus

$$a^\varepsilon{}_{ii}(x) = a_{ii}(x_1/\varepsilon) \text{ and } b^\varepsilon{}_{ii} = b_{ii}(x_1/\varepsilon)$$

where $a_{ii}(y)$ and $b_{ii}(y)$ are Y- periodic functions defined on a reference interval Y, say, $[0,1]$, for $i = 1, 2$. In this case (cf. Murat [12])

$$A_o = \begin{bmatrix} (m(1/a_{11}))^{-1} & 0 \\ 0 & m(a_{22}) \end{bmatrix}$$

where $m(a) = \int_0^1 a(y)dy$ stands for the mean of the function a on $Y = [0,1]$. Then $B^\#$ is given by (cf. Kesavan and Saint Jean Paulin [8])

$$B^\# = \begin{bmatrix} \frac{m(b_{11}/a_{11}^2)}{(m(1/a_{11}))^2} & 0 \\ 0 & m(b_{22}) \end{bmatrix}. \square$$

4 Perforated Domains

As already indicated in the introduction, the main difficulty in the theory of homogenization of problems posed over perforated domains is that the functions are defined over different domains Ω_ε (*viz.* Ω minus the perforations). Thus to study the convergence of the solutions, we need to first extend these functions suitably on Ω.

There are two kinds of boundary conditions one could consider on the boundaries of the holes. The first is the Dirichlet boundary condition $u_\varepsilon = 0$ on the boundaries of the holes. In this case, we have the trivial extension by zero over the holes and we denote the extended function by $\tilde{u}_\varepsilon$. However, there are other kinds of difficulties in this case. The reader is referred to Cioranescu and Murat [4] for a nice exposition.

The other case is to consider a Neumann type boundary condition on the boundaries of the holes and a Dirichlet (or any other) condition on the global boundary $\partial\Omega$. In this case we need to have a prolongation operator from $H^1(\Omega_\varepsilon)$ to $H^1_o(\Omega)$. Once we have this, we can fairly easily imitate the results of homogenization. In the case of a periodic distribution of holes, an extension operator has been explicitly constructed by Cioranescu and Saint Jean Paulin [5]. (In the case of periodic holes, Allaire [1] has also shown that one can successfully use the notion of two - scale convergence together with the trivial extension by zero to get homogenization results in some cases.) For the general case, Briane *et al* [3] have defined the notion of H_o- convergence extending the notion of H- convergence to perforated domains.

Let $\Omega \subset \mathbb{R}^n$ be a bounded domain and let $S_\varepsilon \subset \Omega$ be a closed subset for each $\varepsilon > 0$. We set $\Omega_\varepsilon = \Omega \backslash S_\varepsilon$. We denote by χ_ε, the characteristic function of the perforated domain Ω_ε.

Following Briane *et al* [3], we say that S_ε is an *admissible family of holes* in Ω if the following conditions are satisfied.
(H1) Every weak-* limit point in $L^\infty(\Omega)$ of $\{\chi_\varepsilon\}$ is positive a.e. in Ω.

(H2) For each $\varepsilon > 0$, there exists an extension operator $P^\varepsilon : V_\varepsilon \rightarrow H^1_o(\Omega)$, where

$$V_\varepsilon = \{u \in H^1(\Omega_\varepsilon) | u = 0 \text{ on } \partial\Omega\},$$

such that

$$P^\varepsilon u|_{\Omega_\varepsilon} = u \text{ and } \|\nabla P^\varepsilon u\|_{2,\Omega} \leq C\|\nabla u\|_{2,\Omega_\varepsilon},$$

where $C > 0$ is a constant independent of ε and $\|.\|_{2,\Omega}$ denotes the norm in $L^2(\Omega)$.

As in the case of H- convergence, we have a H_o- compactness result. Let $A_\varepsilon \in \mathcal{M}(\alpha_m, \alpha_M, \Omega)$ and let $f \in H^{-1}(\Omega)$. Let $P^{\varepsilon *} : H^{-1}(\Omega) \rightarrow V_\varepsilon^*$ be the adjoint of P^ε. Let $w_\varepsilon \in V_\varepsilon$ be the weak solution of the problem

$$\left.\begin{array}{rcll} -\operatorname{div}(A_\varepsilon \nabla w_\varepsilon) &=& P^{\varepsilon *} f & \text{in } \Omega_\varepsilon \\ A_\varepsilon \nabla w_\varepsilon . n_\varepsilon &=& 0 & \text{on } \partial S_\varepsilon \\ w_\varepsilon &=& 0 & \text{on } \partial\Omega. \end{array}\right\} \quad (4.1)$$

Let $P^\varepsilon w_\varepsilon \rightharpoonup w_o$ weakly in $H^1_o(\Omega)$. Then, there exists a matrix A_o such that w_o is the weak solution of the problem

$$\left.\begin{array}{rcll} -\operatorname{div}(A_o \nabla w_o) &=& f & \text{in } \Omega \\ w_o &=& 0 & \text{on } \partial\Omega. \end{array}\right\} \quad (4.2)$$

A_o is the H_o- limit of $\{A_\varepsilon\}$.

In addition to the hypotheses (H1) and (H2) above, we will need the following:

(H3) If $\chi_\varepsilon \rightharpoonup \chi_o$ in $L^\infty(\Omega)$ weak-*, then $\chi_o^{-1} \in L^\infty(\Omega)$.

We denote by $\widetilde{\theta}_\varepsilon$ the trivial extension, by zero over the holes, of a function θ_ε defined on Ω_ε.

We now describe the optimal control problem.

Let $U^\varepsilon_{ad} \subset L^2(\Omega_\varepsilon)$ be a closed convex set for each $\varepsilon > 0$. For $\theta_\varepsilon \in U^\varepsilon_{ad}$, the state is defined as the weak solution in V_ε of the problem

$$\left.\begin{array}{rcll} -\operatorname{div}(A_\varepsilon \nabla u_\varepsilon) &=& f + \theta_\varepsilon & \text{in } \Omega_\varepsilon \\ A_\varepsilon \nabla u_\varepsilon . n_\varepsilon &=& 0 & \text{on } \partial S_\varepsilon \\ u_\varepsilon &=& 0 & \text{on } \partial\Omega \end{array}\right\} \quad (4.3)$$

where $f \in L^2(\Omega)$ is fixed. The cost functional is given by

$$J_\varepsilon(\theta_\varepsilon) = \frac{1}{2}\int_{\Omega_\varepsilon} B_\varepsilon \nabla u_\varepsilon \nabla u_\varepsilon dx + \frac{N}{2}\int_{\Omega_\varepsilon} {\theta_\varepsilon}^2 dx. \quad (4.4)$$

The optimal control $\theta_\varepsilon{}^*$ satisfies

$$J_\varepsilon(\theta_\varepsilon{}^*) = \min_{\theta_\varepsilon \in U^\varepsilon_{ad}} J_\varepsilon(\theta_\varepsilon).$$

As before, we introduce the adjoint state p_ε and prove the following result.

Theorem 4.1 *Assume that hypotheses (H1) and (H2) hold. Let* $A_\varepsilon \in \mathcal{M}(\alpha_m, \alpha_M, \Omega)$ *and* $B_\varepsilon \in \mathcal{M}(\beta_m, \beta_M, \Omega)$. *Let* $\theta_\varepsilon \in L^2(\Omega_\varepsilon)$ *such that* $\widetilde{\theta}_\varepsilon$ *is bounded in* $L^2(\Omega)$. *Let* $f \in L^2(\Omega)$ *be fixed. Let* $(u_\varepsilon, p_\varepsilon) \in V_\varepsilon \times V_\varepsilon$ *be the solution of the system*

$$\left.\begin{array}{rcll} -\operatorname{div}(A_\varepsilon \nabla u_\varepsilon) &=& f + \theta_\varepsilon & \text{in } \Omega_\varepsilon \\ \operatorname{div}({}^t A_\varepsilon \nabla p_\varepsilon - B_\varepsilon \nabla u_\varepsilon) &=& 0 & \text{in } \Omega_\varepsilon \\ A_\varepsilon \nabla u_\varepsilon . n_\varepsilon &=& 0 & \text{on } \partial S_\varepsilon \\ ({}^t A_\varepsilon \nabla p_\varepsilon - B_\varepsilon \nabla u_\varepsilon). n_\varepsilon &=& 0 & \text{on } \partial S_\varepsilon \\ u_\varepsilon = p_\varepsilon &=& 0 & \text{on } \partial\Omega. \end{array}\right\} \tag{4.5}$$

Assume that (for a subsequence), $\widetilde{\theta}_\varepsilon \rightharpoonup \theta_o$ *in* $L^2(\Omega)$ *weakly and that* $P^\varepsilon u_\varepsilon \rightharpoonup u_o$ *and* $P^\varepsilon p_\varepsilon \rightharpoonup p_o$ *in* $H^1_o(\Omega)$ *weakly. Then there exist matrices* A_o *and* $B^\#$ *such that*

$$\left.\begin{array}{rcll} -\operatorname{div}(A_o \nabla u_o) &=& \chi_o f + \theta_o & \text{in } \Omega \\ \operatorname{div}({}^t A_o \nabla p_o - B^\# \nabla u_o) &=& 0 & \text{in } \Omega \\ u_o = p_o &=& 0 & \text{on } \partial\Omega. \end{array}\right\} \tag{4.6}$$

The matrix A_o *is the* H_o*- limit of* $\{A_\varepsilon\}$. *The matrix* $B^\#$ *is symmetric if all the* B_ε *are symmetric; the bilinear form defined on* $H^1_o(\Omega)$ *by* $B^\#$ *is elliptic.*□

Let us return to the control problem. We assume henceforth that all the hypotheses (H1) - (H3) are satisfied. Let U^ε_{ad} be one of the following sets.

$$U^\varepsilon_{ad} = L^2(\Omega_\varepsilon) \tag{4.7}$$

$$U^\varepsilon_{ad} = \{\theta \in L^2(\Omega_\varepsilon) | \theta \geq \psi \text{ a.e. in } \Omega_\varepsilon\} \tag{4.8}$$

$$U^\varepsilon_{ad} = \{\theta \in L^2(\Omega_\varepsilon) | \psi_1 \leq \theta \leq \psi_2 \text{ a.e. in } \Omega_\varepsilon\} \tag{4.9}$$

$$U^\varepsilon_{ad} = \{\theta \in L^2(\Omega_\varepsilon) | \int_{\Omega_\varepsilon} \theta^2 dx \leq 1\} \tag{4.10}$$

where ψ, ψ_1 and ψ_2 are fixed functions in $L^2(\Omega)$. We define correspondingly

$$U_{ad} = L^2(\Omega) \tag{4.11}$$

$$U_{ad} = \{\theta \in L^2(\Omega) | \theta \geq \chi_o \psi \text{ a.e. in } \Omega\} \quad (4.12)$$
$$U_{ad} = \{\theta \in L^2(\Omega) | \chi_o \psi_1 \leq \theta \leq \chi_o \psi_2 \text{ a.e. in } \Omega\} \quad (4.13)$$
$$U_{ad} = \{\theta \in L^2(\Omega) | \int_\Omega (\theta^2/\chi_o) dx \leq 1\} \quad (4.14)$$

We can now describe the "homogenized" optimal control problem. Let $\theta \in U_{ad}$. The state $u \in H_o^1(\Omega)$ is the weak solution of the problem

$$\left.\begin{array}{rcll} -\text{div}(A_o \nabla u) & = & \chi_o f + \theta & \text{in } \Omega \\ u & = & 0 & \text{on } \partial\Omega \end{array}\right\} \quad (4.15)$$

and the cost functional is given by

$$J_o(\theta) = \frac{1}{2} \int_\Omega B^{\#} \nabla u \nabla u dx + \frac{N}{2} \int_\Omega \frac{\theta_o^2}{\chi_o} dx. \quad (4.16)$$

Theorem 4.2 *Let $\theta_\varepsilon{}^*$ be the optimal control for the problem (4.3) - (4.4). Then $\{\widetilde{\theta}_\varepsilon{}^*\}$ is bounded in $L^2(\Omega)$. Let (for a subsequence) $\widetilde{\theta}_\varepsilon{}^* \rightharpoonup \theta_o^*$ weakly in $L^2(\Omega)$. Then θ_o^* is the optimal control for the problem (4.15) - (4.16) where U_{ad}^ε and U_{ad} are given as above. Further,*

$$J_\varepsilon(\theta_\varepsilon{}^*) \to J_o(\theta_o^*).\square$$

Corollary 4.1 *If w_ε is the solution of (4.1) and if $P^\varepsilon w_\varepsilon \rightharpoonup w_o$ in $H_o^1(\Omega)$ weakly, then*

$$\int_{\Omega_\varepsilon} B_\varepsilon \nabla w_\varepsilon \nabla w_\varepsilon dx \to \int_\Omega B^{\#} \nabla w_o \nabla w_o dx.\square$$

Note that contrary to the situation in Section 3, the optimal control $\theta_\varepsilon{}^*$ (or rather its extension $\widetilde{\theta}_\varepsilon{}^*$) will *not* converge strongly in $L^2(\Omega)$ to θ_o^*. However, we have the following corrector type result.

Theorem 4.3 *Let $\theta_\varepsilon{}^*$ and θ_o^* be respectively the optimal controls of the problems (4.3) - (4.4) and (4.15) - (4.16). Then*

$$\widetilde{\theta}_\varepsilon{}^* - \frac{\chi_\varepsilon}{\chi_o} \theta_o^* \to 0$$

strongly in $L^2(\Omega)$.$\square$

The proofs of all the results mentioned in this section can be found in Kesavan and Saint Jean Paulin [9].

References

[1] Allaire, G. Homogenization and two-scale convergence, *SIAM J. Math. Anal.*, **23**, No. 6, pp. 1482 - 1518, 1992.

[2] Bensoussan, A., Lions, J. L. and Papanicolaou, G. *Asymptotic Analysis for Periodic Structures*, North Holland, Amsterdam, 1978.

[3] Briane, M., Damlamian, A. and Donato, P. H- convergence for perforated domains, in *Nonlinear Partial Differential Equations and their Applications, Collège de France Seminar*, Vol. XIII, Pitman Research Notes in Mathematics, 1996.

[4] Cioranescu, D. and Murat, F. A strange term coming from nowhere, Preprint A 94003, Publications du Laboratoire d' Analyse Numérique, Université de Paris VI, Paris, 1994.

[5] Cioranescu, D. and Saint Jean Paulin, J. Homogenization in open sets with holes, *J. Math. Anal. Appl.*, **71**, pp. 590 - 607, 1979.

[6] Dal Maso, G. *An Introduction to* Γ*-Convergence*, Birkhäuser, Boston, 1993.

[7] De Giorgi, E. G - operators and Γ - convergence, *Proceedings of the International Congress of Mathematicians, Warsaw, 1983*, PWN Polish Scientific Publishers and North - Holland, Amsterdam, 1984.

[8] Kesavan, S. and Saint Jean Paulin, J. Homogenization of an optimal control problem, *SIAM J. Control*, to appear.

[9] Kesavan, S. and Saint Jean Paulin, J. Optimal control on perforated domains, to appear.

[10] Kesavan, S. and Vanninathan, M. L' homogénéisation d'un problème de contrôle optimal, *C.R.A.S., Série A*, t **285**, pp. 441 - 444, 1977.

[11] Lions, J. L. *Optimal Control of Systems Governed by Partial Differential Equations*, Springer-Verlag, 1971.

[12] Murat, F. H- convergence, Mimeographed Notes, Séminaire d'Analyse Fonctionnelle et Numérique, Université d'Alger, 1977-78.

[13] Nguetseng, G. A general convergence result for a functional related to the theory of homogenization, *SIAM J. Math. Anal.*, **20**, pp. 608 - 623, 1989.

[14] Sanchez - Palencia, E. *Nonhomogeneous Media and Vibration Theory*, Lecture Notes in Physics, **127**, Springer - Verlag, New York, 1980.

[15] Spagnolo, G. Convergence in energy for elliptic operators, in *Numerical Solutions of Partial Differential Equations III, SYNSPADE, 1975*, B. Hubbard (ed.), Academic Press, New York, 1976.

The Institute of Mathematical Sciences,
C. I. T. Campus, Taramani,
Chennai - 600 113, INDIA.

P N SRIKANTH

On periodic motions of two dimensional lattices

$ 1.INTRODUCTION

In [5] one dimensional lattices of N particles with nearest-neighbour interaction was considered and the existence of infinitely many T-periodic solutions was proved. The motivation for such a study was to see how to adopt the recent advances in Critical Point Theory in the context of such equations. The results obtained cover more general potentials then the traditional Toda potential.
In this note we consider a two dimensional generalization. The motivation for the lattice problem considered comes from Crystal Theory (see [3]).

$ 2.DESCRIPTION OF THE PROBLEM

We consider an array of $N \times M$ particles with nearest-neighbour central and noncentral forces. Rigid boundary conditions are used in long dimension. Periodic boundary con-

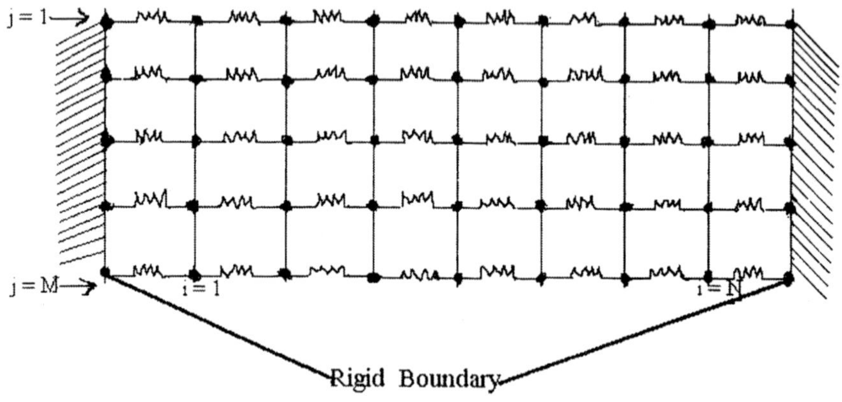

ditions are used in short dimension. The two dimensional lattice may be considered to be a tubular lattice because of the Periodic boundary condition. The equations of motion are

$$(1a) \qquad \frac{d^2xij}{dt^2} = -\frac{\partial\Phi}{\partial xij} \qquad i = 1, 2, ----, N$$

$$(1b) \qquad \frac{d^2yij}{dt^2} = -\frac{\partial\Phi}{\partial yij} \qquad j = 1, 2, ---, M.$$

where Φ is the total potential energy of the system, assumed to be a function of coordinates only. The simplifying assumption.

$$\Phi = \Phi_x + \Phi_y$$

where Φ_x is a function of x_{ij} is only and Φ_y is a function of y_{ij} is makes (1a) and (1b) independent and we need to solve only one of the equations.
Φ_x is sum of terms over nearest-neigbour pairs.

$$\Phi_x = \Sigma_{ij} V_c(x_{i,j} - x_{i-1,j}) + \Sigma_{i,T} V_{nc}(x_{i,j} - x_{i,j-1})$$

The terms V_c and V_{nc} relate to central and noncentral forces i.e. interactions parallel and transverse to the line of centers between pair of particles. In the equations we consider we omit interactions with external systems.

$ 3.MAIN RESULT

In the context of what has been described in $2, we consider the following problem:

$$(2) \qquad \begin{cases} \frac{d^2x_{ij}}{dt^2} = [V'(x_{i-1,j} - x_{i,j}) - V'[x_{i,j} - x_i + i, j)] \\ \quad +[V'(x_{i,j-1} - x_{i,j}) - V'(x_{i,j} - x_{i,j} + 1)] \\ with \quad x_{0,j} = x_{N+1,j} = 0 \qquad j = 1, ...; M \\ and \quad x_{i,1} = x_{i,M+1} \qquad i = 1, ...; N. \end{cases}$$

which correspond to rigid boundary conditions on the long dimension and periodic boundary conditions in the short dimension. We have taken $V_c \equiv V_{nc} \equiv V$ as the potential in the context of what has been described in $2. For the purposes of this note we specialize V further, $V(t) = \frac{a}{b}e^{-bt} + at$ with $a > 0$, $b > 0$, the Toda type potential. See remark at the end of this section for general potentials. We now state and give some idea of the result.

Theorem 1. Under the stated hypothesis on V, for each $T > 0$, the system(3) admits infinitely many T-periodic solutions.
Proof of this Theorem is based on minimax method. We consider the associated energy functional.

$$J(x) = \frac{1}{2}\int_0^T |x|^2\,dt - \Sigma_{i=1}^{N+1} \quad \Sigma_{j=1}^{M} \quad \int_0^T [V(x_{i-1,j} - x_{i,j}) + V(x_{i,j-1} - x_{i,j}]$$

On the Sobolev space $H^1(S_T, \mathbb{R}_P^{N(M+1)})$ of T-periodic L^2- functions whose weak derivatives are also in L^2. A vector $x \in \mathbb{R}_P^{N(M+1)}$ is written
$x = (x_{i,j})_{i=1,}^{N} \ {}_{j=1}^{M+1}$ with $x_{i,1} = x_{i,M+1}$. Note $\mathbb{R}_P^{N(M+1)}$ can be identified with $\mathbb{R}^{NM}$.
In contrast with the one-dimensional periodic lattice (See[5]) the functional here is not invariant under translations. This is somewhat surprising but not unexpected. This is essentially due to the rigid boundary conditions in the long dimensions.

We employ the following Theorem due to Benci [1] to prove our result (See also [2] and [4]).

Theorem 2: Assume that $\Phi : \mathbb{R}^N \to \mathbb{R}$ is C^{-1}.

Let $J : H^1 \equiv H^1(S_T^1, \mathbb{R}^N) \to \mathbb{R}$ be given by

$$J(x) = \frac{1}{2}\int_0^T |\dot{x}|^2\,dt - \int_0^T \Phi(x)dt,$$

assume j satiesfies Palais- Smale condition on H^1(See below). Furthermore, assume that

$$\lim_{|x|\to\infty} \frac{\int_0^T \Phi(x)}{|x|^2} = +\infty \quad \mathit{for}\ x \in H^1,$$

where $|x|$ denotes the L^∞- norm. Then

$$\ddot{x} = -\Phi'(x)$$

has infinitely many T-periodic solutions.
Remark 1: $j : E \to \mathbb{R}$, E *a* Banach space is said to satisfy the Palais - Smale condition (P.S. Condition) if for every sequence $(u_n) cE$ with $|j(u_n)|$ bounded and $j'(u_n) \to 0$ in E' (the dual) there exists a convergent subsequence (u_{nk}) in E.

Proof of Theorem (A Sketch):
Define $\Phi : \mathbb{R}_{\mathrm{P}}^{\mathrm{N(M+1)}} \to \mathbb{R}$ by

$$\Phi(x) = \Sigma_{i=1}^{N+1} \quad \Sigma_{j=1}^{M} [V(x_{i-1,j} - x_{i,j}) + V(x_{i,j-1} - x_{i,j})]$$

Note $\mathbb{R}_{\mathrm{P}}^{\mathrm{N(M+1)}}$ can be identified with $\mathbb{R}^{\mathrm{NM}}$. Assuming.

$$(3) \qquad V(t) = \frac{a}{b} e^{-bt} + at \qquad a > 0, b > 0 \qquad we \quad claim,$$

Lemma: There exists $R > 0$ such that

$$(4) \qquad < \Phi'(x),\ x > \geq (2 + \frac{\delta}{2}) \Phi(x) \forall x \in \mathbb{R}^{\mathrm{NM}} \text{with} \mid \mathrm{x} \mid > \mathbb{R},$$

for some $\delta > 0$.
Proof (a Sketch): Given (3) it is easy to see

$$(5) \qquad V'(t) \cdot (t) \geq (2 + \delta) V(t) \qquad \forall t < -t_0$$

and

$$(6) \qquad \mid V'(t) \mid \leq c|t|^{1+\delta_0} + d \qquad \forall t \geq t_0$$

for some $\delta > 0$, $\delta_0 > 0$ and $t_0 \gg 0$. For $x = (x_{i,j})_{i=1, j=1}^{N \quad M+1}$ with $x_{i,1} = x_{i,M+1}$ or with identification as $x \in \mathbb{R}^{\mathrm{NM}}$, we write

$$(7) \qquad \Phi(x) = \Sigma_{i=1}^{N+1} \quad \Sigma_{j=1}^{M} [V(x_{i-1,j} - x_{i,j})^-) + V[(x_{i,j-1} - x_{i,j})^-)]$$

$$+ \Sigma_{i=1}^{N+1} \quad \Sigma_{j=1}^{M} [V(x_{i-i,j} - x_{i,j})^+) + V((x_{i,j-1} - x_{i,j})^+)]$$

We write (6) as

$$\Phi(x) =: \phi^-(x) + \phi^+(x)$$

It is easy to see,

$$< \Phi'(x), x > =: < (\Phi^-)' x, x > + < (\Phi^+)' x, x > .$$

We further prove (we omit the details) that there exists numerals $R > 0$ and $C > 0$ such that.

$$(8) \qquad < (\Phi^{-1})'(x), x > \geq (2 + \delta) \Phi^-(x) - c \qquad \forall \mid x \mid \geq \mathbb{R}$$

$$(9) \qquad \Phi(x) \geq \alpha \mid x \mid^{2+\delta} - C \ for \ some \ \alpha > 0, \qquad \forall \mid x \mid \geq \mathbb{R}.$$

and

$$(10) \qquad < (\Phi^+)'(x), x > \geq -c \mid x \mid^{2+\delta_0} \quad \forall \mid x \mid \geq \mathbb{R}.$$

from which the Lemma follows. It is very standard to show that if Φ satisfies (4) then (P-S) holds and hypothesis of Theorem 2 are satisfied. Hence Theorem 1 follows from Theorem 2.

Remark 2: A more general result with details of the proof will appear elsewhere . Also we can handle the case of N being infinity.

REFERENCES

1. Benci, V. A geometrical index for the group S^1 and some applications to the study of periodic solutions of ordinary differential equation, Comm. pure Appl. Math. 34 (1981), 393-432.

2. Nirenberg, L., Variational methods in Nonlinear problems, Lect. Notes in Mathematics 1365, M.Giaquinta (ed.), Springer-Verlog, 1987.

3. Payton, D.N., Rich, R. and Visscher, W.M., Lattice Thermal Conductivity in Disordered Harmonic and Anharmonic Crystal Models, Phys.Rev.160(1967) 706-711.

4. Rabinowitz, P.H., Periodic solutions of Hamiltanian Systems, Comm pure Appl. Math 31 (1978)157-184.

5. Ruf, B. and Srikanth P.N., On Periodic Motions of Lattices of Toda Type via Critical Point Theory, Arch. Rat.Mech.Anal.126(1994) 369-385.

TIFR Centre

Bangalore, India

ADIMURTHI

Uniqueness of positive solutions of Dirichlet problem with exponential nonlinearity

Introduction

Recently there has been considerable interest in the uniqueness of solutions of the problem

$$\begin{aligned} -\Delta u &= f(u) \quad \text{in } B \\ u &> 0 \\ u &= 0 \quad \text{on } \partial B \end{aligned} \tag{1.1}$$

where $B \subset \boldsymbol{R}^n$ is the ball of radius one and $f = \overline{\boldsymbol{R}}^+ \to \overline{\boldsymbol{R}}^+$ is a C^1 function with $f(0) = 0$. When the nonlinearity $f(u) = u^p + \lambda u$, $\lambda \geq 0$ and $p > 1$, uniqueness results have been proved for $p \leq (n+2)/(n-2)$ (See [2], [5], [6], [7], [8]). For $p > \frac{n+2}{n-2}$ and $3 \leq n \leq 9$, it has been shown by Budd-Norbury [4] that there exist a $\lambda > 0$ for which (1.1) admits infinitely many solutions.

In view of these results and the Trudinger imbedding for $n = 2$, it is expected that the uniqueness results continue to hold if the nonlinearities f is of the form $f(s) = \lambda s \exp(s^\theta)$ for $0 < \theta \leq 2$ and $\lambda > 0$. In order to substantiate this we prove the following

Theorem 1.1 : Let $\lambda > 0$ and $B \subset \boldsymbol{R}^2$ be the unit ball. Then

$$\begin{aligned} -\Delta u &= \lambda u e^u \quad \textit{in } B \\ u &> 0 \\ u &= 0 \quad \textit{on } \partial B \end{aligned} \tag{1.2}$$

admits at most one solution.

The main ingredient in the proof of uniqueness results for nonlinearities of polynomial growth is the pohozaev's identy. When it comes to $n = 2$ and the non linearity is of exponential growth, then the pohozaev's identity becomes ineffective. In this case one derives an identity from the beautiful analysis developed by

Atkinson-Peletier [3] and this leads to theorem (1.1). Further more the solutions of (1.2) are non degenerate .

2.Proof of theorem(1.1)

Let $f(u) = \lambda u e^u$. Then from Gidas-Ni-Nirenberg, solution u of (1.1) is radial are hence satisfies

$$\begin{cases} -(r^{n-1}u')' = r^{n-1}f(u) & \text{in } (0,1) \\ u > 0 & \text{in } (0,1) \\ u'(0) = u(1) = 0 \end{cases} \tag{2.1}$$

As in [3] we use the following transformation. Let $r = 2e^{-t/2}$, $y(t) = u(2e^{-t/2})$, $T_0 = 2\log 2$, Then y satisfies.

$$\begin{cases} -y'' = f(y)e^{-t} & \text{in } (T_0, \infty) \\ y > 0 \\ y'(\infty) = y(T_0) = 0 \end{cases} \tag{2.2}$$

Let $g(s) = \log f(s)$ and E the energy associated to (2.2) (see[3]) and is given by

$$E(t) = y' - \frac{1}{2}y'^2 g'(y) - e^{g(y)-t} \tag{2.3}$$

then $E' = -\frac{1}{2}g''(y)y'^3$. Since $E(\infty) = 0$, integrating from t *to* ∞ we obtain.

$$y' - \frac{1}{2}y'^2 g'(y) - e^{g(y)-t} = \frac{1}{2}\int_t^\infty g''(y)y'^3 ds.$$

Dividing by y' and integrating the above equations from t *to* ∞ to obtain the following identify

$$g(y(\infty)) + g(y(t)) - \log y'(t) - 2t = \int_t^\infty [y](s)dt \tag{2.4}$$

where we have used the fact that

$$\lim_{t \to \infty}[t + \log y'(t)] = g(y(\infty)) \tag{2.5}$$

$$[y](s) = \frac{1}{y'(s)}\int_s^\infty g''(y(s))y'(s)^3 ds. \tag{2.6}$$

Our main ingredient in the proof of the theorem is the identity (2.4).

Let λ_1 denote the first Dirichlet eigenvalue of $-\Delta$ in B. Then it is easy to see to see that (1.2) admits a solutions if and only if $\lambda \in (0, \lambda_1)$. In order to prove the theorem, It is enough to show that the solutions of (1.2) are non degenarate.

From the bifurcation result of Crandall-Rabinowitz, the solutions of (1.2) are non degenerate near $\lambda = \lambda_1$ (with Morse index one). Hence by continuity if there exist a degenerate solution, then there exist a $\lambda_0 \in (0, \lambda_1)$ such that the linearized problem

$$\begin{aligned} -\Delta w &= \lambda_0(1+u)e^u w \quad \text{in } B \\ w &= 0 \qquad \text{on} \quad \partial B \end{aligned} \tag{2.7}$$

admits a radial solution $w(x) = w(|x|)$ changing sign exactly once in $(0, 1)$ where u is a solution of $(1, 2)$ with $\lambda = \lambda_0$. Use the following transformation, $r = 2e^{-t/2}$, $y(t) = u(r)$, $v(t) = w(r), T_0 = 2 \log 2$, then v satisfies

$$\begin{aligned} -v'' &= \lambda_0(1+y)e^y v \quad \text{in } (T_0, \infty) \\ v(T_0) &= v'(\infty) = 0 \end{aligned} \tag{2.8}$$

and v changes sign exactly once in (T_0, ∞). Now we make the following normalization (which was suggested by Dr.S.L.Yadava), namely by multiplying v by a suitable constant, we can assume that $v'(T_0) = y'(T_0)$.

Step 1: v/y and v'/y' are decreasing functions in (T_0, ∞). Proof of this step follows easily from equations (1.2) and (2.8) and will be omitted.

Step 2: Let $r = y(\infty)$, then $v = \frac{\partial y}{\partial r}$ and is denoted by $\delta(y)$ the variation y with respect to the initial data. Taking the variation in (2.4), we obtain for $t \in (T_0, \infty)$

$$g'(y(\infty))v(\infty) + g'(y(t))v(t) - 2\frac{v'(t)}{y'(t)} = \int_t^\infty \delta[y](s)ds$$

since $\frac{v(T_0)}{y(T_0)} = \frac{v'(T_0)}{y'(T_0)} = 1$, hence letting $t \to T_0$ in the above expression we obtain

$$\begin{aligned}
0 > v(\infty)g'(y(\infty)) - 1 &= \int_{T_0}^{\infty} \frac{1}{y'(s)} \int_s^{\infty} (3y'^2 g''(y)v' + y'^3 g'''(y)v)\,d\varphi\,ds \\
&- \int_{T_0}^{\infty} \frac{v'}{(y'(s))^2} \int_s^{\infty} y'^3 g''(y)\,d\varphi\,ds \\
&= \int_{T_0}^{\infty} \frac{2}{y'(s)} \int_s^{\infty} \frac{(y'v - yv')y'^3}{y^3}\,d\varphi\,ds \\
&+ \int_{T_0}^{\infty} \int_s^{\infty} \frac{y'(\varphi)^3}{y'(s)y(\varphi)^2} \left(\frac{v'(s)}{y'(s)} - \frac{v'(\varphi)}{y'(\varphi)} \right) d\varphi\,ds > 0,
\end{aligned}$$

since from step 1, $y'v - yv' \geq 0$ and $\dfrac{v'(s)}{y'(s)} - \dfrac{v'(\varphi)}{y'(\varphi)} > 0$ for $\varphi > s$. This gives a contradiction and hence the result.

Remark: The uniqueness results for polynomial growth has been extended to q-Laplacians in [2].In a forth coming paper [1] theorem (1.1) is generalized to n-Laplacian as follows:

Theorem 1.2: Let $\lambda > 0$ and $B \subset \mathbb{R}^n$ $(n \geq 2)$ be the unit ball. Then

$$\begin{aligned}
-div(|\nabla u|^{n-2}\nabla u) &= \lambda u^{n-1}e^u \quad \text{in } B \\
u &> 0 \quad \text{in } B \\
u &\in W_0^{1,n}(B)
\end{aligned}$$

admits at most one solution.

Acknowledgement: I would like to thank Dr.S.L.Yadava for having many discussions during the preparation of this paper.

References

1. Adimurthi, uniqueness of positive solutions of Quasilinear Dirichlet problem with exponential non linearity, Preprint, 1996.

2. Adimurthi and S.L.Yadava, An elementary proof of the uniqueness of positive radial solutions of a quasilinear Dirichlet problem, Arch. Rat.Mech. Anal, 127 (1994), 219-229.

3. F.V.Atkinson and L.A.Peletier, Ground states and Dirichlet problem for $-\Delta u = f(u)$ in $\mathbb{R}^2$, Arch.Rat.Mech.Anal. 96 (1988), 147-165.

4. C.Budd and J.Norbury, Semilinear elliptic equations and super critical growth, J.diff.eqns. 68 (1987), 169-197.

5. M.K.Kwong and Y.Li, Uniqueness of radial solutions of semilinear elliptic equations, Trans.Amer.Math.Soc. 333 (1992), 339-363.

6. Z.Liquin, Uniqueness of positive solutions of $\Delta u + u + u^P = 0$ in a finite ball, Comm. Part. Diff.Eqns. 17 (1992), 1141-1164.

7. W.M.Ni and R.D.Nussbaum, Uniqueness and non-uniqueness for positive radial solutions of $\Delta u + f(u, r) = 0$, Comm. Pure Appl. Math. 38 (1985), 171-185.

8. P.N.Srikanth, Uniqueness of solutions of nonlinear Dirichlet problems, Diff. Integ. Eqns. 7(1993), 663-670.

Adimurthi
TIFR Centre
P.B. 1234
BANGALORE-560012
INDIA

L JONES AND A K PANI

A qualocation method for a semilinear second-order two-point boundary value problem[1]

1. Introduction

In this paper, we discuss a qualocation method for the following semilinear two-point boundary value problem

$$\mathcal{L}u := u'' + b(x)u' + c(x)u = f(u), \ x \in [0,1] \tag{1.1}$$

subject to the general linear seperated boundary conditions

$$\lambda_0 u'(0) + \mu_0 u(0) = \alpha, \ \lambda_1 u'(1) + \mu_1 u(1) = \beta, \tag{1.2}$$

where $\lambda_i, \mu_i, \ i = 0, 1, \alpha$ and β are constants.

Given $N \geq 1$, let

$$\Pi_N : 0 = x_0 < x_1 < \cdots < x_N = 1$$

be an arbitrary partition of [0,1], with the property that $h \to 0$ as $N \to \infty$, where

$$I_k = [x_{k-1}, x_k] \text{ with } h_k = x_k - x_{k-1}, \ k = 1, \cdots, N \text{ and } h = \max_{1 \leq k \leq N} h_k.$$

Let

$$S_h = \left\{ v \in C^2[0,1] : v|_{I_k} \in P_3, k = 1, \cdots, N \right\}$$

be a space of cubic splines and let

$$T_h = \left\{ v \in C[0,1] : v|_{I_k} \in P_1, k = 1, \cdots, N \right\}$$

be a space of linear splines.

An H^1-Galerkin procedure for approximating the solution u of (1.1) is defined as a function $\widehat{u} \in S_h$ which satisfies

$$(\mathcal{L}\widehat{u}, \chi) = (f(\widehat{u}), \chi), \ \ \chi \in T_h. \tag{1.3}$$

[1]Supported by NBHM Postdoctoral Fellowship, Department of Atomic Energy, Government of India.

In fact, this can be considered as a Petrov- Galerkin method using a cubic spline trial space and a piece-wise linear test space. For the computation of the Galerkin approximation $\widehat{u}$, the integrals involved in (1.3) are rarely evaluated exactly. Therefore, a discrete version of the H^1-Galerkin procedure in which the discretization is achieved by approximating the integrals either by composite two point Gauss rule or composite Simpson quadrature rule is proposed for the problem (1.1). More precisely, we define the method as follows.
Set

$$\langle v, w\rangle = \mathcal{Q}_h(vw), \tag{1.4}$$

where $\mathcal{Q}_h$ is a fourth-order quadrature rule, in practice either a two-point Gauss rule for the partition Π_N,

$$\mathcal{Q}_h g = \frac{1}{2}\sum_{k=1}^{N} h_k[g(x_{k,1}) + g(x_{k,2})] \tag{1.5}$$

with

$$x_{k,i} = \frac{1}{2}(x_k + x_{k-1}) + (-1)^i \frac{h_k}{2\sqrt{3}} \quad i = 1, 2;$$

or a composite Simpson quadrature rule for the partition Π_N,

$$\mathcal{Q}_h g = \sum_{k=1}^{N} h_k[\frac{1}{6}g(x_{k-1}) + \frac{2}{3}g(x_{k+\frac{1}{2}}) + \frac{1}{6}g(x_k)] \tag{1.6}$$

with $h_{-1} = h_N = 0$ and $x_{k+\frac{1}{2}} = \frac{(x_k+x_{k-1})}{2}$.

The quadrature rule in (1.4) is assumed to have an error bound of the form

$$|\epsilon_h(g)| := |\mathcal{Q}_h(g) - \int_0^1 g| \leq C\sum_{k=1}^{N} h_k^4 \left(\int_{I_k} |v^{(4)}(x)|\right) dx. \tag{1.7}$$

For both the quadrature rules (1.5) and (1.6), the above result follows from the Peano's kernel Theorem (see Davis [4] or Davis et. al. [5]).

The discrete H^1-Galerkin procedure for approximating solution u of (1.1) is now defined as a solution $u^h \in S_h$ of

$$\left\langle \mathcal{L}u^h, \chi\right\rangle = \left\langle f(u^h), \chi\right\rangle, \quad \chi \in T_h \tag{1.8}$$

with u^h satisfying the corresponding boundary conditions (1.2)

The present method may be described as a "qualocation" approximation which is a quadrature based modification of the collocation procedure. Compared to collocation or quadrature method, the size of the system of algebraic equations derived from this scheme is quite small and hence, it has a computational advantage.

Earlier Sloan [8], he has proposed and analysed a qualocation method for boundary-integral equations on smooth curves. Subsequently, Sloan *et.al.* [9], they have extended the result to a linear second order two point boundary value problem and obtained optimal error estimates in $W^{i,p}$ norms for $i = 0, 1, 2$ and $1 \leq p \leq \infty$ with out quasi-uniformity assumption on the finite element space. But their analysis does not yield superconvergence results for the first derivative at the mesh points. In this paper, while considering a semilinear two piont boundary value problem we have derived knot superconvergence of order $O(h^4)$ for the flux at the nodal points. A different proof of superconvergence results at the knots for the mth-order linear two point boundary value problem can be found in Grigorieff and Sloan [7].

A practical advantage of this method over orthogonal spline collocation is that for a given partition there are fewer unknowns, thereby reducing by about half the size of the linear system. An overview of the H^1-Galerkin and related methods is given in Sloan *et.al.* [9].

The organization of this paper is as follows. In Section 2. some preliminaries are discussed which are essential to our error analysis. Section 3. deals with convergence analysis, and related error estaimates are obtained in $W^{i,p}$- norms for $i = 0, 1, 2$ without quasiuniformity assumption on the finite element mesh. In Section 4. existence and uniqueness of the approximate solutions are proved, using a priori estimates and fixed point arguments. Throughout this paper C will denote a positive constant which may depend on u and some of its derivatives.

2. Preliminaries

Without loss of generality, we consider (1.1) subject to the following homogeneous boundary conditions

$$\lambda_0 u'(0) + \mu_0 u(0) = 0, \ \lambda_1 u'(1) + \mu_1 u(1) = 0, \tag{2.1}$$

since the given problem can always be converted to one of this form by a simple transformation. We assume that the coefficients b, c and the right hand side nonlinear function f are sufficiently smooth and are bounded with bounded derivatives. Now using Taylor expansion, we write

$$f(w) = f(u) + f_u(u)(w-u) + \tilde{f}_{uu}(w-u)^2, \tag{2.2}$$

where

$$\tilde{f}_{uu} = \int_0^1 (1-s) f_{uu}(w + s(u-w))\, ds. \tag{2.3}$$

Define

$$\mathcal{M}(u)\phi = \mathcal{L}\phi - f_u(u)\phi \tag{2.4}$$

and its adjoint as

$$\mathcal{M}^*(u)\phi = \mathcal{L}^*\phi - f_u(u)\phi, \tag{2.5}$$

with the adjoint boundary conditions

$$\left.\begin{array}{rcl}\lambda_0\left[\phi'(0)-(b\phi)(0)\right]+\mu_0\phi(0) & = & 0\\ \lambda_1\left[\phi'(1)-(b\phi)(1)\right]+\mu_1\phi(1) & = & 0\end{array}\right\}, \tag{2.6}$$

where $\mathcal{L}^*$, the formal adjoint of $\mathcal{L}$ is given by $\mathcal{L}^*\phi = \phi'' - (b\phi)' + c\phi$. For an open interval I and a non-negative integer m, let

$$\|v\|_{W^{m,p}(I)} = \begin{cases} \left(\sum_{i=0}^{m}\|v^{(i)}\|^p_{L^p(I)}\right)^{1/p}, & if\ 1\le p<\infty,\\ \max\limits_{0\le i\le m}\|v^{(i)}\|_{L^\infty(I)}, & if\ p=\infty.\end{cases}$$

We shall use the following notations through the paper

$$\|v\|_{s,p} = \|v\|_{W^{s,p}(I)},\quad \|v\|_{s,p,k} = \|v\|_{W^{s,p}(I_k)}.$$

Now assume that the operator $\mathcal{M}(u)$ and its adjoint $\mathcal{M}^*(u)$ have the following regularity property i.e., for $p\in[1,\infty]$ and $q\in[1,\infty)$ with $\frac{1}{q}+\frac{1}{p}=1$ there exists a positive constant $C>0$ such that

$$\|\mathcal{M}(u)v\|_p \ge C\|v\|_{2,p} \text{ and } \|\mathcal{M}^*(u)\phi\|_q \ge C\|\phi\|_{2,q}. \tag{2.7}$$

The following interpolation inequality which is proved in the appendix of Sloan et.al., [9] (see also Adams [1]) is used frequently in our analysis

Lemma 2.1 *If $v\in W^{m,p}(E)$ with $p\in[1,\infty]$, then there exists a constant C, independent of v and the length of the interval E, such that, for any d, $0<d\le 1$,*

$$\|v\|_{W^{s,p}(E)} \le C[\, d^{m-s}\|v\|_{W^{m,p}(E)} + d^{-s}\|v\|_{L^p(E)}], 0\le s\le m-1. \tag{2.8}$$

We state the following results without proof which are essential for the convergence analysis. For the proof of *Sobolev imbedding theorem*, one may refer to Adams ([1], page 97).

Lemma 2.2 *Let $\Omega\subset\mathbf{R}^n$ be bounded open set. For $mp>n$, we have*

$$W^{j+m,p}(\Omega)\hookrightarrow C_B^j(\Omega),$$

and for $p=1, m=n$, we have

$$W^{j+m,1}(\Omega)\hookrightarrow C_B^j(\Omega),$$

where $C_B^j(\Omega)=\{u\in C^j(\Omega): D^\alpha u$ is bounded on $\Omega for|\alpha|\le j\}$ is a Banach space with respect to the norm $\|u\| = \max\limits_{0\le|\alpha|\le j}\sup\limits_{x\in\Omega}|D^\alpha u(x)|$.

In particular for I, and $v \in W^{1,p}(I), 1 \leq p \leq \infty$, we have

$$\|v\|_{L^{p'}(I)} \leq \|v\|_{W^{1,p}(I)} \quad 1 \leq p, p' \leq \infty. \tag{2.9}$$

Moreover we shall use the following lemma due to *Ehrling* and *Browder* (see Adams [1], page 79) in our error analysis.

Lemma 2.3 *If $\Omega \subset \mathbf{R}^n$ has the uniform cone property and if $1 \leq p \leq \infty$ then there exists a constant $K = K(m,p,\Omega)$ such that for $0 \leq j \leq m$ and $v \in W^{m,p}(\Omega)$*

$$\|v\|_{W^{j,p}(\Omega)} \leq K \|v\|^{j/m}_{W^{m,p}(\Omega)} \|v\|^{(m-j)/m}_{L^p(\Omega)}.$$

Therefore, for $j = 1, m = 2$ and $\Omega = I_k$ we have

$$\|v\|^2_{1,p,k} \leq K_1 \|v\|_{2,p,k} \|v\|_{0,p,k}. \tag{2.10}$$

3. Convergence Analysis

Let $e = u - u^h$. From (1.1) and (1.8), we have the following error equation in e

$$\langle \mathcal{M}(u)e, \chi \rangle = -\left\langle \tilde{f}_{uu} e^2, \chi \right\rangle, \ \chi \in T_h, \tag{3.1}$$

where $\tilde{f}_{uu} = \int_0^1 (1-s) f_{uu} \left(u^h - se\right) ds$.

In the following lemma, we shall obtain an estimate for $\epsilon_h(\mathcal{M}(u)e\chi)$. The proof technique follows closely that of Sloan *et al.* [9] with appropriate modification due to the presence of a term $f_u e$. So for completeness, we shall present the proof.

Lemma 3.1 *Assume that (1.7) holds. If $e = u - u^h$ then for all $\chi \in T_h$*

$$\begin{aligned} (i) \quad & |\epsilon_h(\mathcal{M}(u)e\chi)| \leq C\left(h\|e\|_{0,p} + h^4\|u\|_{6,p}\right)\|\chi\|_{1,q} \\ (ii) \quad & |\epsilon_h(\mathcal{M}(u)e\chi)| \leq C\left(\|e\|_{0,p} + h^3\|u\|_{5,p} + h^4\|u\|_{6,p}\right)\|\chi\|_{q}. \end{aligned}$$

Here the constant C depends on $\max_{0 \leq k \leq N} \|u\|_{4,\infty,k}$, the boundedness of b, c, f and their derivatives.

Proof. On each subinterval I_k the second and higher order derivatives of the elements in T_h vanish. Thus

$$\begin{aligned} \|(\mathcal{M}(u)e\chi)^{(4)}\|_{1,k} &= \|(\mathcal{M}(u)e)^{(4)}\chi + 4(\mathcal{M}(u)e)^{(3)}(\chi)^{(1)}\|_{1,k} \\ &\leq \|\mathcal{L}e\|_{4,p,k}\|\chi\|_{q,k} + 4\|\mathcal{L}e\|_{3,p,k}\|\chi'\|_{q,k} \qquad (3.2) \\ &+ C(f)\|u\|_{4,\infty,k}\|e\|_{4,p,k}\|\chi\|_{q,k} + C(f)\|u\|_{3,\infty,k}\|e\|_{3,p,k}\|\chi'\|_{q,k} \\ &\leq C\left(\|e\|_{6,p,k}\,\|\chi\|_{q,k} + \|e\|_{5,p,k}\,\|\chi'\|_{q,k}\right) \qquad (3.3) \\ &\leq C\|e\|_{6,p,k}\,\|\chi\|_{1,q,k} \qquad (3.4) \end{aligned}$$

As $u^h \in S_h$, we have

$$\begin{aligned}\|e\|_{6,p,k} &\leq \left(\|e\|_{3,p,k}^p + \|u\|_{6,p,k}^p\right)^{1/p} \leq C\left(\|e\|_{3,p,k} + \|u\|_{6,p,k}\right) \\ &\leq C\left(h_k^{-3}\|e\|_{p,k} + h_k^3\|e\|_{6,p,k}\right) + \|u\|_{6,p,k},\end{aligned}$$

where the second step follows from (2.8) inequality and last step follows from the interpolation inequality (2.8) with $s = 3$ and $m = 6$. Thus for h_k sufficiently small, we have

$$\|e\|_{6,p,k} \leq C\left(h_k^{-3}\|e\|_{0,p,k} + \|u\|_{6,p,k}\right). \tag{3.5}$$

From (1.7), (3.4), it follows that

$$|\epsilon_h(\mathcal{M}(u)e\chi)| \leq C\sum_{k=1}^{N} h_k^4\|(\mathcal{M}(u)e\chi)^{(4)}\|_{1,k} \leq C\sum_{k=1}^{N}\left(h_k\|e\|_{0,p,k} + h_k^4\|u\|_{6,p,k}\right)\|\chi\|_{1,q,k}$$

and with the aid of the Hölder inequality for sums, we obtain (i).
To prove (ii), apply the inverse estimate on the interval I_k

$$\|\chi'\|_{q,k} \leq Ch_k^{-1}\|\chi\|_{q,k}, \quad \chi \in T_h,$$

in (3.3) and use (3.5) to have

$$\begin{aligned}\|(\mathcal{M}(u)e\chi)^{(4)}\|_{1,k} &\leq C\left(\|e\|_{6,p,k} + h_k^{-1}\|e\|_{5,p,k}\right)\|\chi\|_{q,k} \\ &\leq \left(h_k^{-3}\|e\|_{0,p,k} + h_k^{-1}\|u\|_{5,p,k} + \|u\|_{6,p,k}\right)\|\chi\|_{q,k} \\ &\leq \left(h_k^{-4}\|e\|_{p,k} + h_k^{-1}\|u\|_{5,p,k} + \|u\|_{6,p,k}\right)\|\chi\|_{q,k}.\end{aligned}$$

Here, we have used (2.8) in the second last inequality for the term $|\frac{\partial^j u}{\partial t^j}\|_{5,p,k}$ with $m = 6$ and $s = 3$. The result (ii) now follows from (1.7), (3.3). ∎
We now obtain an estmate of the error $\epsilon_h(\tilde{f}_{uu}e^2\chi)$ assocated with the quadrature rules for the nonlinear term of the error equation (3.1). For the rest of this paper, we shall assume that $\|h^h\|_{2,\infty} \leq K^*$ for somr $K^* \geq K$ with $\|u\|_{2,\infty} \leq K$.

Lemma 3.2 *Assume that (1.7) holds. If $e = u - u^h$ then for K^* with $\|u_h\|_{2,\infty}$ and for all $\chi \in T_h$*

$$\begin{aligned}(i) \qquad |\epsilon_h(\tilde{f}_{uu}e^2\chi)| &\leq C\left(h\|e\|_{0,p} + h^4\|u\|_{4,p}\right)\|\chi\|_{1,q} \\ (ii) \qquad |\epsilon_h(\tilde{f}_{uu}e^2\chi)| &\leq C\left(\|e\|_{0,p} + h^3\|u\|_{3,p} + h^4\|u\|_{4,p}\right)\|\chi\|_{q}.\end{aligned}$$

Here the constant C depends on K^*, $\|u\|_{2,\infty}$, the boundedness of b, c, f and their derivatives.

Proof. Using the fact that on each subinterval I_k the second and higher order derivatives of the elements in T_h vanish, we have

$$\|(\tilde{f}_{uu}e^2\chi)^{(4)}\|_{1,k} \leq \|(\tilde{f}_{uu}e^2)^{(4)}\chi\|_{1,k} + 4\|(\tilde{f}_{uu}e^2)^{(3)}(\chi)^{(1)}\|_{1,k}. \tag{3.6}$$

Apply Leibntz's rule to obtain the following,

$$\begin{aligned}(\tilde{f}_{uu}e^2)^{(4)} &= (\tilde{f}_{uu})^{(4)}e^2 + 4(\tilde{f}_{uu})^{(3)}(e^2)^{(1)} + 6(\tilde{f}_{uu})^{(2)}(e^2)^{(2)} + 4(\tilde{f}_{uu})^{(1}(e^2)^{(3)} \\ &+ 4(\tilde{f}_{uu})(e^2)^{(4)} \\ &= (\tilde{f}_{uu})^{(4)}e^2 + 8(\tilde{f}_{uu})^{(3)}\left(ee^{(1)}\right) + 12(\tilde{f}_{uu})^{(2)}\left(e^{(1)}e^{(1)} + ee^{(2)}\right) \\ &+ 8(\tilde{f}_{uu})^{(1)}\left(3e^{(1)}e^{(2)} + ee^{(3)}\right) + 8(\tilde{f}_{uu})\left(2ee^{(4)} + 8e^{(1)}e^{(3)} + e^{(2)}e^{(2)}\right),\end{aligned}$$

where $e^{(j)}$ denotes the jth derivative and use Hölder's inequality for the first integral in (3.6) to have

$$\|(\tilde{f}_{uu}e^2)^{(4)}\chi\|_{1,k} \leq C(K^*)\left[\|e\|_{p,k} + \|e^{(1)}\|_{p,k} + \|e^{(2)}\|_{p,k} + \|e^{(3)}\|_{p,k} + \|e^{(4)}\|_{p,k}\right]\|\chi\|_{q,k}.$$

Now applying Hölder inequality for the sums involving $e-$ terms, we have

$$\|(\tilde{f}_{uu}e^2)^{(4)}\chi\|_{1,k} \leq C(K^*)\|e\|_{4,p,k}\|\chi\|_{q,k}.$$

Similarly for the last term in (3.6), we obtain

$$\|(\tilde{f}_{uu}e^2)^{(3)}\chi^{(1)}\|_{1,k} \leq C(K^*)\|e\|_{3,p,k}\|\chi'\|_{q,k}.$$

Substituting the above two inequalities in (3.6), it follows that

$$\begin{aligned}\|(\tilde{f}_{uu}e^2\chi)^{(4)}\|_{1,k} &\leq C(K^*)\left(\|e\|_{4,p,k}\,\|\chi\|_{q,k} + \|e\|_{3,p,k}\,\|\chi'\|_{q,k}\right) \\ &\leq C(K^*)\|e\|_{4,p,k}\,\|\chi\|_{1,q,k}. \end{aligned} \tag{3.7}$$

As $u^h \in S_h$, we have using (2.8)

$$\begin{aligned}\|e\|_{4,p,k} &\leq C(K^*)\left(\|e\|^p_{3,p,k} + \|u\|^p_{4,p,k}\right)^{1/p} \leq C(K^*)\left(\|e\|_{3,p,k} + \|u\|_{4,p,k}\right) \\ &\leq C(K^*)\left(h_k^{-3}\|e\|_{p,k} + h_k\|e\|_{4,p,k} + \|u\|_{4,p,k}\right),\end{aligned}$$

last step follows from the interpolation inequality (2.8) with $s = 3$ and $m = 4$. Thus, for h_k sufficiently small, we obtain

$$\|e\|_{4,p,k} \leq C(K^*)\left(h_k^{-3}\|e\|_{0,p,k} + \|u\|_{4,p,k}\right). \tag{3.8}$$

From (1.7), (3.4), it follows that

$$\begin{aligned}|\epsilon_h(\tilde{f}_{uu}e^2\chi)| &\leq C(K^*)\sum_{k=1}^{N} h_k^4\|(\tilde{f}_{uu}e^2\chi)^{(4)}\|_{1,k} \\ &\leq C(K^*)\sum_{k=1}^{N}\left(h_k\|e\|_{0,p,k} + h_k^4\|u\|_{4,p,k}\right)\|\chi\|_{1,q,k}\end{aligned}$$

and with the aid of the Hölder inequality for sums, we obtain (i).
To prove (ii), apply the inverse estimate on the interval I_k,

$$\|\chi'\|_{q,k} \leq Ch_k^{-1}\|\chi\|_{q,k}, \;\; \chi \in T_h,$$

in (3.7) and use (3.8) to have

$$\begin{aligned}
\|(\tilde{f}_{uu}e^2\chi)^{(4)}\|_{1,k} &\leq C(K^*)\left(\|e\|_{4,p,k}+h_k^{-1}\|e\|_{3,p,k}\right)\|\chi\|_{q,k}\\
&\leq \left(h_k^{-3}\|e\|_{0,p,k}+h_k^{-1}\|u\|_{3,p,k}+\|u\|_{4,p,k}\right)\|\chi\|_{q,k}\\
&\leq \left(h_k^{-4}\|e\|_{p,k}+h_k^{-1}\|u\|_{5,p,k}+\|u\|_{6,p,k}\right)\|\chi\|_{q,k}.
\end{aligned}$$

Here we have used (2.8) in the second last inequality for the term $|\frac{\partial^j u}{\partial t^j}\|_{3,p,k}$ with $m = 6$ and $s = 3$. The result (ii) now follows from (1.7), (3.7). ∎

Below, we shall prove the main result of this section i.e., the main error estimate of e in $W^{i,p}, i = 0,1,2$ and $1 \leq p \leq \infty$.

Theorem 3.1 *Let u satisfy (1.1) and (2.1). Let the coefficients b, c and the nonlinear function f be sufficiently smooth. Further let the operator $\mathcal{M}(u)$ and its formal adjoint $\mathcal{M}^*(u)$ satisfy the required regularity condition (2.7). Assume that there exists a unique solution $u^h \in S_h$ with $\|u^h\|_{2,\infty} \leq K^*$ for some K^*. Then for $u \in W^{6,p}$ with $1 \leq p \leq \infty$ and for h sufficiently small there is a constant $C = C(K^*)$ such that*

$$\|u-u^h\|_{i,p} \leq Ch^{4-i}\|u\|_{6,p}, i = 0,1,2.$$

Moreover for small h, the constant C can be made independent of K^*.

Proof. Let Ψ be an arbitrary element of L^q, and let $\phi \in W^{2,q}$ be the unique solution of

$$\mathcal{M}^*(u)\phi = \Psi, \ x \in I \tag{3.9}$$

subject to the adjoint boundary conditions (2.6). Using (2.7) the solution ϕ satisfies the following regularity condition

$$\|\phi\|_{2,q} \leq C\|\Psi\|_q.$$

Take L^2-inner product with e and use the error equation (3.1) to have

$$\begin{aligned}
(e, \Psi) &= (e, \mathcal{M}^*(u)\phi) = (\mathcal{M}(u)e, \phi)\\
&= (\mathcal{M}(u)e, (\phi-\chi)) + (\mathcal{M}(u)e, \chi) - \langle\mathcal{M}(u)e, \chi\rangle - \left\langle \tilde{f}_{uu}e^2, \chi\right\rangle\\
&= (\mathcal{M}(u)e, (\phi-\chi)) - \epsilon_h(\mathcal{M}(u)e\chi) - \epsilon_h(\tilde{f}_{uu}e^2\chi) - \left(\tilde{f}_{uu}e^2, \chi\right).
\end{aligned}$$

Choose $\chi = \phi_h$, where ϕ_h is the piece-wise linear interpolant of ϕ. To estimate the last term, we assume temporarily that $\|e\|_{0,\infty} \leq C(K^*)\nu(h)$ with $\nu(h) \to 0$ as $h \to 0$

$$|\left(\tilde{f}_{uu}e^2, \phi_h\right)| \leq C(f)\|u-u_h\|_{0,\infty}\|e\|_{0,p}\|\phi_h\|_{0,q} \leq C\nu(h)\|e\|_{0,p}\|\phi_h\|_{0,q}. \tag{3.10}$$

Substitute this to have

$$\begin{aligned}
|(e, \Psi)| &\leq C\|\eta\|_{2,p}\|\phi-\phi_h\|_q + |\epsilon_h(\mathcal{M}(u)e\phi_h)| + |\epsilon_h(\tilde{f}_{uu}e^2\phi_h)|\\
&+ C\nu(h)\|e\|_{0,p}\|\phi_h\|_{1,q},
\end{aligned}$$

and hence, using Lemma 3.1 (i) ans Lemma 3.2 (i), we find that

$$| (e, \Psi) | \leq C(K^*) \left[h^2 \|e\|_{2,p} \|\phi\|_{2,q} + (h + \nu(h)) \|e\|_{0,p} + h^4 \|u\|_{6,p} + h^4 \|u\|_{4,p}\right] \|\phi_h\|_{1,q}.$$

Since,

$$\|\phi_h\|_{1,q} \leq \|\phi - \phi_h\|_{1,q} + \|\phi\|_{1,q} \leq Ch \|\phi\|_{2,q} + \|\phi\|_{1,q} \leq C \|\phi\|_{2,q}, \tag{3.11}$$

we obtain

$$| (e, \Psi) | \leq C(K^*)(h^2 \|e\|_{2,p} + (h + \nu(h)) \|e\|_{0,p} + h^4 \|u\|_{4,p} + h^4 \|u\|_{6,p}) \|\phi\|_{2,q}.$$

From the regularity condition and the fact that $\Psi \in L^q$ is arbitrary, it follows that

$$\|e\|_p \leq C(K^*)(h^2 \|e\|_{2,p} + (h + \nu(h)) \|e\|_{0,p} + h^4 \|u\|_{6,p}).$$

Choose h sufficiently small so that $(1 - C(h + \nu(h))) > 0$ and hence,

$$\|e\|_p \leq C(h^2 \|e\|_{2,p} + h^4 \|u\|_{6,p}). \tag{3.12}$$

The next step is to estimate $\|e''\|_p$. For this, let us introduce an orthogonal projection P_h from H^2 onto the space T_h with respect to the following form

$$\left((u - P_h u)'', \chi\right) = 0, \quad \forall \chi \in T_h. \tag{3.13}$$

It is shown in De Boor [2] that this L^2 projection is stable in L^p-norm i.e.,

$$\|P_h v\|_{0,p} \leq \|v\|_{0,p}, \quad v \in L^p, 1 \leq p \leq \infty. \tag{3.14}$$

(It is noted that, the elements of T_h have just C^0 continuity, this result holds with no restriction on the mesh (see de Boor [2], Section 4). Further, one may refer Crouzeix and Thomée [3], for a different proof.) Now write

$$\|e''\|_{0,p} = \|(u - u^h)''\|_{0,p} \leq \|(u - P_h u)''\|_{0,p} + \|(P_h u - u^h)''\|_{0,p}, \tag{3.15}$$

and from the stability property (3.14) it follows as in de Boor [2] that

$$\|(u - P_h u)''\|_{0,p} \leq Ch^2 \|u''\|_{2,p} \leq Ch^2 \|u\|_{4,p}. \tag{3.16}$$

Therefore, it remains to estimate the second term on the right hand side of (3.15). From (3.1) and (3.13), we have

$$\begin{aligned}
&\left((P_h u - u^h)'', \chi\right) = \left((u - u^h)'', \chi\right) = \left(e'', \chi\right) \\
&= \left(\mathcal{M}(u)e - be' - ce - f_u(u)e, \chi\right) \\
&= \left(-be' - ce - f_u(u)e, \chi\right) + [(\mathcal{M}(u)e, \chi) - \langle \mathcal{M}(u)e, \chi \rangle] - \left\langle \tilde{f}_{uu} e^2, \chi \right\rangle \\
&= \left(-be' - ce - f_u(u)e, \chi\right) + \epsilon_h(\mathcal{M}(u)e\chi) - \epsilon_h(\tilde{f}_{uu} e^2 \chi) - \left(\tilde{f}_{uu} e^2, \chi\right).
\end{aligned}$$

For the estimation of the last term in the above equation, we use (3.10) to obtain

$$\begin{aligned} |\left(\tilde{f}_{uu}e^2, \chi\right)| &\leq C\|u - u_h\|_{0,\infty}\|e\|_{0,p}\|\chi\|_{0,q} \leq C(K^*)\nu(h)\|e\|_{0,p}\|\chi\|_{0,q} \\ &\leq C(K^*)\nu(h)\|e\|_{1,p}\|\chi\|_{1,q} \end{aligned}$$

From Lemma 3.1 (ii) and Lemma 3.2 (ii) the above inequality, we have

$$|\left((P_h u - u^h)'', \chi\right)| \leq C(K^*)(\nu(h)\|e\|_{1,p} + h^3\|u\|_{5,p} + h^4\|u\|_{6,p})\ \|\chi''\|_{0,q}.$$

Let ψ be an arbitrary element of L^q. Then from the definition of P_h and the above inequality, it follows that

$$\begin{aligned} |\left((P_h u - u^h)'', \psi\right)| &= |\left((P_h u - u^h)'', P_h\psi\right)| \\ &\leq C(K^*)(\nu(h)\|e\|_{1,p} + h^3\|u\|_{5,p} + h^4\|u\|_{6,p})\|P_h\psi\|_{0,q} \\ &\leq C(K^*)(\nu(h)\|e\|_{1,p} + h^3\|u\|_{5,p} + h^4\|u\|_{6,p})\|\psi\|_{0,q}. \end{aligned}$$

Here in the last step we have used (3.14) with p replaced by q. Hence,

$$\|(P_h u - u^h)''\|_{0,p} \leq C(h^*)(\nu(h)\|e\|_{1,p} + h^3\|u\|_{5,p} + h^4\|u\|_{6,p}).$$

On combining the above estimate with (3.16), we have from (3.15)

$$\|e''\|_{0,p} \leq C(K^*)(\nu(h)\|e\|_{1,p} + h^2\|u\|_{4,p} + h^3\|u\|_{5,p} + h^4\|u\|_{6,p}).$$

Note that

$$\|e\|_{2,p} \leq \|e\|_{1,p} + \|e''\|_p,$$

and from interpolation inequality (2.8), we have

$$\|e\|_{1,p} \leq C(K^*)(h^{-1}\nu(h)\|e\|_{0,p} + h\|e\|_{2,p}), \tag{3.17}$$

Substituting above estimate and $\|e''\|_{0,p}$ in $\|e\|_{2,p}$, we obtain again for h sufficiently small

$$\|e\|_{2,p} \leq C(h^{-1}\|e\|_{0,p} + h^2\|u\|_{4,p} + h^3\|u\|_{5,p} + h^4\|u\|_{6,p}),$$

and hence, using (3.12) it follows that for h sufficiently small

$$\|e\|_{2,p} \leq C(K^*)(h^2\|u\|_{4,p} + h^3\|u\|_{6,p}) \leq Ch^2\|u\|_{6,p}.$$

Using the above inequality in (3.12), we obtain for h sufficiently small

$$\|e\|_p \leq C(K^*)h^4\|u\|_{6,p}.$$

For i=1 we have from (3.17)

$$\|e\|_{1,p} \leq C(K^*)h^3\|u\|_{6,p}.$$

Finally with $\|u\|_{2,\infty} \leq K$, we obtain for $K^* \geq 2K$ and sufficiently small h

$$\|u^h\|_{2,\infty} \leq \|u - u^h\|_{2,\infty} + \|u\|_{2,\infty} \leq C(K^*)h^2\|u\|_{6,p} + K \leq 2K \leq K^*.$$

Hence the constant C can be chosen independentof K^*. For the temporary assumption, we note that

$$\|u - u_h\|_{L^\infty} \leq Ch^4\|u\|_{6,\infty} = \nu(h).$$

Therefore, as $h \to 0, \nu(h) \to 0$. This completes the proof. ■

Remark. In case we assume quasiuniformity condition on the finite dimensional space S_h, we obtain using global inverse property the following error estimates in $W^{3,p}$-norm,

$$\begin{aligned}
\|e\|_{3,p} &\leq \|u - \Pi_h u\|_{3,p} + \|\Pi_h u - u^h\|_{3,p} \\
&\leq C\left[h\|u\|_{4,p} + h^{-1}\|\Pi_h u - u^h\|_{2,p}\right] \\
&\leq C\left[h\|u\|_{4,p} + h^{-1}(\|u - \Pi_h u\|_{2,p} + \|e\|_{2,p})\right] \\
&\leq Ch\|u\|_{6,p}
\end{aligned}$$

Here $\Pi_h u$ is an interpolant of u on to S_h and in second step, we have used global inverse inequality. ■

To find an error bound for the derivatives of e at the knot point $\bar{x}$ in $[0,1]$, it is convenient to define the following spaces that generalize $W^{s,p}(I)$. For $s \geq 0$ and for a fixed point $\bar{x} \in [0,1]$, let

$$\tilde{W}^{s,p}(I) = \{u : u|_{(0,\bar{x})} \in W^{s,p}(0,\bar{x}), u|_{(\bar{x},1)} \in W^{s,p}(\bar{x},1)\},$$

and let

$$|||u|||_{s,p} = \left(\|u\|^p_{W^{s,p}(0,\bar{x})} + \|u\|^p_{W^{s,p}(\bar{x},1)}\right)^{1/p}$$

be a norm on $\tilde{W}^{s,p}(I)$. For $s = 0$, $\tilde{W}^{s,p}(I) = L^p$. Below, we shall prove superconvergence result for the flux at the knot point $\bar{x} \in [0,1]$.

Theorem 3.2 *With the assumptions of Theorem 3.1, the following superconvergene result*

$$|e'(\bar{x})| \leq Ch^4\|u\|_{6,p}, \quad 1 \leq p \leq \infty,$$

holds for the flux at the nodal point $\bar{x} \in [0,1]$.

Proof. For the estimation of flux at the knot point $\bar{x} \in [0,1]$, let $\phi \in \tilde{W}^{2,q}$ with $1/p + 1/q = 1$ be a solution of

$$\begin{cases}
\mathcal{M}^*(u)\phi = 0, & x \in I\backslash\{\bar{x}\} \\
\left[\phi'(\bar{x}_+) - \phi'(\bar{x}_-)\right] = b(\bar{x}) & \\
[\phi(\bar{x}_+) - \phi(\bar{x}_-)] = 1 &
\end{cases}$$

with adjoint boundary condition (2.6). Multiplying by e and integrating by parts with respect to x twice we have using boundary conditions (2.1) and (2.6)

$$e'(\bar{x}) = (\mathcal{M}(u)e, \phi) - e(\bar{x})[\phi'(\bar{x}_+) - \phi'(\bar{x}_-)] + b(\bar{x})e(\bar{x})[\phi(\bar{x}_+) - \phi(\bar{x}_-)].$$

Thus, from the above jump conditions we have

$$\begin{aligned} e'(\overline{x}) &= (\mathcal{M}(u)e,\, \phi) = (\mathcal{M}(u)e,\, (\phi-\chi)) - \epsilon_h(\mathcal{M}(u)e\chi) - \epsilon_h(\tilde{f}_{uu}e^2\chi) \\ &- \left(\tilde{f}_{uu}e^2,\, \chi\right). \end{aligned}$$

Now choose $\chi = \phi_h$, where ϕ_h is the piece-wise linear interpolant of ϕ. Then sustitute this to have

$$| e'(\overline{x}) | \leq C[\|e\|_{2,p}\|\phi - \phi_h\|_q + | \epsilon_h(\mathcal{M}(u)e\phi_h) | + | \epsilon_h(\tilde{f}_{uu}e^2\chi) | \, C(K^*)h^2\|e\|_{0,p}\|\phi_h\|_{1,q}.$$

Using the Lemma 3.1 (i) and Lemma 3.2 with $1 \leq p \leq \infty$ and $1/p + 1/q = 1$, we obtain

$$| e'(\overline{x}) | \leq C[h^2\|e\|_{2,p} + (h\|e\|_{0,p} + h^4\|u\|_{6,p})]\||\phi\||_{2,q}.$$

Again the regularity condition on ϕ and Theorem 3.1 complete the rest of proof. ■

4. Existence and Uniqueness of Approximate Solution

In this section, we shall show existence of an unique solution u^h satisfying (1.8) with $\|u^h\|_{2,p} \leq K^*$, for some $K^* > 0$ and for h sufficiently small. We shall demonstrate this by adopting the analysis of Douglas [6].

From the equation (3.1), we say that $w \in S_h$ is an approximation of u if and only if w satisfies the following equation

$$\langle \mathcal{M}(u)(u-w),\, \chi \rangle = -\left\langle \tilde{f}_{uu}(u-w)^2,\, \chi \right\rangle,\ \chi \in T_h, \tag{4.1}$$

with $\tilde{f}_{uu}$ depending on u and w, i.e., $\tilde{f}_{uu} = \int_0^1 (1-s) f_{uu}(w - s(u-w))\, ds$. Thus, to prove existence, it is sufficient to seek a fixed point for the map $\mathcal{F} : w \mapsto z \in S_h$ defined by the equation

$$\langle \mathcal{M}(u)(u-z),\, \chi \rangle = -\left\langle \tilde{f}_{uu}(u-w)^2,\, \chi \right\rangle,\ \chi \in T_h \tag{4.2}$$

with $\tilde{f}_{uu}$ depending on u and w.

Let us first verify that the map $\mathcal{F}$ is well defined, *i.e.*, to check that the system of linear equations (4.2) has a unique solution $z \in S_h$ for a given u and w. For this purpose, it is enough to check that for $u \equiv 0$ and $w \equiv 0$, the equation

$$\langle \mathcal{M}(u)(u-z),\, \chi \rangle = 0\ \chi \in T_h,$$

has identically zero solution, *i.e.*, $z \equiv 0$. From the arguments of Sloan et.al., 1993 with $u \equiv 0$ that for h sufficiently small, we have $\|z\|_{2,p} \leq 0$ and hence $z \equiv 0$. As existence follows from uniqueness for sufficiently small h, well defindedness of $\mathcal{F}$ is established.

Below, we shall discuss existence of an approximate solution in a neighbourhood of u.

Theorem 4.1 *Let u be smooth solution of (1.1) and (2.1). If $\delta > 0$, then for h sufficiently small there exists an approximate solution $u^h \in S_h$ of (1.8) lying inside the ball $\|u - u^h\|_{2,p} \leq \delta$.*

Proof. To show the existence of u^h to (1.8), it is enough to apply Brouwer fixed point arguments to the map $\mathcal{F}$. Therefore, let us first check that $\mathcal{F}$ maps a ball onto itself for sufficiently small h. Now, let Ψ be an arbitrary element of L^q, and let $\phi \in W^{2,q}$ be the unique solution of the auxilliary problem (3.9) with adjoint boundary conditions (2.6). Take L^2-inner product with $(u - z)$ and use (4.2) to have

$$\begin{aligned}((u-z), \Psi) &= (\mathcal{M}(u)(u-z), (\phi - \chi)) - \epsilon_h(\mathcal{M}(u)(u-z)\chi) \\ &- \epsilon_h(\tilde{f}_{uu}(u-w)^2\chi) - \left(\tilde{f}_{uu}(u-w)^2, \chi\right).\end{aligned}$$

Choose $\chi = \phi_h$, where ϕ_h is the piece-wise linear interpolant of ϕ. For the estimation of the last term, use (3.10) with $(u - w)$ in place of e, to have

$$\left|\left(\tilde{f}_{uu}(u-w)^2, \phi_h\right)\right| \leq C\nu(h)\|u-w\|_{0,p}\|\phi_h\|_{1,q}. \tag{4.3}$$

With appropriate modification of the arguments in Theroem 3.1 (*i.e.*, replacing e by $(u - z)$ on the left hand side and by $(u - w)$ on the right hand side), we obtain for sufficiently small h

$$\|u-z\|_p \leq C\left(h^2\|u-z\|_{2,p} + \nu(h)\|u-w\|_{0,p} + h^4\|u\|_{6,p}\right). \tag{4.4}$$

For the estimation of $\|(u-z)''\|_p$. we shall again use the orthogonal projection P_h as defined in (3.13) and obtain

$$\|(u-z)''\|_{0,p} \leq \|(u-P_hu)''\|_{0,p} + \|(P_hu-z)''\|_{0,p}.$$

Since the estimate of the first term on the right hand side follows from (3.14), it remains to estimate the second term. From (4.2), we have

$$\begin{aligned}&\left((P_hu-z)'', \chi\right) = \left((u-z)'', \chi\right) \\ &= (\mathcal{M}(u)(u-z) - b(u-z)' - c(u-z) - f_u(u)(u-z), \chi) \\ &= \left(-b(u-z)' - c(u-z) - f_u(u)(u-z), \chi\right) + \epsilon_h(\mathcal{M}(u)(u-z)\chi) \\ &-\epsilon_h(\tilde{f}_{uu}(u-w)^2\chi) - \left(\tilde{f}_{uu}(u-w)^2, \chi\right).\end{aligned}$$

From Lemma 3.1 (ii), Lemma 3.2 and (4.3), we have

$$\left|\left((P_hu-z)'', \chi\right)\right| \leq C(\|u-z\|_{1,p} + (\nu(h)+h^2)\|u-w\|_{0,p} + h^3\|u\|_{5,p} + h^4\|u\|_{6,p})\ \|\chi\|_{0,q}.$$

Let ψ be an arbitrary element of L^q. Then from the definition of P_h and the above inequality, it follows that

$$\begin{aligned}&\left|\left((P_hu-z)'', \psi\right)\right| = \left|\left((P_hu-z)'', P_h\psi\right)\right| \\ &\leq C(\|u-z\|_{1,p} + (\nu(h)+h^2)\|u-w\|_{0,p} + h^3\|u\|_{5,p} + h^4\|u\|_{6,p})\|P_h\psi\|_{0,q} \\ &\leq C(\|u-z\|_{1,p} + (\nu(h)+h^2)\|u-w\|_{0,p} + h^3\|u\|_{5,p} + h^4\|u\|_{6,p})\|\psi\|_{0,q}\end{aligned}$$

where in the last step we have used (3.14) with p replaced by q. Similarly, for the estimation of $\|u-z\|_{2,p}$ we obtain

$$\begin{aligned}\|u-z\|_{2,p} &\leq C\left(\|u-z\|_{1,p}+\|(u-z)''\|_{0,p}\right)\\ &\leq C\left(h^{-1}\|u-z\|_{0,p}+h\|u-z\|_{2,p}+\|(u-z)''\|_{0,p}\right)\end{aligned}$$

Using the estimates of $\|(u-z)''\|_{0,p}$ and $\|u-z\|_{0,p}$ in (4.4), we have for small h

$$\|u-z\|_{0,p} \leq C\left((1+h^2)h\|u-w\|_{0,p}+h^4\|u\|_{6,p}\right).$$

For $0<\delta<1$, let $\|u-w\|_{2,p}\leq\delta$, then

$$\|u-z\|_{2,p} \leq C\left((\nu(h)+h^2)h\delta^2+h^4\|u\|_{6,p}\right).$$

Now for h sufficiently small, the left hand side can be made smaller that δ, and hence

$$\|u-z\|_{2,p}\leq\delta.$$

Therefore, we conclude that the map $\mathcal{F}$ maps a δ-ball into itself. Further it is not difficult to check that the map $\mathcal{F}$ is contionous on the finite dimensional subspace S_h. Now an application of Brouwer fixed point theorem implies existence of a solution $u_h \in S_h$ of (3.1) lying in the ball $\|u-u_h\|\leq\delta$. Thus we have proved that if $u\in W^{6,p}$ is a solution of (1.1) and if $\delta>0$, then for h sufficiently small there exists an approximate solution $u^h\in S_h$ of (1.8) lying inside the ball $\|u-u^h\|_{2,p}\leq\delta$. ■

In the following we shall prove uniqueness result.

Theorem 4.2 *Let u be a smooth solution of (1.1) and (2.1). Let K be an arbitrary positive number. Then for h sufficiently small, there is a unique approximate solution u^h of (1.8) such that $u^h\in S_h\cap\{\|u-u^h\|_{2,p}\leq K\}$.*

Proof. Assume that there are two solutions $u_i^h, i=1,2$ of (1.8), i.e., each u_i satisfies

$$\left\langle\mathcal{M}(u)(u-u_i^h),\,\chi\right\rangle=-\left\langle\tilde{f}_{i,uu}(u-u_i^h)^2,\,\chi\right\rangle,\ \chi\in T_h, i=1,2, \tag{4.5}$$

where $\tilde{f}_{i,uu}=\int_0^1(1-s)f_{uu}\left(u_i^h-s(u-u_i^h)\right)ds$. Let $\theta=u_1^h-u_2^h$ and $\zeta_i=u-u_i^h$. Then from (4.5), it now follows that

$$\begin{aligned}\langle\mathcal{M}(u)\theta,\,\chi\rangle &=\left\langle\tilde{f}_{1,uu}(\zeta_1)^2,\,\chi\right\rangle-\left\langle\tilde{f}_{2,uu}(\zeta_2)^2,\,\chi\right\rangle\\ &=\left\langle\tilde{f}_{1,uu}(\zeta_1^2-\zeta_2^2),\,\chi\right\rangle-\left\langle(\tilde{f}_{2,uu}-\tilde{f}_{1,uu})(\zeta_2)^2,\,\chi\right\rangle\left\langle\tilde{f}_{1,uu}\theta(\zeta_1+\zeta_2),\,\chi\right\rangle\\ &-\left\langle(\tilde{f}_{2,uu}-\tilde{f}_{1,uu})(\zeta_2)^2,\,\chi\right\rangle.\end{aligned}$$

To obtain an estimation for θ, we shall once more repeat the arguments given in the proof of Theorem 3.1. Let Ψ be an arbitrary element of L^q, and let $\phi \in W^{2,q}$ be the unique solution of the auxilliary problem (3.9). Take L^2-inner product with θ and use (4.5) to have

$$\begin{aligned}(\theta,\,\Psi) &= (\mathcal{M}(u)\theta,\,(\phi-\chi)) - \epsilon_h(\mathcal{M}(u)\theta\phi) \\ &+ \left\langle \tilde{f}_{1,uu}\theta(\zeta_1+\zeta_2),\,\chi\right\rangle - \left\langle (\tilde{f}_{2,uu}-\tilde{f}_{1,uu})(\zeta_2)^2,\,\chi\right\rangle \qquad (4.6)\end{aligned}$$

Choose $\chi = \phi_h$, where ϕ_h is the piece-wise linear interpolant of ϕ. For the estimation of the second term, using the fact that $\theta \in S_h$, the fourth and higher order derivatives of ϕ_h and θ vanish on each subinterval I_k, we have

$$\begin{aligned}\|(\mathcal{M}(u)\theta\phi_h)^{(4)}\|_{1,k} &= \|(\mathcal{M}(u)\theta)^{(4)}\phi_h + 4(\mathcal{M}(u)\theta^{(3)}(\phi_h)^{(1)}\|_{1,k} \\ &\le C\left(\|\mathcal{L}\theta\|_{3,p,k}\|\phi_h\|_{q,k} + \|\mathcal{L}\theta\|_{3,p,k}\|\phi_h'\|_{q,k}\right) \qquad (4.7) \\ &\le C\|\theta\|_{3,p,k}\,\|\phi_h\|_{1,q,k}. \qquad (4.8)\end{aligned}$$

From (1.7), it follows that

$$\begin{aligned}|\epsilon_h(\mathcal{M}(u)\theta\phi_h)| &\le C\sum_{k=1}^{N} h_k^4\|(\mathcal{M}(u)\theta\phi_h)^{(4)}\|_{1,k} \\ &\le C\sum_{k=1}^{N} h_k^4\|\theta\|_{3,p,k}\,\|\phi_h\|_{1,q,k} \qquad (4.9) \\ &\le Ch^3\|\theta\|_{2,p}\|\phi_h\|_{1,q}.\end{aligned}$$

Since f is sufficiently smooth, we note that

$$\begin{aligned}&\|\tilde{f}_{2,uu} - \tilde{f}_{1,uu}\|_{L^\infty(I_k)} \\ &= \int_0^1 (1-s)\|f_{uu}(u_2^h - s(u-u_2^h)) - f_{uu}(u_1^h - s(u-u_1^h))\|_{L^\infty(I_k)}ds \\ &\le C\|\theta\|_{L^\infty(I_k)} \le C\|\theta\|_{1,p,k}.\end{aligned}$$

Therefore using the inverse estimate $\|\theta\|_{L^\infty(I_k)} \le h_k^{-1/2}\|\theta\|_{1,p,k}$, the last two terms of (4.6) can be estimated as

$$\begin{aligned}&\left\langle \tilde{f}_{1,uu}\theta(\zeta_1+\zeta_2),\,\phi_h\right\rangle - \left\langle (\tilde{f}_{2,uu}-\tilde{f}_{1,uu})(\zeta_2)^2,\,\phi_h\right\rangle \qquad (4.10) \\ &\le C\sum_{k=1}^{N} h_k\left(\|\theta\|_{L^\infty(I_k)} + \|\tilde{f}_{2,uu}-\tilde{f}_{1,uu}\|_{L^\infty(I_k)}\right)\|\phi_h\|_{L^\infty(I_k)} \\ &\le C\sum_{k=1}^{N} h_k^{1/2}\left(\|\theta\|_{1,p,k} + \|\tilde{f}_{2,uu}-\tilde{f}_{1,uu}\|_{1,p,k}\right)\|\phi_h\|_{1,p,k} \\ &\le C\sum_{k=1}^{N} h_k^{1/2}\|\theta\|_{1,p,k}\|\phi_h\|_{1,p,k} \le Ch^{1/2}\|\theta\|_{1,p}\|\phi_h\|_{1,p}. \qquad (4.11)\end{aligned}$$

Following the analysis in the proof of Theorem 3.1 (replacing e by θ) and using interpolation inequality (2.8) for $\|\theta\|_{1,p}$ we obtain with $0 < \delta < 1$

$$\begin{aligned}\|\theta\|_p &\leq C\left(h^2\|\theta\|_{2,p} + h^3\|\theta\|_{2,p} + h^{1/2}\|\theta\|_{1,p}\right)\\ &\leq C\left(h^2\|\theta\|_{2,p} + h^3\|\theta\|_{2,p} + h^{1/2-\delta}\|\theta\|_{0,p} + h^{1/2+\delta}\|\theta\|_{2,p}\right).\end{aligned}$$

For h sufficiently small, it now yields

$$\|\theta\|_p \leq C\left(h^2 + h^3 + h^{1/2+\delta}\right)\|\theta\|_{2,p}. \tag{4.12}$$

The next step is to estimate $\|\theta''\|_p$. From (4.2), we have

$$\begin{aligned}\left(\theta'', \chi\right) &= \left(\mathcal{M}(u)\theta - b\theta' - c\theta - f_u(u)\theta, \chi\right)\\ &= \left(-b\theta' - c\theta - f_u(u)\theta, \chi\right) + \epsilon_h(\mathcal{M}(u)\theta\chi) + \left\langle \tilde{f}_{1,uu}\theta(\zeta_1 + \zeta_2), \chi\right\rangle\\ &- \left\langle (\tilde{f}_{2,uu} - \tilde{f}_{1,uu})(\zeta_2)^2, \chi\right\rangle\end{aligned}$$

We now use the inverse estimate $\|\chi'\|_{q,k} \leq Ch_k^{-1}\|\chi\|_{q,k}$, $\chi \in T_h$, in (4.9) and (4.11), further combine with the above inequality to obtain

$$\left|\left(\theta'', \chi\right)\right| \leq C(\|\theta\|_{1,p} + h^2\|\theta\|_{2,p})\,\|\chi\|_{0,q}.$$

Let ψ be an arbitrary element of L^q. Then from the definition of P_h and the above inequality, it follows that

$$\left|\left(\theta'', \psi\right)\right| = \left|\left(\theta'', (P_h\psi)\right)\right| \leq C(\|\theta\|_{1,p} + h^2\|\theta\|_{2,p})\,\|P_h\chi\|_q.$$

An use of stability result (3.14) with p replaced by q yields

$$\|\theta''\| \leq C(\|\theta\|_{1,p} + h^2\|\theta\|_{2,p}).$$

Note that for $0 < \delta < 1$

$$\begin{aligned}\|\theta\|_{2,p} &\leq \|\theta\|_{1,p} + \|\theta''\|_p \leq C(\|\theta\|_{1,p} + h^2\|\theta\|_{2,p})\\ &\leq C\left(h^{-\delta}\|\theta\|_{0,p} + (h^\delta + h^2)\|\theta\|_{2,p}\right)\end{aligned}$$

Again for small h

$$\|\theta\|_{2,p} \leq Ch^{-\delta}\|\theta\|_{0,p} \tag{4.13}$$

Use (4.13) in (4.12) to have

$$\|\theta\|_p \leq C\left(h^2 + h^3 + h^{1/2+\delta}\right)\|\theta\|_{2,p} \leq C\left(h^{2-\delta} + h^{3-\delta} + h^{1/2}\right)\|\theta\|_{0,p}.$$

For h sufficiently small, $\left(1 - C(h^{2-\delta} + h^{3-\delta} + h^{1/2})\right) > 0$, and hence, $\|\theta\|_p \leq 0$. Therefore $\theta \equiv 0$ for h sufficiently small and this completes the rest of the proof. ■

References

[1] Adams, R.A., *Sobolev Spaces*, New York: Academic Press, 1975.

[2] de Boor, C., A bound on the L^∞-norm of the L^2- approximation by splines in terms of a global mesh ratio, *Math. Comput.*, **30**, 1976, pp 765-771.

[3] Crouzeix, M and Thomée, V., The Stability in L_p and W_p^1 of the L_2-Projection onto Finite Element Function Spaces, *Math. Comp.*, **48**, 1987, pp. 521- 532.

[4] Davis, P.J., *Interpolation and Approximation*, Blaisdell, Waltham, Mass, 1963.

[5] Davis, P.J., & Rabinowitz, P., *Methods of Numerical Integration*, New York: Academic Press, 1975.

[6] Douglas, J.Jr., H^1-Galerkin methods for a nonlinear Dirichlet problem, in the *Proc. Conf. on Mathematical Aspects of Finite Element Methods*, Lecture Notes in Mathematics. Vol. 606, Berlin: Springer-Verlag, pp 64- 86, 1977.

[7] Grigorieff, R.D., & Sloan, I.H., Spline Petrov-Galerkin methods with quadrature, Applied Mathematics Report AMR 94/4, 1994, School of Mathematics, University of New South Wales, Australia.

[8] Sloan, I. H., A quadrature based approach to improve the collocation method., *Numer. Math.*, **54**, 1988, pp 41-56.

[9] Sloan, I. H., Tran, D., & Fairweather, G., A fourth order cubic spline method for linear second order two point boundary value problems, *IMA J. Numer. Anal.*, **13**, 1993, pp 591-607.

L. Jones and A.K. Pani
Deparment of Mathematics
Indian Institute of Technology
PO IIT Powai, Bombay - 400 076
INDIA.
jones@ganit.math.iitb.ernet.in
akp@ganit.math.iitb.ernet.in

W OETTLI AND D SCHLÄGER

Generalized vectorial equilibria and generalized monotonicity

1. Introduction

Let A be a convex set in some real topological vector space X, and let $f\colon A\times A \to \mathbb{R}$ be a given function such that $f(x,x) \geq 0$ on A. The scalar equilibrium problem consists in finding

$$\overline{x} \in A \text{ such that } f(\overline{x},y) \geq 0 \ \forall y \in A. \tag{EP}$$

This problem has many interesting applications [3], [7]. An important special case is the variational inequality problem. In the latter one is given an operator $\Phi\colon A \to X^*$, and one has to find

$$\overline{x} \in A \text{ such that } \langle \Phi(\overline{x}), y-\overline{x}\rangle \geq 0 \ \forall y \in A. \tag{VI}$$

In connection with problem (EP) the function $f(\cdot,\cdot)$ is called pseudomonotone [2] iff, for all $x, y \in A$,

$$f(x,y) \geq 0 \implies f(y,x) \leq 0.$$

Monotonicity permits to weaken the topological requirements needed to establish a solution of (EP), as compared with the non-monotone case. It is possible [10] to subsume the monotone and the non-monotone case under a single heading, by introducing another function $g\colon A \times A \to \mathbb{R}$, and defining f to be g-monotone iff, for all $x, y \in A$,

$$f(x,y) \geq 0 \implies g(x,y) \leq 0.$$

For $g(x,y) := -f(x,y)$, this is trivially satisfied. On the other hand, for $g(x,y) := f(y,x)$ we obtain pseudomonotonicity of f (of course, other choices of g are possible). Below we apply this line of approach to the multivalued vectorial equilibrium problem [1]. Here $\mathbb{R}$ is replaced by some topological vector space Z. We are given multivalued mappings $f\colon A \times A \rightrightarrows Z$ and $C\colon A \rightrightarrows Z$, and we want to find

$$\overline{x} \in A \text{ such that } f(\overline{x},y) \not\subseteq C(\overline{x}) \ \forall y \in A.$$

As particular case we consider the multivalued variational inequality problem [9] of finding

$$\overline{x} \in A \text{ such that } \langle \Phi(\overline{x}), y-\overline{x}\rangle \not\subseteq C(\overline{x}) \ \forall y \in A,$$

where $\Phi\colon A \rightrightarrows L(X,Z)$. We give several existence results for these problems. We admit even the possibility that f is defined on the product of two different sets $A \times B$, which are linked to each other by a multivalued mapping $T\colon B \rightrightarrows A$.

2. Results

We assume the following setting:

X is a real locally convex topological vector space;

Y is a real topological vector space;

$A \subseteq X$ and $B \subseteq Y$ are nonempty and convex;

$T\colon B \rightrightarrows A$ is an upper semicontinuous mapping with nonempty, closed, convex values.

Our starting point is the following result, which is a slight generalization of [1], Theorem 1.

Lemma 1. *Let A be compact and let $F, G \subseteq A \times B$ have the following properties:*

(i) *for all $y \in B$ and $x \in T(y)$, $(x, y) \in F$;*

(ii) *for all $y \in B$, $\{x \in A \mid (x, y) \notin G\}$ is open in A;*

(iii) *for all $x \in A$, $\{y \in B \mid (x, y) \notin F\}$ is convex;*

(iv) *$F \subseteq G$;*

(v) *for every $x \in A$, $(x, y) \in G \;\forall\, y \in B$ implies $(x, y) \in F \;\forall\, y \in B$.*

Then there exists $\overline{x} \in A$ such that $(\overline{x}, y) \in F$ for all $y \in B$.

Proof. In view of (v) it suffices to find $\overline{x} \in A$ such that $(\overline{x}, y) \in G$ for all $y \in B$. If such an $\overline{x}$ does not exist, then the compact set A is covered by the sets

$$V(y) := \{x \in A \mid (x, y) \notin G\}, \quad y \in B,$$

which are open in A by (ii). Let $V(y_1), \ldots, V(y_n)$ be a finite subcover of A, and let $\beta_1, \ldots, \beta_n$ be a continuous partition of unity subordinate to this cover. The functions $\beta_i\colon A \to \mathbb{R}$ are nonnegative and continuous with $\sum_{i=1}^n \beta_i(x) = 1$ for all $x \in A$, and $x \in V(y_i)$ whenever $\beta_i(x) > 0$. Then $p(x) := \sum_{i=1}^n \beta_i(x) y_i$ defines a continuous function $p\colon A \to B$. The mapping $T(p(\cdot))\colon A \rightrightarrows A$ is upper semicontinuous with nonempty, closed, convex values, hence has a fixed point $x^0 \in T(p(x^0))$. Let $I := \{i \mid \beta_i(x^0) > 0\}$; then, for all $i \in I$, $x^0 \in V(y_i)$, hence $(x^0, y_i) \notin G$, hence $(x^0, y_i) \notin F$ by (iv). Then from $p(x^0) \in \operatorname{conv}\{y_i \mid i \in I\}$ and (iii) it follows that $(x^0, p(x^0)) \notin F$, which contradicts (i). □

Let Z be another real topological vector space. Let $f, g\colon A \times B \rightrightarrows Z$ and $C, D\colon A \rightrightarrows Z$ be multivalued mappings. By f^{-1} we denote the upper inverse of f, i.e.,

$$f^{-1}(S) := \{(x, y) \in A \times B \mid f(x, y) \subseteq S\}$$

for all $S \subseteq Z$. Likewise, g^{-1} denotes the upper inverse of g. Letting

$$(x, y) \in F :\Longleftrightarrow (x, y) \notin f^{-1}(C(x)) \Longleftrightarrow f(x, y) \not\subseteq C(x),$$
$$(x, y) \in G :\Longleftrightarrow (x, y) \notin g^{-1}(D(x)) \Longleftrightarrow g(x, y) \not\subseteq D(x),$$

we obtain from Lemma 1 immediately

Theorem 1. *Assume that A is compact and*

(i) *for all $y \in B$ and $x \in T(y)$, $(x, y) \notin f^{-1}(C(x))$;*

(ii) *for all $y \in B$, $\{x \in A \mid (x, y) \in g^{-1}(D(x))\}$ is open in A;*

(iii) *for all $x \in A$, $\{y \in B \mid (x, y) \in f^{-1}(C(x))\}$ is convex;*

(iv) *for all $x \in A$, $y \in B$, $(x, y) \notin f^{-1}(C(x))$ implies $(x, y) \notin g^{-1}(D(x))$;*

(v) *for all $x \in A$, $(x, y) \notin g^{-1}(D(x))\ \forall y \in B$ implies $(x, y) \notin f^{-1}(C(x))\ \forall y \in B$.*

Then there exists $\overline{x} \in A$ such that $(\overline{x}, y) \notin f^{-1}(C(\overline{x}))$ for all $y \in B$.

Note that condition (iv) may be considered as an abstract monotonicity requirement.

Let us single out two particular cases of Theorem 1. First we let $g(x, y) := f(x, y)$, $D(x) := C(x)$. Then (iv) and (v) are obviously satisfied, and spelling out the upper inverse we obtain from Theorem 1

Theorem 2. *Assume that A is compact and*

(i) *for all $y \in B$ and $x \in T(y)$, $f(x, y) \not\subseteq C(x)$;*

(ii) *for all $y \in B$, $\{x \in A \mid f(x, y) \subseteq C(x)\}$ is open in A;*

(iii) *for all $x \in A$, $\{y \in B \mid f(x, y) \subseteq C(x)\}$ is convex.*

Then there exists $\overline{x} \in A$ such that $f(\overline{x}, y) \not\subseteq C(\overline{x})$ for all $y \in B$.

Remark 1. In Theorem 2, compactness of A may be replaced by the following coercivity hypothesis:

(iv) there exist a nonempty, compact set $A^0 \subseteq A$ and a compact, convex set $B^0 \subseteq B$ such that, for every $x \in A \setminus A^0$, there exists $y \in B^0$ with $f(x, y) \subseteq C(x)$;

(v) for every compact, convex set $B' \subseteq B$ there exists a compact, convex set $A' \subseteq A$ with $T(B') \subseteq A'$.

To prove this, let $y_1, \ldots, y_n \in B$, and $B' := \operatorname{conv}(B^0 \cup \{y_1, \ldots, y_n\})$. B' is compact, since B^0 is compact and convex. By (v) there exists a compact, convex set $A' \subseteq A$ with $T(B') \subseteq A'$. It is easily verified that $T|_{B'}$ and $f|_{A' \times B'}$ satisfy the conditions of Theorem 2. Hence there exists $x' \in A'$ such that $f(x', y) \not\subseteq C(x')$ for all $y \in B'$. Then from $B^0 \subseteq B'$ and (iv) follows $x' \in A^0$, and from $y_i \in B'$ follows $f(x', y_i) \not\subseteq C(x')$, $i = 1, \ldots, n$. Hence every finite subfamily of the family of sets

$$W(y) := \{x \in A \mid f(x, y) \not\subseteq C(x)\}, \quad y \in B,$$

which are closed in A by (ii), has nonempty intersection over A^0. Since A^0 is compact, there exists $\overline{x} \in A^0$ belonging to all $W(y)$, $y \in B$, which is the claimed result. □

Theorem 2 contains as a special case (namely when f is single-valued, $B = A$, and with a simpler coercivity condition) Lemma 2.1 from [8].

To consider another particular case of Theorem 1, we assume that $Y = X$, $B = A$, $T = id$, X Hausdorff (so that id has closed values). We let $g(x, y) := f(y, x)$. Then, spelling out the upper inverses, we obtain from Theorem 1

Theorem 3. *Assume that A is compact and*

(i) *for all $y \in A$, $f(y, y) \nsubseteq C(y)$;*

(ii) *for all $y \in A$, $\{\xi \in A \mid f(y, \xi) \subseteq D(\xi)\}$ is open in A;*

(iii) *for all $x \in A$, $\{\xi \in A \mid f(x, \xi) \subseteq C(x)\}$ is convex;*

(iv) *for all $x, y \in A$, $f(x, y) \nsubseteq C(x)$ implies $f(y, x) \nsubseteq D(x)$;*

(v) *for all $x, y \in A$ and $u \in]x, y[$, if $f(u, x) \nsubseteq D(x)$ and $f(u, y) \subseteq C(x)$ then $f(u, \xi) \nsubseteq \operatorname{cl} D(x)$ for all $\xi \in]x, y]$;*

(vi) *for all $x, y \in A$, $\{\xi \in [x, y] \mid f(\xi, y) \subseteq C(x)\}$ is open in $[x, y]$;*

(vii) *for all $x, y \in A$, $f(y, y) \subseteq \operatorname{cl} D(x)$.*

Then there exists $\overline{x} \in A$ such that $f(\overline{x}, y) \nsubseteq C(\overline{x})$ for all $y \in A$.

Proof. It only remains to verify condition (v) of Theorem 1, which reads now:

$$\text{for all } x \in A,\ f(y, x) \nsubseteq D(x)\ \forall\, y \in A \text{ implies } f(x, y) \nsubseteq C(x)\ \forall\, y \in A.$$

To this end, let $x \in A$ with $f(y, x) \nsubseteq D(x)$ for all $y \in A$. Assume, for contradiction, that $f(x, y) \subseteq C(x)$ for some $y \in A$. By (i), $y \neq x$. From $f(x, y) \subseteq C(x)$ and (vi) there exists $u \in]x, y[$ such that $f(u, y) \subseteq C(x)$. Since $f(u, x) \nsubseteq D(x)$, we obtain from (v) that $f(u, u) \nsubseteq \operatorname{cl} D(x)$, contradicting (vii). □

Remark 2. In Theorem 3, we can replace compactness of A by the following coercivity condition:

(viii) there exist a nonempty, compact set $A^0 \subseteq A$ and a compact, convex set $B^0 \subseteq A$ such that, for every $x \in A \setminus A^0$, there exists $y \in B^0$ with $f(x, y) \subseteq C(x)$.

To prove this, let $y_1, \ldots, y_n \in A$, and $A' := \operatorname{conv}(B^0 \cup \{y_1, \ldots, y_n\})$. A' is compact, since B^0 is compact and convex. It is easily verified that $f|_{A' \times A'}$ satisfies the conditions of Theorem 3. Hence there exists $x' \in A'$ such that $f(x', y) \nsubseteq C(x')$ for all $y \in A'$. Then from $B^0 \subseteq A'$ and (viii) follows $x' \in A^0$, and from $y_i \in A'$ and (iv) follows $f(y_i, x') \nsubseteq D(x')$, $i = 1, \ldots, n$. Hence every finite subfamily of the family of sets

$$W(y) := \{x \in A \mid f(y, x) \nsubseteq D(x)\}, \quad y \in A,$$

which are closed in A by (ii), has nonempty intersection over A^0. Since A^0 is compact, there exists $\overline{x} \in A^0$ belonging to all $W(y)$, i.e., $f(y, \overline{x}) \nsubseteq D(\overline{x})$ for all $y \in A$. From this follows, as shown in the proof of Theorem 3, that $f(\overline{x}, y) \nsubseteq C(\overline{x})$ for all $y \in A$, the claimed result. □

3. Examples

We analyze further the conditions of Theorem 2 and 3, and apply the results to the multivalued vectorial equilibrium problem.

We turn first to Theorem 2. Condition (ii) is fulfilled if, for every $y \in B$, the mapping $f(\cdot, y)$ is upper semicontinuous with nonempty, compact values, and $C(\cdot)$ has open graph in $A \times Z$. To see this, let $R := \{x \in A \mid f(x,y) \subseteq C(x)\}$, and let $\{x_i\}$ be a net in $A \setminus R$ which converges to $x \in A$. For every i, there exists $z_i \in f(x_i, y)$ with $z_i \notin C(x_i)$. Since $f(x,y)$ is compact and $f(\cdot,y)$ is upper semicontinuous, a subnet of $\{z_i\}$ converges to some $z \in f(x,y)$. Then $z \notin C(x)$, since $C(\cdot)$ has open graph. Hence $x \in A \setminus R$, and R is open in A.

If $C(x) = C$ is open and independent of x, then (ii) is satisfied if $f(\cdot, y)$ is upper semicontinuous. Indeed, in this case R is open, being the upper inverse of an open set under an upper semicontinuous mapping.

Condition (iii), namely the convexity of the set $\{y \in B \mid f(x,y) \subseteq C(x)\}$, is fulfilled if $C(x)$ is convex and

$$f(x, \alpha v + (1-\alpha)w) \subseteq \alpha f(x,v) + (1-\alpha) f(x,w) \quad \forall \alpha \in [0,1]. \tag{1}$$

If $C(x)$ is an open, convex cone, then (iii) is satisfied if

$$f(x, \alpha v + (1-\alpha)w) \subseteq \alpha f(x,v) + (1-\alpha) f(x,w) + \operatorname{cl} C(x) \quad \forall \alpha \in [0,1]$$

(since $\operatorname{int} C(x) + \operatorname{cl} C(x) \subseteq \operatorname{int} C(x)$).

We apply now Theorem 2 and the observations made above to the variational inequality problem. Let $Y = X$ be Hausdorff, $B = A$, and $T = id$. Let $L(X, Z)$ be a given subspace of the space of all linear operators from X into Z. For $\phi \in L(X,Z)$ we write $\langle \phi, x \rangle := \phi(x)$, and for $\Phi \subseteq L(X,Z)$ we write $\langle \Phi, x \rangle := \{\langle \phi, x \rangle \mid \phi \in \Phi\}$. Then from Theorem 2 and Remark 1 we obtain

Theorem 4. *Let $\Phi \colon A \rightrightarrows L(X,Z)$ and $C \colon A \rightrightarrows Z$ be such that $\Phi(y) \neq \emptyset$ for all $y \in A$, and*

(i) *for all $y \in A$, $0 \notin C(y)$;*

(ii) *for all $y \in A$, $\langle \Phi(\cdot), y - \cdot \rangle$ is upper semicontinuous on A with compact values; C has open graph in $A \times Z$;*

(iii) *for all $x \in A$, $C(x)$ is convex;*

(iv) *there exist a nonempty, compact set $A^0 \subseteq A$ and a compact, convex set $B^0 \subseteq A$ such that, for every $x \in A \setminus A^0$, there exists $y \in B^0$ with $\langle \Phi(x), y - x \rangle \subseteq C(x)$.*

Then there exists $\overline{x} \in A^0$ such that $\langle \Phi(\overline{x}), y - \overline{x} \rangle \not\subseteq C(\overline{x})$ for all $y \in A$.

Proof. Let $f(x,y) := \langle \Phi(x), y - x \rangle$. Then $\Phi(y) \neq \emptyset$ and $0 \notin C(y)$ yield $f(y,y) = \{0\} \not\subseteq C(y)$; moreover, f satisfies (1). Thus, by the above observations, the conditions of Theorem 2 are fulfilled. Condition (v) of Remark 1 is trivially satisfied, since $T = id$. □

Now we apply Theorem 3. We have to assume that $L(X, Z)$ is topologized in such a way that $\langle \cdot, \cdot \rangle$ is continuous on $L(X, Z) \times A$.

Theorem 5. *Let $\Phi: A \rightrightarrows L(X, Z)$ and $C, D: A \rightrightarrows Z$ be such that $\Phi(y)$ is compact and nonempty for all $y \in A$, and*

(i) *for all $y \in A$, $0 \notin C(y)$;*

(ii) *D has open graph in $A \times Z$;*

(iii) *for all $x \in A$, $C(x)$ is convex;*

(iv) *for all $x, y \in A$, $\langle \Phi(x), y - x \rangle \not\subseteq C(x)$ implies $\langle \Phi(y), x - y \rangle \not\subseteq D(x)$;*

(v) *for all $x \in A$, if $a \in Z \setminus D(x)$ and $b \in C(x)$, then $c \notin \operatorname{cl} D(x)$ for all $c \in]a, b]$;*

(vi) *for all $x, y \in A$, $C(x)$ is open and $\langle \Phi(\cdot), y - x \rangle$ is upper semicontinuous on $[x, y]$;*

(vii) *for all $x \in A$, $0 \in \operatorname{cl} D(x)$;*

(viii) *there exist a nonempty, compact set $A^0 \subseteq A$ and a compact, convex set $B^0 \subseteq A$ such that, for every $x \in A \setminus A^0$, there exists $y \in B^0$ with $\langle \Phi(x), y - x \rangle \subseteq C(x)$.*

Then there exists $\overline{x} \in A^0$ such that $\langle \Phi(\overline{x}), y - \overline{x} \rangle \not\subseteq C(\overline{x})$ for all $y \in A$.

Proof. Let $f(x, y) := \langle \Phi(x), y - x \rangle$. We shall verify the conditions of Theorem 3 and Remark 2: (i), (iv), (vii), and (viii) are obviously fulfilled; (iii) holds, since f satisfies (1).

(ii): We have to show that $S := \{\xi \in A \mid \langle \Phi(y), \xi - y \rangle \not\subseteq D(\xi)\}$ is closed in A. Let $\{\xi_i\}$ be a net in S converging to $\xi \in A$. For every i, there exists $\phi_i \in \Phi(y)$ with $\langle \phi_i, \xi_i - y \rangle \notin D(\xi_i)$. Since $\Phi(y)$ is compact, a subnet $\{\phi_j\}$ converges to some $\phi \in \Phi(y)$. By the continuity of $\langle \cdot, \cdot \rangle$ on $L(X, Z) \times A$, $\langle \phi_j, \xi_j \rangle \to \langle \phi, \xi \rangle$ and $\langle \phi_j, y \rangle \to \langle \phi, y \rangle$, so $\langle \phi_j, \xi_j - y \rangle \to \langle \phi, \xi - y \rangle$. Since D has open graph, we obtain $\langle \phi, \xi - y \rangle \notin D(\xi)$, hence $\xi \in S$.

(v): Suppose that $\langle \Phi(u), x - u \rangle \not\subseteq D(x)$ and $\langle \Phi(u), y - u \rangle \subseteq C(x)$. There exists $\phi \in \Phi(u)$ with $a := \langle \phi, x - u \rangle \notin D(x)$, and $b := \langle \phi, y - u \rangle \in C(x)$. For arbitrary $\xi = \lambda x + (1 - \lambda) y \in]x, y]$, let $c := \lambda a + (1 - \lambda) b$. Then $c = \langle \phi, \xi - u \rangle$ and $c \notin \operatorname{cl} D(x)$, hence $\langle \Phi(u), \xi - u \rangle \not\subseteq \operatorname{cl} D(x)$.

(vi): We have to show that $R := \{\xi \in [x, y] \mid \langle \Phi(\xi), y - \xi \rangle \not\subseteq C(x)\}$ is closed in $[x, y]$. Let $\{\xi_i\}$ be a net in R converging to $\xi \in [x, y]$; we may assume $\xi \neq y$, since $y \in R$ from (i), and we may assume $\xi_i \neq y$ for all i as well. Thus $y - \xi = \lambda(y - x)$ with $\lambda \neq 0$, $y - \xi_i = \lambda_i (y - x)$ with $\lambda_i \neq 0$, and $\lambda_i \to \lambda$. For every i, there exists $z_i \in \langle \Phi(\xi_i), y - \xi_i \rangle$ with $z_i \notin C(x)$. Then $u_i := \lambda_i^{-1} z_i \in \langle \Phi(\xi_i), y - x \rangle$. Since $\langle \Phi(\cdot), y - x \rangle$ is upper semicontinuous at ξ and $\langle \Phi(\xi), y - x \rangle$ is nonempty and compact, a subnet $\{u_j\}$ converges to some $u \in \langle \Phi(\xi), y - x \rangle$. Then $z_j = \lambda_j u_j \to \lambda u =: z \in \langle \Phi(\xi), y - \xi \rangle$. Since $C(x)$ is open and $z_j \notin C(x)$, we obtain $z \notin C(x)$, hence $\xi \in R$. □

From Theorem 5 we can obtain by suitable specialization Theorem 3.1 of [9]. The details are as follows: Let X and Z be normed spaces. Let $L(X, Z)$ be the space of all continuous linear operators from X into Z. $L(X, Z)$ is equipped with the norm topology. Let $A \subseteq X$ be bounded. We endow A and Z with their respective weak topologies, which we indicate by the notation A_{w} and Z_{w}. Then $\langle \cdot, \cdot \rangle$ is continuous from $L(X, Z) \times A_{\mathrm{w}}$ into Z_{w}. To see this, let $\{(\phi_i, x_i)\}$ be a net in $L(X, Z) \times A$ and $(\phi, x) \in L(X, Z) \times A$ such that $\|\phi_i - \phi\| \to 0$ and $x^*(x_i) \to x^*(x)$ for all $x^* \in X^*$. Then, for each $z^* \in Z^*$,

$$|z^*(\langle \phi_i - \phi, x_i \rangle)| \leq \|z^*\| \cdot \|\phi_i - \phi\| \cdot \|x_i\| \to 0,$$

since $\|x_i\|$ remains bounded; thus

$$z^*(\langle \phi_i, x_i \rangle) = z^*(\langle \phi_i - \phi, x_i \rangle) + z^*(\langle \phi, x_i \rangle) \to z^*(\langle \phi, x \rangle),$$

since $z^* \circ \phi \in X^*$.

Now for all $x \in A$ let $P(x) \subseteq Z$ be a convex cone such that $P(x) \neq Z$ and, with respect to the norm topology on Z, $P(x)$ is closed and $\operatorname{int} P(x) \neq \emptyset$. We let

$$C(x) := -\operatorname{int} P(x), \quad D(x) := \operatorname{int} P(x).$$

Observe that $P(x)$, since convex and closed, is weakly closed. This implies that

$$\text{w-cl}\, D(x) = P(x),$$

since $P(x) = \operatorname{cl} P(x) = \operatorname{cl} \operatorname{int} P(x) \subseteq \text{w-cl}\operatorname{int} P(x) \subseteq \text{w-cl}\, P(x) = P(x)$.

With this choice of $C(\cdot)$ and $D(\cdot)$, conditions (i) and (iii) of Theorem 5 are clearly satisfied. The same holds for condition (vii), which reads now $0 \in P(x)$. Also condition (v) is fulfilled, which requests now, for every $x \in A$, that

$$\text{if } a \notin \operatorname{int} P(x) \text{ and } b \in -\operatorname{int} P(x), \text{ then } c \notin P(x) \text{ for all } c \in \left]a, b\right].$$

To see this, let a and b be as hypothesized, and assume that $c := \lambda a + (1-\lambda) b \in P(x)$ for some $0 \leq \lambda < 1$. Then $c - (1-\lambda) b \in P(x) + \operatorname{int} P(x) = \operatorname{int} P(x)$, contradicting $\lambda a \notin \operatorname{int} P(x)$.

Condition (ii) requires that $D(\cdot)$ has open graph in $A_{\mathrm{w}} \times Z_{\mathrm{w}}$. Since $C(\cdot) = -D(\cdot)$, this implies that $C(x)$ is weakly open; so the latter requirement can be omitted from condition (vi). Altogether, we have only to enforce conditions (ii), (iv), (viii), and the remaining part of condition (vi). This leaves us with the following result.

Theorem 6. *Let X, Z be normed spaces. Let A be bounded. Suppose that mappings $\Phi\colon A \rightrightarrows L(X, Z)$ and $P\colon A \rightrightarrows Z$ are given such that, for all $x \in A$, $\Phi(x)$ is nonempty and compact, and $P(x) \neq Z$ is a closed, convex cone with* $\operatorname{int} P(x) \neq \emptyset$. *Assume that* (ii)′ *the mapping* $\operatorname{int} P(\cdot)$ *has open graph in* $A_w \times Z_w$;

(iv)′ *for all* $x, y \in A$, $\langle \Phi(x), y - x\rangle \not\subseteq -\operatorname{int} P(x)$ *implies* $\langle \Phi(y), x - y\rangle \not\subseteq \operatorname{int} P(x)$;

(vi)′ *for all* $x, y \in A$, $\langle \Phi(\cdot), y - x\rangle$ *is upper semicontinuous from* $[x, y]$ *into* Z_w;

(viii)′ *there exist a nonempty, weakly compact set* $A^0 \subseteq A$ *and a weakly compact, convex set* $B^0 \subseteq A$ *such that, for every* $x \in A \setminus A^0$, *there exists* $y \in B^0$ *with* $\langle \Phi(x), y - x\rangle \subseteq -\operatorname{int} P(x)$.

Then there exists $\overline{x} \in A^0$ *such that* $\langle \Phi(\overline{x}), y - \overline{x}\rangle \not\subseteq -\operatorname{int} P(\overline{x})$ *for all* $y \in A$.

Proof. We apply Theorem 5 with $C(x) := -\operatorname{int} P(x)$, $D(x) := \operatorname{int} P(x)$, providing A and Z with their weak topologies. The details have already been given in the preceding discussion. □

In [9], X was supposed to be a Banach space, and A was supposed to be weakly compact. We recall that if X is a normed space and $A \subseteq X$ is weakly compact, then A is bounded ([4], Prop. 15.7).

Let us return to the conclusion of Theorem 4 or 5, namely the existence of $\overline{x} \in A$ such that

$$\langle \Phi(\overline{x}), y - \overline{x}\rangle \not\subseteq C(\overline{x}) \text{ for all } y \in A. \tag{2}$$

If $Z = \mathbb{R}$ and $C(\overline{x}) := \{z \in \mathbb{R} \mid z < 0\}$, then usually one wants to obtain a stronger result, namely the existence of

$$\overline{\phi} \in \Phi(\overline{x}) \text{ such that } \langle \overline{\phi}, y - \overline{x}\rangle \notin C(\overline{x}) \text{ for all } y \in A. \tag{3}$$

The next theorem shows that it is possible to deduce (3) from (2), if $C(\overline{x})$ and $Z \setminus C(\overline{x})$ are both convex.

Theorem 7. *Let* $A \subseteq X$ *be convex. Let* $\Phi \subseteq L(X, Z)$ *be convex and compact. Let* $C \subseteq Z$ *be convex and open. Assume that* $Z \setminus C$ *is convex as well. Suppose that* $\langle \cdot, \cdot \rangle$ *is continuous on* $\Phi \times A$. *Let* $\overline{x} \in A$ *be such that* $\langle \Phi, y - \overline{x}\rangle \not\subseteq C$ *for all* $y \in A$. *Then there exists* $\overline{\phi} \in \Phi$ *such that* $\langle \overline{\phi}, y - \overline{x}\rangle \notin C$ *for all* $y \in A$.

Proof. We apply Theorem 2. Define $T\colon A \rightrightarrows \Phi$ by

$$T(y) := \{\phi \in \Phi \mid \langle \phi, y - \overline{x}\rangle \notin C\}.$$

Define $f\colon \Phi \times A \to Z$ by $f(\phi, y) := \langle \phi, y - \overline{x}\rangle$. Then, by hypothesis, $T(y) \neq \emptyset$ for all $y \in A$. T has convex, closed values. T is upper semicontinuous, since it has closed graph and Φ is compact. Furthermore: (i) for all $y \in A$ and $\phi \in T(y)$, $f(\phi, y) \notin C$; (ii) for all $y \in A$, $\{\phi \in \Phi \mid f(\phi, y) \in C\}$ is open in Φ; (iii) for all $\phi \in \Phi$, $\{y \in A \mid f(\phi, y) \in C\}$ is convex. Hence all assumptions of Theorem 2 are satisfied, and so there exists $\overline{\phi} \in \Phi$ such that $f(\overline{\phi}, y) \notin C$ for all $y \in A$, the claimed result. □

For the benefit of those readers who are primarily interested in the scalar-valued case we conclude with an extension of Fan's classical result [6]. The space X is assumed to be Hausdorff.

Theorem 8. *Let $A \subseteq X$ and $B \subseteq Y$ be convex. Let $T\colon Y \to X$ be linear and continuous such that $T(B)$ is a closed subset of A. Let $f, g\colon A \times B \to \mathbb{R}$ have the following properties:*

(i) *for all $y \in B$, $f(T(y), y) \geq 0$;*

(ii) *for all $y \in B$, $\{x \in A \,|\, g(x, y) > 0\}$ is open in A;*

(iii) *for all $x \in A$, $\{y \in B \,|\, f(x, y) < 0\}$ is convex;*

(iv) *for all $x \in A$ and $y \in B$, $f(x, y) \geq 0$ implies $g(x, y) \leq 0$;*

(v) *for all $u \neq v$ in B, $g(T(u), y) \leq 0 \;\forall\, y \in \,]u, v]$ implies $f(T(u), y) \geq 0 \;\forall\, y \in \,]u, v]$;*

(vi) *there exist a nonempty, compact set $A^0 \subseteq A$ and a compact, convex set $B^0 \subseteq B$ such that, for every $x \in A \setminus A^0$, there exists $y \in B^0$ with $f(x, y) < 0$.*

Then there exists $\overline{x} \in A^0$ such that $f(\overline{x}, y) \geq 0$ for all $y \in B$.

Proof. For arbitrary $y_1, \ldots, y_n \in B$, the sets $B' := \operatorname{conv}(B^0 \cup \{y_1, \ldots, y_n\})$ and $A' := T(B')$ are compact and convex. Observe that condition (v) together with (i) has the following effect: if $\overline{x} \in A'$ is such that $g(\overline{x}, y) \leq 0$ for all $y \in B'$, then $f(\overline{x}, y) \geq 0$ for all $y \in B'$. Hence, setting $C(x) := \,]-\infty, 0[$, $D(x) := \,]0, \infty[$, the restrictions of T, f, g to A' and B' satisfy the assumptions of Theorem 1.

Thus, there exists $x' \in A'$ such that $f(x', y) \geq 0$ for all $y \in B'$. Then $x' \in T(B)$, $x' \in A^0$ by (vi), and $g(x', y_k) \leq 0$ for all k by (iv). From the compactness of $A^0 \cap T(B)$ and (ii) we conclude that there exists $\overline{x} \in A^0 \cap T(B)$ such that $g(\overline{x}, y) \leq 0$ for all $y \in B$. Using (v) and (i) again, we obtain $f(\overline{x}, y) \geq 0$ for all $y \in B$. □

Extending a definition used in [10], we say that the function f is g-monotone, iff it satisfies (iv), and f is maximal g-monotone, iff it satisfies (iv) and (v). Note that if $g(x, y) = -f(x, y)$, then f is maximal g-monotone, i.e., (iv) and (v) constitute no restriction at all. On the other hand, if $A = B$ and f is pseudomonotone in the sense that, for all $x, y \in A$,

$$f(x, y) \geq 0 \implies f(y, x) \leq 0,$$

then choosing $g(x, y) := f(y, x)$ we see that f is g-monotone.

References

[1] Q. H. Ansari, W. Oettli, and D. Schläger. A generalization of vectorial equilibria. *Math. Methods Operations Res.* **46** (2), 1997 (to appear). Extended Abstract in Du et al. [5], 181–185.

[2] M. Bianchi, N. Hadjisavvas, and S. Schaible. Vector equilibrium problems with generalized monotone bifunctions. *J. Optimization Theory Appl.* **92**, 1997, 531–546.

[3] E. Blum and W. Oettli. From optimization and variational inequalities to equilibrium problems. *The Mathematics Student* **63**, 1994, 123–145.

[4] A. Brown and C. Pearcy. *Introduction to Operator Theory I. Elements of Functional Analysis.* Springer, New York, 1977.

[5] D.-Z. Du, X.-S. Zhang, and K. Cheng, editors. *Operations Research and Its Applications.* World Scientific Co., Beijing, 1996.

[6] K. Fan. A minimax inequality and applications. In O. Shisha, editor. *Inequalities 3.* Academic Press, New York, 1972, 103–113.

[7] T. Husain and E. Tarafdar. Simultaneous variational inequalities, minimization problems and related results. *Math. Japonica* **39**, 1994, 221–231.

[8] G. M. Lee, D. S. Kim, and B. S. Lee. On noncooperative vector equilibrium. *Indian J. Pure Appl. Math.* **27** (8), 1996, 735–739.

[9] K. L. Lin, D. P. Yang, and J. C. Yao. Generalized vector variational inequalities. In Du et al. [5], 78–84.

[10] W. Oettli. A remark on vectorial equilibria and generalized monotonicity. *Acta Math. Vietnamica*, 1997 (to appear).

W. Oettli, D. Schläger
Lehrstuhl für Mathematik VII
Universität Mannheim
68131 Mannheim
Germany

H BEN-EL-MECHAIEKH

Existence theorems for set valued maps

1. Introduction

The purpose of this paper is to discuss possible directions for relaxing compactness or convexity hypotheses in some nonlinear existence results that can be formulated as fixed point problems. The results presented here are part of an attempt to outline an elementary fixed point theory that would address broad classes of convex as well as non-convex maps defined on spaces lacking convexity or compactness (see [Be2], [BD], [BOT], [BCFL], [BId], [BIs], [BK1], [BK2], [GGK], [GL]).

After a brief description of the various concepts of maps and spaces studied in this paper, we formulate in Section 2 the general coincidence principle that garantees the existence of a coincidence point between a set-valued map that can be approximated by continuous functions and a set-valued map whose inverse admits a continuous selection. There is here a certain duality that garantees the matching of both maps: the first map has closed graph (upper semicontinuous type) as in Cellina's approximation theorem ([Ce]) and the second one is lower semicontinuous (more particularly, is of open graph type) as in Michael's selection theorem ([M]). This duality, together with topological properties of compactness and strong conditions of connectedness as well as some regularity of the maps force the existence of coincidence points.

We then derive Kakutani-Fan type topological fixed point theorems for approachable maps defined on extension spaces or on spaces with abstract convexity structures and a Browder-Fan type fixed point theorem for convex valued set-valued maps with open fibers defined on non-compact domains as in [BDG].

Section 3 is devoted to applications of this principle (in its analytical form) to the problems of existence of equilibria for upper hemicontinuous set-valued maps with convex values defined on non-compact spaces and to set-valued maps defined on compact neighborhood retracts and

I wish to thank professors George Isac and Wojciech Kryszewski for stimulating discussions. Thanks are extended to the Natural Sciences and Engineering Research Council of Canada for its support under grant OGP0042422.

subject to tangency conditions expressed in terms of new concepts of tangent cones to neighborhood retracts.

Section 4 discusses a typical application to control problems where maps with non-convex values appear in a natural way as Poincaré operators of (i.e. solution sets to) differential inclusions. This application follows from a general multivalued quasi-variational inequality.

2. A coincidence principle for set-valued maps

We formulate a simple and very general coincidence principle between an upper semicontinuous set-valued map whose graph can be approximated by continuous functions and a set-valued map whose inverse admits a continuous selection. First, let us recall some background concepts and facts.

A set-valued map Φ from a set X into a set Y is a map whose values are elements of $\mathcal{P}(Y)$, the power set of Y. The map $\Phi^{-1} : Y \longrightarrow \mathcal{P}(X)$ whose values are the fibers of Φ, $\Phi^{-1}(y) = \{x \in X : y \in \Phi(x)\}, y \in Y$, is the *inverse* of Φ.

Assume that X and Y are topological spaces (all topological spaces considered here are assumed to be Hausdorff). A *selection* for a map $\Psi : X \longrightarrow \mathcal{P}(Y)$ is a single-valued map $f : X \longrightarrow Y$ verifying $f(x) \in \Psi(x)$ for all $x \in X$. Denote by $\mathbf{s}(\Psi)$ the set of all continuous selections for Ψ.

Definition 2.1. *The map Ψ is said to be selectionable if and only if $\mathbf{s}(\Psi)$ is non-empty.*

The class of selectionable maps from X into Y is denoted by $\mathcal{S}(X, Y)$ ($\mathcal{S}(X)$ stands for $\mathcal{S}(X, X)$).

The most fundamental selectionability result is Michael's theorem ([M]). It garantees the existence of a continuous selection for a lower semicontinous map Ψ with closed convex values from a paracompact topological space into a Fréchet space. If the codomain is a non-complete convex subset of a topological vector space, a stronger regularity condition on Ψ, namely the openess of the fibers, implies the existence of a continuous selection for Ψ without the closedness of values (see [BDG]). This result was implicit in F. Browder's proof of the Browder-Fan fixed point theorem ([Bro]).

One might ask if such results hold without convexity assumptions. The answer is a partial yes. A prospective way to address this question is to consider topological (or metric) spaces with abstract convexity structures. We refer to the papers [BCFL], [Bi], [Ho1], and [Ho2] for results of Michael's type along these lines. Removing convexity in Browder's argument comes at a certain price: stronger regularity of the mapping. By way of illustration, let us mention that if X is a finite polyhedron, Y a compact ANR, and $\Psi : X \longrightarrow \mathcal{P}(Y)$ has contractible values and open graph (a stronger lower semicontinuity), then $\Psi \in \mathcal{S}(X,Y)$ ([Be2]). In fact this result holds true when X is a locally finite simplicial complex and Y an absolute retract ([MC]) with the weaker assumption of openess of values and fibers ([Be2]). No much more is known about Browder-type results for maps with contractible values. It is important to bring to the reader's attention the following limitation: the self-map of the Hilbert cube I^∞, $\Psi(x) = I^\infty \backslash \{x\}$, has open graph and contractible values but cannot be in $\mathcal{S}(I^\infty)$ otherwise it would have a fixed point by the generalized Brouwer's theorem of Borsuk. This is impossible.

Assume now that X and Y are endowed with compatible uniformities $\mathcal{U}$ and $\mathcal{V}$ respectively. Given $U \in \mathcal{U}$ and $V \in \mathcal{V}$, a $(U,V)-$*approximative selection* for a map $\Phi : X \longrightarrow \mathcal{P}(Y)$ is a single-valued map $s : X \longrightarrow Y$ verifying $s(x) \in V[\Phi(U[x])]$ for all $x \in X$. Denote by $\mathbf{a}(\Phi; U, V)$ the set of all continuous $(U,V)-$approximative selections for Φ.

Definition 2.2. *(see* [BD], [BOT]*) The map Φ is said to be approachable if and only if it is upper semicontinuous and $\mathbf{a}(\Phi; U, V)$ is non-empty for all $U \in \mathcal{U}$ and $V \in \mathcal{V}$. The class of approachable maps from X into Y is denoted by $\mathcal{A}(X,Y)$ ($\mathcal{A}(X)$ denotes $\mathcal{A}(X,X)$).*

The class $\mathcal{A}$ of approachable maps is quite broad. The approximation theorem of Cellina referred to in the introduction asserts that if X is a paracompact topological space equipped with a compatible uniformity $\mathcal{U}$ and Y is a convex subset of a locally convex topological vector space, then every upper semicontinuous map $\Phi : X \longrightarrow \mathcal{P}(Y)$ with non-empty convex values is approachable ([Ce]). Also, if X and Y are two absolute neighborhood retracts (ANR's for short) with X compact, then every upper semicontinuous map $\Phi : X \longrightarrow \mathcal{P}(Y)$ with compact contractible values is approachable. More generally (see [Be2] and compare with [GGK]), if X is an approximative absolute neighborhood extension space for compact spaces, and Y is a uniform space, then every upper semicontinuous map $\Phi : X \longrightarrow \mathcal{P}(Y)$ with non-empty compact $\infty-$proximally connected values in Y is approachable. It is shown in [BOT] that if X is an ANR, $(Y, \mathcal{V})$ is a uniform space, and

$\Phi : X \longrightarrow \mathcal{P}(Y)$ is an upper semicontinuous map with non-empty values such that the restriction $\Phi|P$ of Φ to any finite polyhedron $P \subset X$ is approachable. Then the restriction $\Phi|K$ of Φ to any compact subset K of X is approachable.

We start with the following general coincidence principle.

Theorem 2.1. *Let $(X,\mathcal{U})$ and $(Y,\mathcal{V})$ be two uniform spaces and let $\Phi, \Psi : X \longrightarrow \mathcal{P}(Y)$ be two maps such that $\Phi \in \mathcal{A}(X,Y)$ and has non-empty compact values, $\Psi^{-1} \in \mathcal{S}(Y,X)$. Assume that either one of X or Y is compact and has the fixed point property for continuous maps. Then Φ and Ψ have a coincidence, that is $\Phi(x_0) \cap \Psi(x_0) \neq \emptyset$ for some $x_0 \in X$.*

Proof. Assume that Y is compact and has the fixed point property for continuous maps, the other case being similar. One invokes the stability of the class $\mathcal{A}$ under composition products (see [Be2]) to infer that given $f \in \mathbf{s}(\Psi^{-1})$, the composite map $\Phi \circ f : Y \xrightarrow{f} X \xrightarrow{\Phi} \mathcal{P}(Y)$ is in $\mathcal{A}(Y)$. By hypothesis, each continuous map $s \circ f$ where $s \in \mathbf{a}(\Phi; U, V), U \in \mathcal{U}, V \in \mathcal{V}$, has a fixed point in Y which is an approximative fixed point for $\Phi \circ f$. One then obtain a net of approximative fixed points which must converge due to the compactness of Y. Since the map $\Phi \circ f$ is upper semicontinuous and has closed values, the limit of this net is necessarily a fixed point of $\Phi \circ f$, and therefore a coincidence point for Φ and Ψ. ∎

Note that if $X = Y$ and Ψ is the identity on X, we are in presence of a fixed point for $\Phi, x_0 \in \Phi(x_0)$. This is the object of the following

Corollary 2.2. *Let $\Phi \in \mathcal{A}(X)$ be a map with non-empty closed values. Assume that either one of the following conditions holds:*

(i) X is a convex subset of a locally convex topological vector space and Φ is compact ([Be2]);

(ii) X is a compact approximate extension space for compact spaces ([BD]);

(iii) X is a "convex" in a generalized sense subset of a uniform space E equipped with an abstract structure of local convexity ([BCFL]).

Then Φ has a fixed point.

If both maps have convex values and the underlying sets are convex and compact, the principle goes back to Browder ([Bro]) and find its motivation in the work of Ky Fan (see [BDG], [F2]).

Corollary 2.3. *Let X be a convex subset of a topological vector space, $\widetilde{X}$ be a nonempty subset of X, Y be a convex subset of a topological vector space. Let $\Phi : X \to \mathcal{P}(Y)$ be a map satisfying the following properties:*

(i) given any finite subset N of $\widetilde{X}$, the restriction of Φ to the convex hull co(N) of N is upper semicontinuous;

(ii) Φ has nonempty closed convex values.

Let $\Psi : X \to \mathcal{P}(Y), \tilde{\Psi} : \widetilde{X} \to \mathcal{P}(Y)$ be two maps satisfying the properties:

(iii) $\emptyset \neq \tilde{\Psi}^{-1}(y) \subseteq \Psi^{-1}(y)$ for all $y \in \overline{\Phi(X)}$;

(iv) $\tilde{\Psi}(x)$ is open in X for all $x \in \widetilde{X}$;

(v) $\Psi^{-1}(y)$ is convex for all $y \in \overline{\Phi(X)}$.

If the map Φ is compact, then Φ and Ψ have a coincidence.

Proof. A simple partition of unity argument implies that the restriction of Ψ^{-1} to the compact set $K = \overline{\Phi(X)}$ admits a continuous selection with values in the convex hull $C = co(N)$ of a finite subset N of $\widetilde{X}$. The restriction of Φ to the compact convex subset C of X verifies the hypotheses of Theorem 2.1. The conclusion follows. ∎

We deduce in a standard way a fixed point theorem for maps with convex values and open fibers. This formulation is essentially due to [BDG].

Corollary 2.4. *Let X be a convex subset of a topological vector space, $\Psi : X \to \mathcal{P}(X)$ a map satisfying the following properties:*

(i) $\Psi^{-1}(y)$ is open in X for all $y \in Y$;

(ii) $\Psi(x)$ is convex nonempty for all $x \in X$;

(iii) there exists a compact subset K of X such that for any finite subset N of X there exists a compact convex subset C_N of X containing N such that $\Psi(x) \cap C_N \neq \emptyset$ for all $x \in C_N \setminus K$.

Then there exists $x_0 \in X$ such that $x_0 \in \Psi(x_0)$.

Proof. It suffices to apply Corollary 2.3 to $Y = \widetilde{X} = X$ and $\Phi = 1_X, \tilde{\Psi} = \Psi$. The restriction $\Psi^{-1}|K$ has a continuous selection f with values in the convex hull $co(N)$ of a finite subset N of X. Let C_N be a compact convex subset of X containing $co(N)$ and let $\Psi_N : C_N \to \mathcal{P}(C_N)$ be the map defined by:

$$\Psi_N^{-1}(y) = \Psi^{-1}(y) \cap C_N, y \in Y.$$

We claim that Ψ_N satisfies all conditions of Corollary 2.3 (only the surjectivity of Ψ_N onto C_N needs to be verified). Hence, the maps Ψ_N and Φ_N have a coincidence point which is a coincidence point for Φ and Ψ. ∎

Remarks. (1) The Browder-Fan fixed point theorem (see [Bro]) corresponds to the case where X is compact, $K = \emptyset$, and $C_N = X$ for any N.

(2) Compactness conditions similar to (iii) were considered in [F2], [BP], [Pa], [L], and other papers. It is clearly a coercivity condition that compensates full compactness of the space X by a control outside the compact subset K of X.

3. On zeros of non-self maps

In this section, we discuss two possible directions in which one can extend to non-compact or non-convex domains equilibrium theorems for non-self maps. The reader is referred to [BK1] and [BK2] for details and related results.

Equilibrium theorems provide sufficient conditions for the existence of an equilibrium (or a zero) for a given map Φ, that is a solution to the inclusion $0 \in \Phi(x)$ required to belong to a certain constraint set X. Many important problems in nonlinear analysis can be reduced to equilibrium problems (for example the problem of existence of critical points for convex or nonsmooth functions, the problem of existence of stationary solutions to differential inclusions, etc...). We discuss a first equilibrium theorem for the perturbation of an upper hemicontinuous map by closed convex processes defined on a closed convex subset of nromed space. The lack of compactness of the domain is compensated by tangency conditions of the Ky Fan type on a compact subset of the domain and outside of it. These tangency conditions are closely related to the concept of viability domains of differential inclusions. In our second result the convexity of the domain is dropped. The set X is assumed to be a compact neighborhood retract of a normed space with nontrivial Euler characteristic. From the point of view of fixed point theory, the nontriviality of the Euler characteristic is crucial (indeed, spheres with odd dimensions have trivial Euler characteristic and admit nontrivial vector fields). The tangency conditions here are expressed in terms of a new concept of retraction tangent cone. This tangent cone coincides with the usual tangent cone of convex analysis for locally convex sets and with the tangent cone of non-smooth analysis for tangentially regular sets.

3.1. **Relaxation of Compactness.** For a given subset X of a normed space E, denote by $T_X(x)$ the cone $\overline{\bigcup_{t>0}\frac{1}{t}(X-x)}$. Let $\Phi : X \longrightarrow \mathcal{P}(F)$ be a map with non-empty values in another normed space F and let Λ be a map from X into the normed space $\Lambda(E,F)$ of all closed convex processes from E into F.

Definition 3.1. *Given a subset X_1 of X and a subset X_2 of E, we say that the map Φ satisfies the condition of Ky Fan on (X_1, X_2) with respect to Λ if and only if:*

$$\overline{\Lambda(x)(-T_{X_2}(x))} \cap \Phi(x) \neq \emptyset \text{ for all } x \in X_1. \tag{3.1}$$

Weaker forms of this tangency condition were considered by many authors ([A1], [Bel], [F2], [Pa] and others). In the case where $E = F$ and $\Lambda(x) \equiv -Id_E$, and $X = X_1 = X_2$, (3.1) implies that the map Φ is *inward* in the sense of Ky Fan (in the terminology of viability theory, it is said that X is a *viability domain* for Φ.)

Let $N_X(x)$ be the set $\{\varphi \in E^*; \langle\varphi, x\rangle \geq \sup_{v\in X}\langle\varphi, v\rangle\}$. If X is convex in E and $x \in X$, the cone $T_X(x)$ defined above is the *tangent cone* to X at x and $N_X(x)$ is the negative polar cone of $T_X(x)$, called the *normal cone* to X at x. In this case, condition (3.1) can be expressed in terms of the normal cone as follows: if the map Φ satisfies the condition of Ky Fan on (X_1, X_2) with respect to Λ then:

$$\inf_{\psi\in F^*, \Lambda(x)^*(\psi)\in N_{K_2}(x)} \sup_{y\in\Phi(x)} \langle\psi, y\rangle \geq 0 \text{ for all } x \in X_1, \tag{3.2}$$

where, for a given convex process $\Lambda : E \longrightarrow \mathcal{P}(F), \Lambda^* : F^* \longrightarrow \mathcal{P}(E^*)$ is the transpose convex process of Λ defined by:

$$\varphi \in \Lambda^*(\psi) \Leftrightarrow \langle\varphi, x\rangle \leq \langle\psi, y\rangle \text{ for all } x \in E \text{ and all } y \in \Lambda(x).$$

The converse holds true whenever the values of Φ are compact.

In finite dimensional spaces, condition (3.1) is necessary for the solvability of the following differential inclusion:

$$\Lambda(x'(t)) \subset \Phi(x(t)), t \in [0, T]. \tag{3.3}$$

where $\Lambda : \mathbf{R}^n \to \mathcal{P}(\mathbf{R}^m)$ is a linear process and $\Phi : \mathbf{R}^n \to \mathcal{P}(\mathbf{R}^m)$ is an upper hemicontinuous[1] map with closed convex values. More precisely, we obtain after a straightforward computation the following result.

Proposition 3.1. *Let X be a subset of $\mathbf{R}^n$ containing the initial point $x(0) = x_0$. If $x(.)$ is a solution of problem (3.3) satisfying the following condition:*

(3.4) *for each $0 < T$, there exists $t \in (0, T]$ such that $x(t) \in X$.*

Then $\overline{\Lambda(T_X(x_0))} \cap \Phi(x_0) \neq \emptyset$, that is $-\Phi$ satisfies the condition of Ky Fan on $(\{x_0\}, X)$ with respect to Λ.

Remark. If for any $x_0 \in X$, there exists $T > 0$ and a trajectory $x(.)$ on $[0, T)$ of the differential inclusion (3.3) viable in X (that is $x(t) \in X$ for all $t \in [0, T)$) and starting at x_0, then (3.4) is obviously satisfied. Therefore $-\Phi$ satisfies the condition of Ky Fan on (X, X) with respect to Λ.

We are ready to state our first equilibrium theorem.

Theorem 3.2. *Let X be a convex subset of a normed space E, F be a normed space, $\Phi : X \longrightarrow \mathcal{P}(F)$ be an upper hemicontinuous map with closed convex values, and $\Lambda : X \longrightarrow \Lambda(E, F)$ be a continuous mapping satisfying the following boundedness condition:*

(i) there exists a constant $M > 0$ such that for each $x \in X$ and each $u \in E$ with $\|u\| = 1$, there exists $v \in \Lambda(x)(u)$ with $\|v\| \leq M$.

Assume that there exists a compact subset K of X such that:

(ii) for each finite subset N of X, there exists a compact convex subset C_N of X containing N such that Φ satisfies the condition of Ky Fan on $(C_N \setminus K, C_N)$ with respect to Λ;

(ii) Φ satisfies the condition of Ky Fan on (K, X) with respect to Λ.

Then,
(A) Φ has an equilibrium in X,
(B) for every $x_0 \in X$, the map $\Phi(.) + \Lambda(.)(. - x_0)$ has an equilibrium in X.

[1]A map Φ is said to be upper hemicontinuous if, for any linear form ψ, the real function $x \longmapsto \sup_{y \in \Phi(x)} \langle \psi, y \rangle$ is upper semicontinuous. The graph of an upper hemicontinuous map with closed convex values is closed. Weakly upper semicontinuous maps are upper hemicontinuous.

Remarks. (1) With suitable changes in the hypotheses, this theorem remains valid when E is a topological vector space, F is a topological vector space having separating dual and X is paracompact; (the paracompactness of X could be dropped by an argument similar to that of [BP]).

(2) In the case where for every $x \in K$, $\Lambda(x) \equiv L$ is a linear operator from E into F, and $C_N = C$ is the same set for all finite subsets N of K, this result can be found in [Bel]. In this case (ii) is equivalent to the boundary condition:

$$\Phi \text{ satisfies the condition of Ky Fan on } (K \cap \partial X, X) \text{ with respect to } L,$$

and (B) garantees the surjectivity of the perturbation $\Phi + L$ onto $L(X)$.

(3) In the case where X is compact and $\Lambda(x) = L(x)$ is a linear operator, this result is precisely the solvability theorem in [AF]. By putting $K = \emptyset$ and $C_N = X$ for any N, (i) and (ii) reduce to:

$$(i)' \ \Phi \text{ satisfies the condition of Ky Fan on } (X, X) \text{ with respect to } L.$$

(4) A particular case of our theorem involving the concept of $c-$compactness of [L] is due to [BP]. Recall that a subset C of X is said to be $c-$ *compact* in X if for each finite subset N of X there exists a compact convex subset C_N of X such that $C \cup N \subset C_N$. Condition (i) in Theorem 3.2 may be replaced by:

(i)" There exists a $c-$compact subset C of X such that $\overline{\Lambda(-T_{co(x,C)}(x))} \cap \Phi(x) \neq \emptyset$ for every $x \in X \setminus K$.

As an immediate consequence of Theorem 3.2 and of the fact that every bounded subset of a finite dimensional normed space is $c-$compact, we show that existence of viable solutions in a non-compact domain implies the existence of a *stationary solution* (or *rest point*) of inclusion (3.3). More precisely, we have:

Corollary 3.3. *Let X be a closed convex subset of $\mathbf{R}^n$, K a compact subset of X, C a bounded subset of X, $\Lambda : \mathbf{R}^n \longrightarrow \mathcal{P}(\mathbf{R}^m)$ a strict linear process, and $\Phi : K \longrightarrow \mathcal{P}(\mathbf{R}^m)$ be an upper hemicontinuous map with closed convex values.*

Assume that the following properties are satisfied:

(i) given any $x_0 \in K$, there exists $T > 0$ and a solution $x(.)$ of the differential inclusion (3.3) on $[0, T)$ starting at x_0 such that for any $T' \in (0, T]$ there exists $t \in (0, T']$ such that $x(t) \in X$;

(ii) given any $x_0 \in X \setminus K$, there exists $T > 0$ and a solution $x(.)$ of the differential inclusion (3.3) on $[0, T)$ starting at x_0 such that for any $T' \in (0, T]$ there exists $t \in (0, T']$ such that $x(t) \in conv(x_0, C)$;

Then (3.3) has a stationary solution in X.

Remarks. (1) Conditions (i)-(ii) state that if a trajectory of the differential inclusion (3.3) starts in K then it must first enter X, and if it starts in $X \setminus K$ then it is first attracted by C.

(2) When X is a compact convex viability domain of Φ, $\mathbf{R}^n = \mathbf{R}^m$ and $\Lambda = Id_{\mathbf{R}^n}$, then (i) and (ii) are obviously satisfied with $K = C = X$. This corresponds to the equilibrium theorem of [AF].

The proof of Theorem 3.2 is based on the following more general version of a remarkable theorem of Ky Fan ([F2]).

Theorem 3.4. ([Bel]) *Let X be a convex subset in a topological vector space, Y a subset in $\{\varphi : X \longrightarrow \mathbf{R} : \varphi$ is upper semicontinuous and quasiconcave$\}$, and Γ a map in $\mathcal{S}(X, Y)$. Assume that there exists a compact subset K of X such that for each finite subset N of X there exists a compact convex subset C_N of X containing N such that:*

for every $x \in C_N \setminus K$ and every $\varphi \in \Gamma(x), \varphi(x) < \max_{u \in C_N} \varphi(u)$.

Then there exist $x_0 \in K$ and $\varphi_0 \in \Gamma(x_0)$ such that $\varphi_0(x_0) = \max_{u \in X} \varphi_0(u)$.

Proof. Define $f : X \times X \longrightarrow \mathbf{R} \cup \{+\infty\}$ by putting:

$$f(x, y) := s(x)(y) - s(x)(x) \text{ for all } (x, y) \in X \times X.$$

The function f is lower semicontinuous in the first variable and quasiconcave in the second one. Given any finite subset N of X and any $x \in C_N \setminus K$, choose $y \in C_N$ so that $s(x)(y) = \max_{u \in C_N} s(x)(u)$. By hypothesis, $s(x)(x) < s(x)(y)$, that is $f(x, y) > 0$. Consider now the map $\Psi : X \longrightarrow \mathcal{P}(Y)$ defined by:

$$\Psi(x) = \{y \in X : f(x, y) > 0\}, x \in X.$$

The map Ψ satisfies all hypotheses of Corollary 2.4. Since $f(y, y) = 0$ for all $y \in X$, Ψ cannot have a fixed point. Hence, for some $x_0 \in X$, $\Psi(x_0)$ must be empty, that is $f(x_0, y) \leq 0$ for all $y \in X$. The proof is complete with $\varphi_0 := s(x_0)$. ∎

Theorem 3.4 implies the following convenient analytic nonlinear alternative.

Corollary 3.5. *Let X be a convex subset of a normed space E, F be a normed space, $\Lambda : X \longrightarrow \Lambda(E, F)$ be a continuous mapping, and $f : X \times F^* \longrightarrow \mathbf{R} \cup \{+\infty\}$ be a real function satisfying the following conditions:*

(i) for each $\psi \in F^*$, the function $f(., \psi)$ is upper semicontinuous on X;

(ii) for each $x \in X$, the function $f(x, .)$ is quasiconvex on F^*.

Assume that there exist a compact subset K of X such that for each finite subset N of X there exists a compact convex subset C_N of X containing N such that:

(iii) for every $x \in C_N \setminus K$ and every $\psi \in F^*, f(x, \psi) \geq 0$ provided that $\Lambda(x)^*(\psi) \cap N_{C_N}(x) \neq \emptyset$.

Then one of the following conditions is satisfied:

(1) there exists $\widehat{x} \in X$ such that $f(\widehat{x}, \psi) \geq 0$ for all $\psi \in F^*$; or

(2) there exists $(x_0, \psi_0) \in K \times F^*$ such that $\Lambda(x_0)^*(\psi_0) \cap N_X(x_0) \neq \emptyset$ and $f(x_0, \psi_0) < 0$.

Proof of Theorem 3.2. (A) We apply Corollary 3.5 to $f(x, \psi) := \sup_{y \in \Phi(x)} \langle \psi, y \rangle$. Since Φ satisfies the condition of Ky Fan on (K, X) with respect to Λ, (3.2) implies that $f(x, \psi) \geq 0$ for all $x \in K$ and all $\psi \in F^*$ such that $\Lambda(x)^*(\psi) \cap N_K(x) \neq \emptyset$. That is, possibillity (2) of Corollary 3.5 fails. Therefore, there exists $\widehat{x} \in X$ such that $\sup_{y \in \Phi(\widehat{x})} \langle \psi, y \rangle \geq 0$ for all $\psi \in F^*$. The fact that $\widehat{x}$ is an equilibrium for Φ follows from the characterization of the closed convex hull in terms of the support function.

(B) Given $x_0 \in X$, define the map $\Psi : K \longrightarrow \mathcal{P}(F)$ by putting:

$$\Psi(x) := \Phi(x) + \Lambda(x)(x - x_0), x \in X,$$

and apply part (A) after checking that all hypotheses are satisfied by Ψ.

Note that Theorem 3.2 can be expressed as a coincidence theorem between two upper hemicontinuous maps. In the case where $E = F$ and one of the maps is the inclusion $K \hookrightarrow E$, we would then obtain a fixed point theorem for inward or outward maps with respect to a continuous family of convex processes that generalizes Ky Fan fixed point theorem for inward or outward maps [F2].

Open Problem. Is the inwardness condition (3.1) sufficient for the solvability of inclusion (3.3)? In other words, do we have a Nagumo type theorem for (3.3)? For the treatment of the classical case ($\Lambda = -Id$, plus compactness assumptions), the reader is referred to the book of Aubin [A2].

3.2. **Relaxation of Convexity.** One of the most natural as well as convenient class of spaces to be considered in extending fixed point results to non-convex sets is the class of absolute retracts, and more generally that of absolute neighborhood retracts. Recall that a compact subset X of a normed space E is said to be *a compact (neighborhood) retract* of E if there exists a mapping $r : E \longrightarrow X$ ($r : U \longrightarrow K$, where U is an open neighborhood of X) such that $r(x) = x$ for all $x \in X$; (the retraction r being not unique).

It is well-known that every continuous self-mapping f of a compact retract X has a fixed point; while if X is a compact neighborhood retract, then f has a fixed point provided the Lefschetz number $\lambda(f)$ of f is non-zero. In particular, assuming that the Euler characteristic $\chi(X) \neq 0$ (which holds true if K is a retract), it follows that a map from X into itself that is homotopic to the identity has a fixed point.

In what follows we shall assume that:

$$(A)\begin{cases} X \text{ is a compact neighborhood retract of a Banach space } E \\ \text{with a given retraction } r : U \longrightarrow X \text{ defined on a neighborhood } U \text{ of } X. \end{cases}$$

Since X is compact, we may assume with no loss of generality that U is a uniform neighborhood $B(X, \delta) = \{x \in E : d(x, X) < \delta\}$ of X.

Let us mention that the class of sets satisfying (A) is quite large. For instance, if X is compact convex then X satisfies (A) in addition of having non-trivial Euler characteristic. Furthermore, since the property of being a compact neighborhood retract, and the Euler characteristic, are both topological invariants, it follows that any subset Y of E homeomorphic to X satisfies (A) and has non-trivial Euler characteristic, provided X has both properties. For an extensive study of retracts the reader is referred to the monographs by Borsuk [Bo] and Hu [?].

The following two examples show that tangency conditions involving the Bouligand cone $T_X(x) := \{v \in E : \liminf_{t\downarrow 0^+} \frac{d(x+tv;X)}{t} = 0\}$ or the Clarke cone $C_X(x) := \{v \in E : \liminf_{y\to x, y\in K, t\downarrow 0^+} \frac{d(x+tv;X)}{t} = 0\}$ (which are equivalent for tangentially regular sets) are inadequate for the study of equilibria in large classes of non-convex sets.

Examples. The set $X = X_1 \cup X_2$ where:

$$\begin{aligned} X_1 &= \{(x,y) \in \mathbf{R}^2 : (x-1)^2 + y^2 = 1\}, \\ X_2 &= \{(x,y) \in \mathbf{R}^2 : (x+1)^2 + y^2 = 1\} \end{aligned}$$

is a compact neighborhood retract of $\mathbf{R}^2$ with non-trivial Euler characteristic. Let $f : X \longrightarrow \mathbf{R}^2$ be given by the formula:

$$f(x,y) = \begin{cases} (y, 1-x) & \text{if } (x,y) \in X_1, \\ (-y, 1+x) & \text{if } (x,y) \in X_2. \end{cases}$$

Clearly for each $x \in X, f(x,y) \in T_X(x,y) = C_X(x,y)$; but f has no zeros.

Being homotopy equivalent to the set X above, The set $Y = Y_1 \cup Y_2$ where:

$$\begin{aligned} Y_1 &= \{(x,y) \in \mathbf{R}^2 : \frac{1}{2} \leq (x-1)^2 + y^2 \leq 1\}, \\ Y_2 &= \{(x,y) \in \mathbf{R}^2 : \frac{1}{2} \leq (x+1)^2 + y^2 \leq 1\}. \end{aligned}$$

is also a compact neighborhood retract of $\mathbf{R}^2$ with non-trivial Euler characteristic. Furthermore, $T_Y(x,y) = C_Y(x,y)$ for $(x,y) \neq (0,0)$ and $T_Y(0,0) = \mathbf{R}^2 \supset \{0\} \times \mathbf{R} = C_Y(0,0)$. One can easily extend the function f above to a function $g : Y \longrightarrow \mathbf{R}^2$ having no zeros and satisfying tangency conditions with Clarke's cone.

These observations motivated us to define a new notion of *retraction normal cone* $N_X^r(x)$ to a compact neighborhood retract X.

We denote by D^* the closed unit ball in the dual E^* of E equipped with the strong topology. Given $X \subset E, x \in X, \epsilon > 0$, we put:

$$D_X(x,\epsilon) := \{y \in X : \|x - y\| \leq \epsilon\}, \quad B_X(x,\epsilon) := \{y \in X : \|x - y\| < \epsilon\}.$$

Let $X \subset E$ be a set satisfying (A). Then there holds:

$$(A') \quad \exists \eta > 0, \eta < \frac{1}{2}\delta, \forall x \in B(X,\eta), \|x - r(x)\| < \eta.$$

Definition 3.2. *For any $x \in X$ and $0 < \epsilon < \eta$, let:*

$$M_X^r(x;\epsilon) := \{\varphi \in D^* : x \in r[\overline{conv}\{y \in D_X(x,\epsilon) : \sup_{z \in \overline{B_X(x,\epsilon)}} \langle \varphi, z - y \rangle \leq 0\}]\}.$$

In other words, $\varphi \in M_X^r(x;\epsilon)$ if and only if x is a fixed point for the map $X \ni x \longmapsto r[\overline{conv}\{G_\epsilon(x,\varphi)\}]$ where $G_\epsilon : X \times D^* \longrightarrow \mathcal{P}(X)$ is the compact-valued map given by:

$$G_\epsilon(x,\varphi) := \{y \in D_X(x,\epsilon) : \sup_{z \in \overline{B_X(x,\epsilon)}} \langle \varphi, z - y \rangle \leq 0\}.$$

Example. Assume that $x \in X$ and that X is locally convex at x, that is there is a neighborhood V of x such that $V \cap X$ is convex. Then, for sufficiently small $\epsilon > 0, (\epsilon < \eta)$, we have:

$$M_X^r(x;\epsilon) = D^* \cap N_X(x),$$

where $N_X(x) = \{\varphi \in E^* : \langle \varphi, v \rangle \leq 0 \text{ for all } v \in T_X(x)\}$ is the normal cone of convex analysis to X at x.

In particular, if $x \in int(X)$, then for small ϵ, $M_X^r(x) = \{0\}$.

Remark. The sets $M_X^r(x;\epsilon)$ are not cones but they are starshaped around zero.

We are ready now to define the retraction normal cone.

Definition 3.3. *Assume that X satisfies (A). The retraction normal pre-cone at $x \in X$ is defined to be the set:*

$$M_X^r(x) := \limsup_{\substack{y \to x, y \in X \\ \epsilon \downarrow 0}} M(y;\epsilon).$$

Definition 3.4. *The retraction normal cone at $x \in X$ is the cone $N_X^r(x)$ spanned by $M_X^r(x)$.*

The following properties are important.

Proposition 3.6. *(i) $N_X^r(x)$ is a closed cone in E^*; (ii) the maps $M_X^r, N_X^r : X \longrightarrow \mathcal{P}(E^*)$ have closed graphs.*

Definition 3.5. *The retraction tangent cone at $x \in X$ is defined by the formula:*

$$T_X^r(x) := [N_X^r(x)]^- = \{v \in E : \langle p, v \rangle \leq 0, \forall p \in N_X^r(x)\}.$$

As expected, $T_X^r(x)$ is a closed convex cone in E^*. Moreover, the map $T_X^r : X \longrightarrow \mathcal{P}(E^*)$ is lower semicontinuous provided the normal cone $N_X^r(x)$ is convex for each $x \in X$ (which is not true in general).

Example. Assume that X is locally convex at $x \in X$ Then $N^r_X(x) = N_X(x)$ and $T^r_X(x) = T_X(x) = \overline{\bigcup_{t>0} \frac{1}{t}(X - x)}$ the tangent cone of convex analysis.

It is important to note that these notions of retraction cones depend upon the choice of the retraction r and on that of the norm in E.

The proof presented here of our second equilibrium theorem is based on the following generalization of the Fan-Kakutani fixed point theorem. A proof based on a Lefschetz type fixed point theorem is presented in our detailed paper [BK2].

The basic tool for our second equilibrium theorem is the following consequence of Corollary 2.2 (ii).

Proposition 3.7. *Assume that X satisfies (A) and that $\chi(X) \neq 0$ and let $\Psi \in \mathcal{A}(X)$ be a map with non-empty closed values such that, for all $x \in X$, $\Psi(x) \subset B(x, 2\eta)$, where η is given by (A'). Then Ψ has a fixed point.*

The idea of the proof is as follows. Consider a sequence $\epsilon_n \downarrow 0^+$ so that $2\epsilon_n + 2\eta < \delta$ for all n. Every ϵ_n–approximative selection s_{ϵ_n} of Ψ is homotopic to the identity (keep in mind (A') for the construction of the homotopy). Hence, s_{ϵ_n} has a fixed point x_n which is an approximative fixed point for Ψ. The compactness of X together with the closedness of the graph of Ψ end the argument.

We are ready to state and prove our second main theorem.

Theorem 3.8. *Assume that X satisfies (A) and that $\chi(X) \neq 0$. If $\Phi : X \longrightarrow \mathcal{P}(X)$ is upper hemicontinuous with closed convex values and satisfies the tangency condition:*

$$\Phi(x) \cap T^r_X(x) \neq \emptyset, \text{ for all } x \in X. \tag{3.5}$$

Then Φ has an equilibrium.

Proof. Observe first that (3.5) implies:

$$\text{for all } x \in X \text{ and all } \varphi \in N^r_X(x), \inf_{y \in \Phi(x)} \langle \varphi, y \rangle \leq 0. \tag{3.6}$$

Suppose now that $0 \notin \Phi(x)$ for each $x \in X$. By the Hahn-Banach separation theorem, for each $x \in X$ there is $\varphi_x \in E^*$ such that $\inf\{\langle \varphi_x, y \rangle : y \in \Phi(x)\} > 0$, i.e. $\sup_{y \in \Phi(x)} \langle -\varphi_x, y \rangle < 0$. Since Φ is upper hemicontinuous, the set:

$$U(x) := \{z \in X : \sup_{y \in \Phi(z)} \langle -\varphi_x, y \rangle < 0\},$$

is an open neighborhood of x, and the collection $\mathcal{U} = \{U(x)\}_{x\in X}$ constitutes an open covering of X. Let $\{\lambda_x\}_{x\in X}$ be a locally finite partition of unity subordinated to $\mathcal{U}$. Let us define a continuous mapping $f : X \longrightarrow E^*$ by the formula:

$$f(z) := \sum_{x\in X} \lambda_x(z)\varphi_x, \text{ for } z \in X.$$

Then, for any $z \in X$, we have $\sup_{y\in\Phi(z)}\langle -f(z), y\rangle < 0$, hence, $f(z) \neq 0$. Indeed, let $\{x_1, ..., x_k\} = \{x \in X : z \in U(x)\}$. Then $f(z) = \sum_{i=1}^{k} \lambda_i(z)\varphi_i$ where $\lambda_i = \lambda_{x_i}, \varphi_i = \varphi_{x_i}, i = 1, ..., k$. Since $z \in U(x_i)$, it follows that $\sup_{y\in\Phi(z)}\langle -\varphi_i, y\rangle < 0$ and we are done.

We prove the existence of an element $\overline{x} \in X$ such that $f(\overline{x}) \in N_X^r(\overline{x})$. This together with condition (3.6) will yield to a contradiction.

Let $n \in \mathbf{N}$ be such that $0 < \frac{1}{n} < \eta$, where η is given by (A'). Consider the map $\Psi_n : X \longrightarrow \mathcal{P}(X)$ defined by:

$$\Psi_n(x) := r[\overline{conv}\{G_{\frac{1}{n}}(x, \frac{f(x)}{\|f(x)\|})\}], \text{ for } x \in X.$$

One easily verifies that $\Psi_n(x) \subset B(x, 2\eta)$ for all $x \in X$ and all n. Since the map Ψ_n is the composition of the continuous mapping r and of the approachable map $\overline{conv}\{G_{\frac{1}{n}}(., \frac{f(.)}{\|f(.)\|})\}$, it follows from Proposition 3.7 that there is $x_n \in \Psi_n(x_n)$. In other words $\frac{f(x_n)}{\|f(x_n)\|} \in M_X^r(x_n; \frac{1}{n})$. The compactness of X implies (without loss of generality) that $x_n \underset{n\to\infty}{\longrightarrow} \overline{x} \in X$. Hence $f(x_n) \to f(\overline{x}) \in N_X^r(\overline{x})$. ∎

Remark. In case X is convex and compact we recover the classical convex case due to Cornet ([Co]).

We leave it to the reader to state a fixed point formulation of Theorem 3.8 subject to inwardness (or outwardness) conditions.

4. A general variational inequality

We present in this section a very general variational like inequality that extends known results to non-convex settings that naturally arise in control theory. We start with the abstract existence result.

Theorem 4.1. *Let C be a non-empty compact convex subset in a locally convex space E, and let Y be a non-empty complete convex subset of a locally convex space F. Let $\Phi : C \longrightarrow \mathcal{P}(Y)$ be a map with non-empty compact values such that one of the following conditions is satisfied:*

(i)$_1$ $\Phi \in \mathcal{A}(C, Y)$, or

(i)$_2$ the restriction $\Phi | conv\{N\} \in \mathcal{A}(conv\{N\}, Y)$ for each finite subset N of C.

Let $\Psi : C \longrightarrow \mathcal{P}(C)$ be a continuous map with non-empty compact convex values, and let $\varphi : C \times Y \times C \longrightarrow \mathbf{R} \cup \{\pm\infty\}$ be a continuous extended proper real function satisfying:

(ii) $\forall (x, y) \in C \times Y$, the function $\varphi(x, y, .)$ is quasiconvex on C.

Then the problem (4.1) below has a solution.

$$(4.1\ ) \qquad \begin{aligned} \exists x_0 &\in \Psi(x_0), \exists y_0 \in \Phi(x_0) \text{ such that} \\ \varphi(x_0, y_0, x) &\geq \varphi(x_0, y_0, x_0), \forall x \in \Psi(x_0). \end{aligned}$$

Proof. Define the marginal map $M_{\varphi,\Psi} : C \times Y \longrightarrow \mathcal{P}(C)$ by putting:

(4.2)
$$M_{\varphi,\Psi}(x, y) := \{u \in \Psi(x); \varphi(x, y, u) = \inf_{z \in \Psi(x)} \varphi(x, y, z)\}, (x, y) \in C \times Y.$$

The compactness of the values of Ψ, together with the continuity of φ, implies that $M_{\varphi,\Psi}$ has non-empty compact values. The convexity of the values of Ψ, together with (ii), implies that $M_{\varphi,\Psi}$ has convex values. We verify that $M_{\varphi,\Psi}$ is upper semicontinuous. To do this, observe that $M_{\varphi,\Psi}(x, y) = \Psi(x) \cap \tilde{M}_{\varphi,\Psi}(x, y)$ where $\tilde{M}_{\varphi,\Psi}(x, y) := \{u \in C; \varphi(x, y, u) = \inf_{z \in \Psi(x)} \varphi(x, y, z)\}$. Since Ψ is upper semicontinuous and has compact values, it suffices to verify that the graph of $\tilde{M}_{\varphi,\Psi}$ is closed. To do this, let $(x_\alpha, y_\alpha, u_\alpha)_\alpha$ be a net in $graph(\tilde{M}_{\varphi,\Psi})$ converging to $(x, y, u) \in C \times Y \times C$. Then,

$$\begin{aligned} \varphi(x, y, u) &= \lim_\alpha \varphi(x_\alpha, y_\alpha, u_\alpha) = \lim_\alpha \inf_{z \in \Psi(x_\alpha)} \varphi(x_\alpha, y_\alpha, z) \\ &= \limsup_\alpha \inf_{z \in \Psi(x_\alpha)} \varphi(x_\alpha, y_\alpha, z) \leq \inf_{z \in \Psi(x)} \varphi(x, y, z), \end{aligned}$$

where the inequality above follows from the upper semicontinuity of the marginal function $\inf_{z \in \Psi(.)} \varphi(., ., z)$ (this follows from the facts that φ is lower semicontinuous as a real function and that Ψ is lower semicontinuous as a map). Hence, $(x, y, u) \in graph(\tilde{M}_{\varphi,\Psi})$.

Now, since Y is convex and complete, $\overline{conv}\Phi(C)$ is a convex compact subset of Y. Since the product $C\times\overline{conv}\Phi(C)$ is compact, the restriction of the map $M_{\varphi,\Psi}$ to $C\times\overline{conv}\Phi(C)$ is approachable. Define a map $\Gamma : C\times\overline{conv}\Phi(C) \longrightarrow \mathcal{P}(C\times\overline{conv}\Phi(C))$ by putting:

$$\Gamma(x,y) := M_{\varphi,\Psi}(x,y)\times\Phi(x), (x,y)\in C\times\overline{conv}\Phi(C).$$

Being the product of compact-valued approachable maps, the map Γ is also approachable and has non-empty compact values. All conditions of Theorem 2 are thus satisfied. Therefore, Γ has a fixed point $(x_0,y_0)\in\Gamma(x_0,y_0)$, that is, $x_0\in\Psi(x_0)$, $y_0\in\Phi(x_0)$ and $\varphi(x_0,y_0,x_0)\leq\varphi(x_0,y_0,x), \forall x\in\Psi(x_0)$. ∎

Remarks. (1) If for all $x\in X$ with $x\in\Psi(x)$, for all $y\in\Phi(x)$ one has $\varphi(x,y,x)\geq 0$, then the inequality in (4.1) becomes $\varphi(x_0,y_0,x)\geq 0$, for all $x\in\Psi(x_0)$.

(2) If $\Psi(x)=C$ for all $x\in C$, the continuity assumptions on φ can be slightly relaxed to: φ is lower semicontinuous and $\varphi(.,.,u)$ is upper semicontinuous.

Theorem 4.1 can be expressed in purely topological terms as follows:

Theorem 4.2. *Let X be an acyclic compact ANR, and let Y be an ANR. Let $\Phi\in\mathcal{A}(X,Y)$ be a map with non-empty compact values. Let $\Psi : X\longrightarrow\mathcal{P}(X)$ be a continuous map with non-empty compact values, and let $\varphi : X\times Y\times X\longrightarrow \mathbf{R}\cup\{\pm\infty\}$ be a continuous extended proper real function. Assume that for any finite polyhedron P contained in $X\times Y$, the restriction of the marginal map $M_{\varphi,\Psi}$ defined by (4.2) to P is in $\mathcal{A}(P,X)$. Then problem (4.1) has a solution.*

4.1. **An application to a control problem.** A classical result of Aronszajn asserts that the solution set of the Cauchy problem with continuous right hand side $y'=f(t,y), y(0)=x$, is an R_δ set (i.e. a countable intersection of a decreasing sequence of compact contractible spaces), hence ∞−proximally connected in the space of continuous functions (see for instance [Be2], [GGK] and references there). This qualitative property of solution sets was extended by many authors to differential inclusions. Let X be a non-empty subset of $\mathbf{R}^n$ and $F:[0,T]\times X\longrightarrow\mathcal{P}(\mathbf{R}^n)$ be a map with non-empty compact values. Denote by $S_F(x;X)$ ($S_F(x)$ for $S_F(x;\mathbf{R}^n)$) the sets of Carathéodory solutions viable in X (i.e. $y\in S(x;X)$ if and only if $y(t)\in X,\forall t\in$

$[0,T]$) of the Cauchy problem with initial value x:

$$(4.3)\qquad \begin{cases} y'(t) \in F(t, y(t)) & a.e. \text{ in } [0,T], \\ y(0) = x. \end{cases}$$

Let C be a non-empty subset of X. Assume that for any given $x \in C$ there corresponds a subset $\Psi(x) \subset C$ of *possible return points.* Starting at an arbitrary point $x \in C$, we travel along a trajectory y of problem (4.3). We then follow a return path to a point $z \in \Psi(x)$. Assume that a *cost* $\varphi(x, y, z)$ is associated to this journey (for instance, $\varphi(x, y, z)$ could be the sum of an *attack cost* $\varphi_1(x, y)$ and a *retreat cost* $\varphi_2(y(T), z)$). We are interested in the problem:

$$(4.4)\qquad \begin{cases} \text{Find } x_0 \in C, x_0 \in \Psi(x_0), y_0 \in S_F(x_0; X), \text{ such that} \\ \varphi(x_0, y_0, x_0) = \inf_{z\in\Psi(x_0)} \varphi(x_0, y_0, z). \end{cases}$$

Let us recall that F is said to be a *Carathéodory map* if the following conditions are satisfied: (i) F has convex values; (ii) $y \longmapsto F(t, y)$ is upper semicontinuous a.e. $t \in [0,T]$; (iii) $\forall y \in X, t \longmapsto F(t, y)$ is measurable; (iv) $\sup\{\|v\|; v \in F(t, y), y \in X\} < \mu(t)$ where $\mu : [0,T] \longrightarrow [0, +\infty)$ is an integrable function. Let us also recall that a non-empty closed subset X of $\mathbf{R}^n$ is said to be a *proximate retract* ([Pl]) if there exists an open neighborhood U of X in $\mathbf{R}^n$ and a continuous mapping $r : U \longrightarrow X$ such that the following two conditions are satisfied: (i) $r(x) = x, \forall x \in X$; (ii) $\|r(x) - x\| = d(x, X) = \inf_{u\in X} \|x - u\|, \forall x \in U$. Any closed convex subset of $\mathbf{R}^n$, and any C^2-submanifold of $\mathbf{R}^n$ is a proximate retract ([Pl]).

Let $\mathcal{C}([0,T], \mathbf{R}^n)$ denote the space of continuous vector fields on $[0,T]$.

Theorem 4.3. *Assume that C is convex compact, Ψ is continuous with non-empty convex compact values, φ is continuous on $C \times \mathcal{C}([0,T], \mathbf{R}^n) \times C$, and φ is quasiconvex with respect to the return variable z. Then problem (4.4) has a solution provided X is a proximate retract and F is a Carathéodory map satisfying the tangency condition*

$$(4.5)\qquad F(t, y) \cap T_X(y) \neq \emptyset, \forall (t, y) \in [0,T] \times X,$$

where $T_X(y) := \{v \in \mathbf{R}^n; \liminf_{t\downarrow 0^+} \frac{d(y+tv, X)}{t} = 0\}$ is the Bouligand contingent cone to X at y.

Proof. In view of Theorem 1.1 in [Pl], the map $\Phi : X \longrightarrow \mathcal{P}(\mathcal{C}([0,T], X))$ defined by $\Phi(x) := S_F(x; X), x \in X$, has R_δ values. Moreover, one can show that Φ is upper semicontinuous. Indeed, the map F extends to a map $\hat{F} : [0,T] \times \mathbf{R}^n \longrightarrow \mathcal{P}(\mathbf{R}^n)$ in such a way that $S_{\hat{F}}(x;X) = \Phi(x)$ (see [Pl] for details). One then invokes well-known results to obtain that the solution set map $S_{\hat{F}} : \mathbf{R}^n \longrightarrow \mathcal{P}(\mathcal{C}([0,T],\mathbf{R}^n))$ is upper semicontinuous. Being a closed-graph multiselection of $S_{\hat{F}}$, the map $S^X_{\hat{F}}$: $X \longrightarrow \mathcal{P}(\mathcal{C}([0,T],X))$ given by $S^X_{\hat{F}}(x) := S_{\hat{F}}(x;X) = \Phi(x), x \in X$, is also upper semicontinuous. Consequently, Φ is approachable. The conclusion immediately follows from Theorem 4.1. ∎

Remarks. (1) Observe that in case X is an open subset of $\mathbf{R}^n$ then $T_X(y) = \mathbf{R}^n$, the inwardness condition (4.5) being automatically satisfied.

(2) Theorem 4.2 also holds for differential inclusions of order k in a Banach space.

When F is lower semicontinuous, the map $S_F(.)$ is generally neither upper semicontinuous or lower semicontinuous, nor are its values always closed. However, $S_F(.)$ admits an upper semicontinuous multiselection Φ with compact values ([Bre]) This fact leads to a lower semicontinuous version of the preceding theorem.

Theorem 4.4. *Assume that C is a closed disk $D(u_0,b)$ and that X is an open subset of $\mathbf{R}^n$ containing the closed disk $D(u_0,b+LT)$, where $b,L>0$. Assume also that Ψ and φ are as in Theorem 4.2 If F is lower semicontinuous with values in the open ball $B(0,L)$, then problem (4.4) has a solution.*

References

[A1] Aubin J. P., *Analyse fonctionnelle non-linéaire et applications à l'équilibre économique,* Ann. Sc. Math. Québec **II** (1978), 5-47.

[A2] Aubin J. P., *Viability theory,* Birkhäuser, Boston, 1991.

[AF] Aubin J. P. and H. Frankowska, *Set-valued analysis,* Birkhäuser, Boston, 1990.

[Be1] Ben-El-Mechaiekh H., *Zeros for set-valued maps with non-compact domains,* Math. Reports Acad. Sc. Canada **XII** (1990), 125-130.

[Be2] Ben-El-Mechaiekh H., *Continuous approximations of set-valued maps and fixed points,* Rapport de recherche **1820**, Centre de recherche mathématique, Université de Montréal, 1992.

[BCFL] Ben-El-Mechaiekh H., Chebbi S., Florenzano M. and J.V. Llinares, *Abstract convexity and fixed points,* preprint 1997.

[BD] Ben-El-Mechaiekh H. and P. Deguire, *Approachability and fixed points for non-convex set-valued maps,* J. Math. Anal. Appl. **170** (1992), 477-500.

[BDG] Ben-El-Mechaiekh H., Deguire P., Granas G., *Une alternative non linéaire en analyse convexe et applications,* C.R. Acad. Sc. Paris **295** (1982), 257-259; *Points fixes et coincidences pour les applications multivoques I, (applications de Ky Fan),* Ibidem **295** (1982), 337-340; *Points fixes et coincidences pour les applications multivoques II, (applications de type Φ et Φ^*),* Ibidem **295** (1982), 381-384.

[BK1] Ben-El-Mechaiekh H. and W. Kryszewski, *Equilibrium for perturbations of upper hemicontinuous set-valued maps by convex processes,* Georgian Math. J. **3** (1996), 201-215.

[BK2] Ben-El-Mechaiekh H. and W. Kryszewski, *Equilibrium for set-valued maps in nonconvex domains,* Trans. Amer. math. Soc., in print.

[BId] Ben-El-Mechaiekh H. and A. Idzik, *A Leray-Schauder type theorem for approximable maps,* Proc. Amer. Math. Soc. **122** (1994), 105-109.

[BIs] Ben-El-Mechaiekh H. and G. Isac, *Generalized multivalued variational inequalities,* in "Analysis and Topology", C. Andreian Cazacu, O. Letho and Th. M. Rassias Eds., World Scientific, Singapore, 1997.

[BOT] Ben-El-Mechaiekh H., Oudadess M. and J. F. Tounkara, *Approximations of multifunctions on uniform spaces and fixed points,* in "Topological vector spaces, algebras and related areas", A.T. Lau and I. Tweddle Eds., 239-250, Pitman Research Notes in Math. Series **316**, Longman, New York, 1994.

[Bi] Bielawski R., *Simplicial convexity and its applications,* J. Math. Anal. Appl. **127** (1987), 155-171.

[Bo] Borsuk K., *Theory of retracts,* Monografle Matematyczne 44, Warszawa, 1967.

[BP] Bae J. S. and S. Park, *Existence of maximizable quasiconcave functions on convex spaces,* J. Korean Math. Soc. **28** (1991), 285-292.

[Bre] Bressan A., *On the qualitative theory of lower semicontinuous differential inclusions,* J. Differential Equations **77** (1989), 379-371.

[Bro] Browder F., *The fixed point theory of multivalued mappings in topological vector spaces,* Mah. Ann., **177** (1968), 283-301.

[Ce] Cellina A., *A theorem on the approximation of compact valued mappings,* Atti Accad. Naz. Lincei Rend. Cl. Sci. Fis. Mat. Natur. (8) **47** (1969), 429-433.

[Co] Cornet B., *Paris avec handicap et théorèmes de surjectivité de correspondances,* C. R. Acad. Sc. Paris **281** (1975), 479-482.

[F1] Fan K., *A minimax inequality and applications,* in "Inequalities III", O. Shisha Ed., 103-113, Academic Press, New York, 1972.

[F2] Fan K., *Some properties of convex sets related to fixed point theorems,* Math. Ann. **266** (1984), 519-537.

[GGK] Górniewicz L., Granas A. and W. Kryszweski, *On the homotopy method in the fixed point index theory of multivalued mappings of compact absolute neighborhood retracts,* J. Math. Anal. Appl. **161** (1991), 457-473.

[GL] Górniewicz L. and M. Lassonde, *On approximable multi-valued mappings,*

[Ho1] Horvath C. D., *Contractibility and generalized convexity,* J. Math. Anal. Appl. **156** (1991), 341-357.

[Ho2] Horvath C. D., *Extension and selection theorems in topological spaces with a generalized convexity structure,* Annales de la faculté des sciences de Toulouse **2**, No. 2 (1993), 253-269.

[Hu] Hu S-T., *Theory of Retracts,* Wayne State Univ. Press, Detroit, 1965.

[K] Kryszewski W., *Topological and approximation methods in degree theory of set-valued maps,* Diss. Math., to appear.

[L] Lassonde M., *On the use of KKM set-valued maps in fixed point theory and related topics*, J. math. Anal. Appl. **97** (1983), 151-201.

[LR] Lasry J. M. and R. Robert, *Analyse nonlinéaire multivoque*, Cahiers de Math. de la Décision **7611**, Université de Paris-Dauphine, Paris, 1976.

[Pa] Park S., *Generalizations of Ky Fan's matching theorems and their applications,* J. Korean Math. Soc. **28** (1991), 275-283.

[Pl] Plaskacz S., *On the solution sets of differential inclusions,* Boll. U. M. I. (7) **6-A** (1992), 387-394.

[M] Michael E., *Continuous selections,* Annals Math. **63** (1956), 361-382.

[MC] McClendon, *Note on a selection theorem of Mas-Colell,* J. Math. Anal. Appl. **77** (1980), 326-327.

[R] Rockafellar R. T., *Clarke's tangent cones and boundaries of closed sets in* $\mathbf{R}^n$, Nonlinear Analysis **3** (1979), 145-154.

[V] Van Mill J., *Infinite dimensional topology,* North Holland, Amsterdam, 1989.

Department of Mathematics, Brock University, St. Catharines, Ontario, L2S 3A1, Canada

E-mail address: hmechaie@abacus.ac.brocku.ca

Q H ANSARI AND A H SIDDIQI

A generalized vector variational-like inequality and optimization over an efficient set

1. INTRODUCTION

The concept of vector variational inequality (abbreviated, VVI) was introduced by Giannessi [11] in 1980, in the finite dimensional Euclidean space with further applications. Since then, it has been studied and generalized in many different directions in a general setting in [1-8,14-18,24,26,29-31,34]. Chen and Cheng [5] have explained that the (VVI) expresses the necessary condition for an efficient solution to the vector optimization problem. Chen and Yang [8] have proved the equivalence and the existence of the vector complementarity problem, the vector variational inequality problem (VVIP), the extremum problem, the minimal element problem, and the vector unilateral minimization problem in the setting of Banach spaces. Various connections among (VVI), vector complementarity, fixed point, nonlinear vector optimization, and vector unilateral minimization problems have been further studied by Yang [31]. Yang [29] has also discussed the duality of the (VVI) which is related to the dual problem of vector optimization problems. For duality of scalar variational inequality, see Mosco [19]. The (VVI) for point-to-set maps is known as generalized vector variational inequality (GVVI), has been studied by Chen and Craven [7] and Lee *et al* [17,18]. They have also shown some relations between weak vector minimization, a (GVVI), and the optimization of a utility function over a set of efficient points. The vector variational-like inequality (VVLI) is a generalized form of (VVI), which was introduced and studied by Yang [30] and Siddiqi *et al* [26]. The relation between (VVLI) and generalized multi-objective programming problem has been investigated in [30]. Inspired and motivated by the applications of (GVVI) and (VVLI), Ansari [1-3] has introduced (VVLI) for point-to-set maps, which is known as generalized vector variational-like inequality (GVVLI), and he proved the existence of its solution in the setting of reflexive Banach spaces and Hausdorff topological vector spaces. Very recently, Lee *et al* [15] have further studied (GVVLI) in the setting of locally convex Hausdorff topological vector spaces.

In this paper, we consider the (GVVLI) and obtain some relations between weak vector minimization, a (GVVLI) and the optimization of a utility function over a set of efficient points, by introducing the concepts of η-subgradient and η-subdifferential.

Let Y be a topological vector space. A nonempty subset P of Y is called a *convex cone* if $P+P=P$ and $\alpha P \subset P$, for all $\alpha>0$. P is called a *pointed cone* if P is a cone and $P\cap(-P)=\{0\}$. The *partial order* in Y is denoted by $\leq$ and is defined as $x\leq y$ if and only if $y-x\in P$, for all $x,y\in Y$, and in this case P is called a *positive cone* in Y. Further more, if such a partial order is induced by a convex cone, it is called a *linear order.* An ordered topological vector space is a pair (Y,P) where Y is a topological vector space and P is a pointed convex cone with linear order induced by P. The *weak order* $\not<$ on ordered topological vector space (Y,P) with *int* $P\neq\emptyset$ is defined as $x\not< y$ if and only if $y-x\notin$ *int* P, for all $x,y\in Y$, where *int* P denotes the interior of P.

Let X and Y be two topological vector spaces and K be a nonempty subset of X. Let $F:K\longrightarrow L(X,Y)$ be a map, where $L(X,Y)$ is the space of all continuous linear operators from X into Y. Let $\{C(x):x\in K\}$ be a family of closed pointed convex cone in Y with *int* $C(x)\neq\emptyset$, for all $x\in K$. Then the *vector variational inequality problem* considered by Chen [4], as follows:

(VVIP): Find $x_0\in K$ such that

$$\langle F(x_0),\ y-x_0\rangle \notin -int\ C(x_0), \quad \text{for all} \ \ y\in K, \tag{1.1}$$

where $\langle F(x),y\rangle$ denotes the evaluation of the linear operator $F(x)$ at y and hence $\langle F(x),y\rangle\in Y$. (1.1) is called *vector variational inequality (VVI).* It has been further studied by Chen and Craven [6], Lai and Yao [14], and Yu and Yao [34] in the setting of Banach spaces with or without monotonicity assumptions.

Let $\eta:K\times K\longrightarrow X$ be a continuous map. Then Siddiqi *et al* [26] have studied the following *vector variational-like inequality problem:*

(VVLIP): Find $x_0\in K$ such that

$$\langle F(x_0),\ \eta(y,x_0)\rangle \notin -int\ C(x_0), \quad \text{for all} \ \ y\in K. \tag{1.2}$$

When $C(x)=P$, for all $x\in K$ and (Y,P) is an ordered topological vector space with weak order, (VVIP) becomes the following problem:

(VVIP)′: Find $x_0\in K$ such that

$$\langle F(x_0),\ y-x_0\rangle \not< 0, \quad \text{for all} \ \ y\in K. \tag{1.1$'$}$$

This problem has been investigated by Chen and Cheng [5], Chen and Yang [8] and Yang [31]. They have also studied the relations among the (VVIP)′, the vector complementarity problem, the fixed point problem and the vector optimization problem.

In this case, the (VVLIP) reduces to the following problem:

(VVLIP)′: Find $x_0 \in K$ such that

$$\langle F(x_0),\ \eta(y,x_0)\rangle \not< 0, \quad \text{for all } \ y \in K. \tag{1.2$'$}$$

Yang [30] has considered the (VVLIP)′ and established its equivalence with a generalized multi-objective programming problem.

Let $F : K \longrightarrow 2^{L(X,Y)}$ be a point-to-set map and $C(x) = P$, for all $x \in K$, then the (VVIP) is called the *generalized vector variational inequality problem* which is defined as follows:

(GVVIP): Find $x_0 \in K$ and $s_0 \in F(x_0)$ such that

$$\langle s_0,\ y - x_0\rangle \notin -int\ P, \quad \text{for all } \ y \in K. \tag{1.3}$$

It has been studied by Chen and Craven [7] and Lee *et al* [17,18] with applications in optimization.

Let $F : K \longrightarrow 2^{L(X,Y)}$ be a point-to-set map, then the (VVLIP) is known as the *generalized vector variational-like inequality problem*, studied by Ansari [1-3] and Lee *et al* [15]:

(GVVLIP): Find $x_0 \in K$ and $s_0 \in F(x_0)$ such that

$$\langle s_0,\ \eta(y,x_0)\rangle \notin -int\ C(x_0), \quad \text{for all } \ y \in K. \tag{1.4}$$

If we take $Y = \mathbb{R}$, $C(x) = \mathbb{R}_+$, for all $x \in K$, then the (VVI), (VVLI), (GVVI) and (GVVLI) reduce to the variational inequality [13], variational-like inequality [20,21,27,32], generalized variational inequality [10,22,23] and the generalized variational-like inequality [25], respectively.

Let (Y, P) be an ordered topological vector space. Then in this paper, we consider the following problems:

(GVVLIP)′: Find $x_0 \in K$ and $s_0 \in F(x_0)$ such that

$$\langle s_0,\ \eta(y,x_0)\rangle \notin -int\ P, \quad \text{for all } \ y \in K. \tag{1.4$'$}$$

Multiobjective Optimization Problem (MOP): Let $h : X \longrightarrow Y$ be a map.

$$WMIN\ h(x) \text{ subject to } x \in K, \tag{1.5}$$

where $WMIN$ denotes the weak minimum [28].

A point x_0 is a *weak minimum*, or *weak efficient point* for problem (1.5) if

$$h(x) - h(x_0) \notin -int\ P, \text{ for all } x \in K. \tag{1.6}$$

The set of all weak minimum points for (MOP) is denoted by E.

2. SOME EXISTENCE RESULTS FOR (GVVLIP)′

Through out in this paper, we denote 2^X the set of all nonempty subsets of X, $conv(A)$, for all $A \subset X$, the convex hull of A and $\partial(A)$, the boundary of A. For K, $B \subset X, int_K(B)$ and $\partial_K(B)$ denote the relative interior and relative boundary of B in K, respectively. The bilinear form $\langle .,. \rangle$ is supposed to be continuous.

We use the following concepts and results.

Definition 2.1 [1]**:** Let X be a topological vector space and (Y, P) be an ordered topological vector space, where P is a closed pointed convex cone in Y with $int\ P \neq \emptyset$. Let $T : X \longrightarrow 2^{L(X,Y)}$ be a point-to-set map and $\eta : K \times K \longrightarrow X$ be a point-to-point map. Then T is called:

(i) $P -$ *monotone* if for all $s \in T(x)$ and $t \in T(y)$,

$$\langle s - t,\ \eta(x, y)\rangle \in P, \text{ for all } x, y \in K;$$

(ii) $\eta -$ *pseudomonotone* if there exists $s \in T(x)$ such that $\langle s,\ \eta(y, x)\rangle \notin -int\ P$ implies that there exists $t \in T(y)$ such that $\langle t,\ \eta(y, x)\rangle \notin -int\ P$, for all $x, y \in K$.

(iii) $V -$ *hemicontinuous* if for any $x, y \in X$, $\alpha > 0$ and $t_\alpha \in T(x + \alpha y)$, there exists $t_0 \in T(x)$ such that for any $z \in X$, $\langle t_\alpha, z\rangle \longrightarrow \langle t_0, z\rangle$ as $\alpha \longrightarrow 0^+$.

Definition 2.2 [3]: A point-to-set map $T : X \longrightarrow 2^Y$ is said to be *uniformly compact near* $x \in X$ (or locally bounded at $x \in X$) if there exists a neighbourhood U of x in X such that $T(U) = \bigcup_{y \in U} T(y)$ is bounded.
If T is uniformly compact near x for any $x \in X$, then T is said to be *uniformly compact* on X.

Definition 2.3: Let K be a convex subset of X. A point-to-set map $T : K \longrightarrow 2^X$ is called *concave* if for any $x, y \in K$, $0 < \alpha < 1$,

$$\alpha T(x) \ + \ (1 - \alpha) T(y) \ \subseteq \ T(\alpha x + (1 - \alpha) y).$$

Now we present the existence theorems for (GVVLIP)$'$.

Theorem 2.1 [1]: Let X be a reflexive Banach space and (Y, P) be an ordered Banach space, where P is a closed pointed convex cone in Y with *int* $P \neq \emptyset$. If $F : X \longrightarrow 2^{L(X,Y)}$ is compact valued, η-pseudomonotone and V-hemicontinuous, and $\eta : K \times K \longrightarrow X$ is continuous and affine such that $\eta(x, x) = 0$, for all $x \in K$, then the (GVVLIP)$'$ is solvable.

When F is not monotone operator, then we have the following result.

Theorem 2.2 [3]: Let K be a nonempty compact convex subset of a Housdorff topological vector space X and (Y, P) be an ordered Hausdorff topological vector space, where P is a closed pointed convex cone with *int* $P \neq \emptyset$. If $F : K \longrightarrow 2^{L(X,Y)}$ is upper semicontinuous and uniformly compact, $\eta : K \times K \longrightarrow X$ is continuous such that $\eta(x, x) = 0$, for all $x \in K$, and the map $x \mapsto \langle s, \ \eta(x, y) \rangle$, for each fixed $y \in K$ and $s \in F(y)$, is affine, then the (GVVLIP)$'$ is solvable.

When K is not necessarily compact, we have the following result of Ansari [3].

Theorem 2.3 [3]: Let K be a nonempty convex subset of a Housdorff topological vector space X and (Y, P) be an ordered Hausdorff topological vector space, where P is a closed pointed convex cone with *int* $P \neq \emptyset$. Assume that

1° $F : K \longrightarrow 2^{L(X,Y)}$ is a point-to-set map;

2° $\eta : K \times K \longrightarrow X$ is a map;

3° there exists a nonempty compact convex subset B of K such that

(i) $F : B \longrightarrow 2^{L(X,Y)}$ is upper semicontinuous and uniformly compact;

(ii) $\eta : B \times B \longrightarrow X$ is continuous such that $\eta(x,x) = 0$, for all $x \in B$;

(iii) the map $x \mapsto \langle s,\ \eta(x,y) \rangle$, for each fixed $y \in B$ and $s \in F(x)$, is affine;

(iv) for any $x \in \partial_K(B)$, there exists an $y \in int_K(B)$ such that for any $s \in F(x)$, $\langle s,\ \eta(y,x) \rangle \in -int\ P$.

Then the (GVVLIP)$'$ is solvable.

3. EQUIVALENCE OF THE (MOP) AND A (GVVLIP)$'$

In this section, we define η-subgradient and η-subdifferential and we see that the (MOP) is equivalent to a (GVVLIP)$'$ under some restrictions.

Definition 3.1 [9]: A nonempty subset K of X is said to be *invex* if for each $x, y \in K$, $\alpha \in (0,1)$, we have $x + \alpha\eta(y,x) \in K$.

From now onward, K is a nonempty invex subset of X, unless otherwise specified.

Definition 3.2: A map $f : K \longrightarrow Y$ is called *P-preinvex function with respect to* η if for all $x, y \in K$ and $\alpha \in (0,1)$, we have

$$\alpha f(y) + (1-\alpha) f(x) - f(x + \alpha\eta(y,x)) \notin -int\ P.$$

A function $f : K \longrightarrow Y$ is called *P-invex with respect to* η if

$$f(y) - f(x) - f'(x)\eta(y,x) \notin -int\ P, \quad \text{for all} \quad x, y \in K,$$

where $f'(x)$ denotes the Frechet derivative of f at x.
If $Y = \mathbb{R}$ and $P = \mathbb{R}_+$ then f is called *invex* [12].

Definition 3.3: A continuous linear operator $v : X \longrightarrow Y$ *i.e.*, $v \in L(X,Y)$ is called $\eta-$ *subgradient* of $f : K \longrightarrow Y$ at $x_0 \in K$ if

$$f(x) - f(x_0) - \langle v,\ \eta(x,x_0) \rangle \notin -int\ P, \quad \text{for all} \quad x \in K.$$

The η-*subdifferential* of f at x_0 is the set $\eta - \partial f(x_0)$ of all η-subgradients of f at x_0.

Remark 3.1:(i) If $X = \mathbb{R}^n$, $Y = \mathbb{R}$ and $P = \mathbb{R}_+$, and $f : K \longrightarrow \mathbb{R}$ is a local Lipschitz function on an open subset K of $\mathbb{R}^n$, then the η-subgradient and η-subdifferential of f at x_0 reduce to the η-subgradient and η-subdifferential, respectively, of f at x_0 in the sence of Yang and Craven [33].
(ii) If $\eta(y,x) = y - x$, then η-subgradient and η-subdifferential of f at x_0 become the weak subgradient and weak subdifferential, respectively, of f at x_0 [7].

Proposition 3.1: If f is P-invex with respect to η at x_0, then $f'(x_0) \in \eta - \partial f(x_0)$.

Proof: If f is P-invex with respect to $\eta(y,x)$ at x_0, then

$$f(x) - f(x_0) - \langle f'(x_0),\ \eta(y,x_0)\rangle \notin \ - int\ P.$$

From the definition of η-subgradient, we have $f'(x_0) \in \eta - \partial f(x_0)$.

Consider the following Generalized Vector Variational-like Inequality Problem: Given nonempty invex set K in X and let $f : K \longrightarrow Y$ be a P-preinvex function with respect to $\eta(y,x)$, find $x_0 \in K$ and $v_0 \in \eta - \partial f(x_0)$ such that

$$\langle v_0,\ \eta(x,x_0)\rangle \notin \ - int\ P, \quad \text{for all } \ x \in K. \tag{3.1}$$

Theorem 3.1: Assume that

1^0. K is an invex set in X with $int\ K \neq \emptyset$;

2^0. P is a pointed convex cone in Y with $int\ P \neq \emptyset$;

3^0. $f : K \longrightarrow Y$ is P-preinvex with respect to $\eta(y,x)$ and continuous at a point $x_0 \in int\ K$.

If x_0 is a weak minimum of (MOP), then x_0 solves (3.1).
Conversely, if (x_0, v_0) solves (3.1), then x_0 is a weak minimum of (MOP).

Proof: Suppose that x_0 is a weak minimum of (MOP) then we heve

$$f(x) - f(x_0) \notin \ -int\ P, \quad \text{for all} \quad x \in K.$$

But for all $x \in K$,

$$\eta - \partial f(x_0) = \{v \in L(X, Y) : f(x) - f(x_0) - \langle v,\ \eta(x, x_0)\rangle \notin -int\ P\}.$$

So x_0 with $v_0 = 0$ satisfies (3.1).
Conversely, let (x_0, v_0) solve (3.1), where $x_0 \in int\ P$ and $v_0 \in \eta - \partial f(x_0)$.
Then, for all $x \in K$,

$$\langle v_0, \eta(x, x_0)\rangle \notin \ -int\ P$$

and

$$f(x) - f(x_0) - \langle v_0, \eta(x, x_0) \notin -int\ P.$$

Combining these, we get

$$f(x) - f(x_0) \notin -int\ P - int\ P \subseteq -int\ P, \quad \text{for all} \quad x \in K.$$

Thus x_0 is a weak minimum of (MOP).

Corollary 3.1: If x_0 is a weak minimum of (MOP) then $0 \in \eta - \partial f(x_0)$.

Definition 3.4:[33] The *η-extended directional derivative* of a function $f : K \longrightarrow Y$ at x in the direction $\eta(y, x)$ is defined as

$$f'_\eta(x)\eta(y, x) = \lim_{\alpha \downarrow 0} \frac{f(x + \alpha\eta(y, x)) - f(x)}{\alpha}.$$

In this case the operator $f'_\eta(x) : X \longrightarrow Y$ which assigns to each $\eta(y, x) \in K$ the vector $f'_\eta(x)\eta(y, x) \in Y$ is called the *η-extended directional derivative* at x_0 in the direction $\eta(y, x)$.

Theorem 3.2: Assume that

1^0. K is a nonempty invex set in X;

2^0. $f : K \longrightarrow Y$ is P-preinvex with respect to η and η-extended directional differentiable at x_0.

Then $x_0 \in K$ is a weak minimum of (MOP) if and only if x_0 and $f'_\eta(x_0)$ solves (3.1).

Proof: Let $x_0 \in K$ be a weak minimum of (MOP), then

$$f(x) - f(x_0) \notin -int\ P, \quad \text{for all} \quad x \in K.$$

Since K is invex, we have $x_0 + \alpha\eta(x, x_0) \in K$, for all $x \in K$ and $\alpha \in (0, 1)$. Then, we have

$$f(x_0 + \alpha\eta(x, x_0)) - f(x_0) \notin -int\ P$$

$$\Longrightarrow \quad \frac{f(x_0 + \alpha\eta(x, x_0)) - f(x_0)}{\alpha} \notin -int\ P.$$

Since f is η-extended directional differentiable at x_0, we have

$$f'_\eta(x_0)\eta(x, x_0) \notin \ -int\ P.$$

$$i.e. \ \langle f'_\eta(x_0),\ \eta(x, x_0)\rangle \notin \ -int\ P.$$

Since f is P-preinvex with respect to η, we have

$$\alpha f(x) + (1 - \alpha)f(x_0) - f(x_0 + \alpha\eta(x, x_0)) \notin -int\ P, \quad \text{for all} \quad \alpha \in (0, 1)$$

$$\Longrightarrow \quad \alpha(f(x) - f(x_0)) - [f(x_0 + \alpha\eta(x, x_0)) - f(x_0)] \notin -int\ P$$

$$\Longrightarrow \quad f(x) - f(x_0) - \frac{f(x_0 + \alpha\eta(x, x_0)) - f(x_0)}{\alpha} \notin -int\ P.$$

Taking $\alpha \longrightarrow 0^+$, we have

$$f(x) - f(x_0) - \langle f'_\eta(x_0),\ \eta(x, x_0)\rangle \notin -int\ P$$

$$\Longrightarrow \quad f'_\eta(x_0) \in \eta - \partial f(x_0).$$

Thus x_0 and $f'_\eta(x_0)$ satisfy (3.1). Converse part follows from the proof of Theorem 3.1.

4. OPTIMIZATION OVER AN EFFICIENT SET

Given nonempty set $K \subset X, f : K \longrightarrow Y$ and $\phi : X \longrightarrow \mathbb{R}$, in this section we consider the following problems:

(I): Multiobjective Optimization Problem (MOP):

$$WMIN\ f(x) \text{ subject to } x \in K.$$

We denote the set of all weak minimum or weak efficient points for this problem by E.

(II): Generalized Vector Variational-like Inequality Problem: Find $x_0 \in K,\ v_0 \in \eta - \partial f(x_0)$ such that

$$\langle v_0,\ \eta(x, x_0)\rangle \notin\ - int\ P, \quad \text{for all} \quad x \in K.$$

We denote the set of optima for this problem by V.

(III): Optimization Problem Over an Efficiet Set:

$$Minimize \quad \phi(x) \quad \text{subject to} \quad x \in E.$$

(IV): $WMIN\ f'_\eta(x_0)\eta(x, x_0)$ subject to $x \in K$.

(V): $WMIN\ [\phi(x), f'_\eta(x_0)\eta(x, x_0)]$ subject to $x \in K$.

(VI): $Minimize \quad \phi(x)$ subject to $x \in V$.

(VII): $Minimize \quad \phi(x)$ subject to $f'_\eta(x_0)\eta(x, x_0) \notin -int\ P, \quad \text{for all} \quad x \in K$.

For (IV), (V) and (VII), f is assumed η-extended directional differentiable. The weak minimization is with respect to the cone P, or $\mathbb{R}_+ \times P$ for (V).

Theorem 4.1: Let K be a nonempty invex subset of X with $int\ K \neq \emptyset$ and $f : X \longrightarrow Y$ be P-preinvex with respect to η and η-extended directional differentiable, with derivative $f'_\eta(x_0)$ at $x_0 \in X$. Then

(a) Problem (I) is equivalent to Problem (IV).

(b) Problem (I) is equivalent to Problem (II).

(c) Problem (III) is equivalent to Problem (VI).

(d) If x_0 is an optimum for Problem (III), then x_0 is an optimum for Problem (V) (So any necessary conditions for an optimum of Problem (V) hold also for Problem (III)).

(e) Problem (III) is equivalent to Problem (VII).

Proof: (a). (I $\Rightarrow$ IV): Let x_0 be a weak minimum of (I), then

$$f(x) - f(x_0) \notin -\,int\ P, \quad \text{for all} \quad x \in K.$$

Since K is invex, we have $x = x_0 + \alpha\eta(y, x_0) \in K$, for all $\alpha \in (0, 1)$ and $y \in K$.
This implies that

$$f(x_0 + \alpha\eta(y, x_0)) - f(x_0) \notin -\,int\ P.$$

Dividing by α, we have

$$\frac{f(x_0 + \alpha\eta(y, x_0)) - f(x_0)}{\alpha} \notin -\,int\ P.$$

Taking $\alpha \longrightarrow 0^+$, and since f is η-extended directional differentiable, we have

$$f'_\eta(x_0)\eta(y, x_0) \notin -\,int\ P.$$

Hence x_0 is a weak minimum for Problem (IV).
(IV $\Rightarrow$I): Since f is P-preinvex, we have from the proof of Theorem 3.2,

$$f(y) - f(x_0) - f'_\eta(x_0)\eta(y, x_0) \notin -\,int\ P, \quad \text{for all} \quad y \in K.$$

Now, since $f'_\eta(x_0)\eta(y, x_0) \notin -\,int\ P$, we have

$$f(y) - f(x_0) \notin -\,int\ P - int\ P \subseteq -int\ P$$

$$\Longrightarrow \quad f(y) - f(x_0) \notin -\,int\ P, \quad \text{for all} \quad y \in K.$$

Hence x_0 is a weak minimum of Problem (I).
(b). It follows from Theorem 3.2.
(c). It also follows from Theorem 3.2, $E = V$.
(d). Let x_0 be an optimum for Problem (III), then

$$\phi(x_0) \leq \phi(x) \text{ and } f'_\eta(x_0)\ \eta(x,x_0) \ \notin \ -int\ P, \quad \text{for all } \ x \in K.$$

which is equivalent to $[\phi(x_0) - \phi(x), f'_\eta(x_0)\eta(x,x_0)] \ \notin \ -int\ [\mathbb{R}_+ \times P]$, for all $x \in K$ which states that x_0 is a weak minimum for (V).
(e). (III $\Rightarrow$ VII): Let x_0 be a minimum for Problem (III), then

$$\phi(x_0) \leq \phi(x) \text{ and } f(x) - f(x_0) \ \notin \ -int\ P, \quad \text{for all } \ x \in K.$$

By a similar argument to that in (a), it follows that

$$\phi(x_0) \leq \phi(x) \text{ and } f'_\eta(x_0)\eta(x,x_0) \ \notin \ -int\ P, \quad \text{for all } \ x \in K.$$

(VII $\Rightarrow$III): If x_0 is a weak minimum for Problem (VII), then we have

$$\phi(x_0) \leq \phi(x) \text{ and } f'_\eta(x_0)\eta(x,x_0) \ \notin \ -int\ P, \quad \text{for all } \ x \in K.$$

Since f is P-preinvex with respect to η, we have from the proof of Theorem 3.2,

$$\phi(x_0) \leq \phi(x) \text{ and } f(x) - f(x_0) - f'_\eta(x_0)\eta(x,x_0) \ \notin \ -int\ P, \quad \text{for all } \ x \in K.$$

Hence,

$$\phi(x_0) \leq \phi(x) \text{ and } f(x) - f(x_0) \ \notin \ -int\ P, \quad \text{for all } \ x \in K.$$

$\Longrightarrow x_0$ is a solution of Problem (III).

References

[1] Ansari QH(1995) On Generalized Vector Variational-like Inequalities. *Ann. Sci. Math. Quebec*19:131-137.

[2] Ansari QH(1997) A Note On Generalized Vector Variational-like Inequatities. *Optimization*:1-9 (forthcoming).

[3] Ansari QH(1997) Extended Generelized Vector Variational-like Inequalities For Nonmonotone Multivalued Maps. *Ann. Sci. Math. Quebec*21: (forthcoming).

[4] Chen GY(1992) Existence of solutions for a Vector Variational Inequality : An Extension of Hartmann-Stampacchia Theorem. *J. Optimi. Theory Appl.*74:445-456.

[5] Chen GY and Cheng GM(1987) Vector variational Inequality and Vector Optimization. *Lecture Notes in Economics and Mathematical Systems,* No. 285:408-416, Springer-Verlag.

[6] Chen GY and Craven BD(1989) Approximate dual and approximate Vector Variational Inequality for Multiobjective Optimization. *J. Austral. Math. Soc.*47:418-423.

[7] Chen GY and Craven BD(1990) A Vector Variational Inequality and Optimization Over an Efficient Set. *Zeitscrift fur Operations Research*3:1-12.

[8] Chen GY and Yang XQ(1990) The Vector Complementarity Problem and its Equivalences with the Weak Minimal Element in Ordered Spaces *J. Math. Anal. Appl.*153:136-158.

[9] Craven BD(1981) Invex Functions and Constrained Local Minima. *Bull. Aust. Math. Soc.*24:357-366.

[10] Fang SC and Peterson EL(1982) Generalized Variational Inequalities. *J. Optimiz. Theory Appl.*38:363-383.

[11] Giannessi F(1980) Theorems of Alternative, Quadratic Programs and Complementarity Problems, in *edited by R.W. Cottle, F. Giannessi and J.L. Lions*:151-186, John Wiley and Sons, Chichester.

[12] Hansan MA(1982) On suffidiency of the Kuhn-Tucker Conditions. *J. Math. Anal. Appl.*80:545-550.

[13] Kinderlehrer D and Stampcchia G(1980) *An Introduction to Variational Inequalities and Their Applications*, Academic Press, New York.

[14] Lai TC and Yao JC(1996) Existence Results For VVIP. *Appl. Math. Lett.*9:17-19.

[15] Lee BS, Lee GM and Kim DS(1996) Generalized Vector Variational-like Inequalities in Locally Convex Hausdorff Topological Vector Spaces. *Preprint.*

[16] Lee GM, Kim DS, Lee BS and Yen DN(1996) Vector Variational Inequality as a tool For Studing Vector Optimization Problems *Preprint.*

[17] Lee GM, Kim DS, Lee BS and Cho SJ(1993) Generalized Vector Variational Inequality and Fuzzy Extension *Appl. Math. Lett.*6:47-51.

[18] Lee GM, Kim DS and Lee BS(1996) Generalized Vector Variational Inequality *Appl. Math. Lett.*9:39-42.

[19] Mosco U(1972) Dual Variational Inequalities *J. Math. Anal. Appl.* 40:202-206.

[20] Noor MA(1995) Nonconvex Functions and Variational Inequalities *J. Optimiz. Theory Appl.*87:615-630.

[21] Parida J, Sahoo M and Kumar A(1989) A variational-like inequality problem *Bull. Austral. Math. Soc.*39:225-231.

[22] Saigal R(1976) Extension of the Generalized Complementarity Problem *Math. Oper. Res.*1:260-266.

[23] Siddiqi AH and Ansari QH(1989) An Iterative Method For Generalized Variational Inequatities *Math. Japonica*34:475-481.

[24] Siddiqi AH, Ansari QH, and Khaliq A(1995) On Vector Variational Inequalities *J. Optimiz. Theory Appl.* 84:171-180.

[25] Siddiqi AH, Ansari QH and Ahmad R(1995) On Generalized Variational-like Inequalities *Indian J. Pure Appl. Math.* 26:1135-1141.

[26] Siddiqi AH, Ansari QH and Ahmad R(1997) On Vector Variational-like Inequalities, To appear in *Indian J. Pure Appl. Math.*

[27] Siddiqi AH, Khaliq A and Ansari QH(1994) On variational-like inequalities *Ann. Sci. Math. Quebec* 18:39-48.

[28] Weir T and Jeyakumar VA(1988) class of nonconvex functions and mathematical programming *Bull. Austral. Math. Soc.*38:177-189.

[29] Yang XQ(1993) Vector Variational Inequality and Its Duality, *Nonlinear Analysis, Theory, Methods Appl.*21:869-877.

[30] Yang XQ(1993) Generalized Convex Functions and Vector Variational Inequalities *J. Optimiz. Theory Appl.*79:563-580.

[31] Yang XQ(1993) Vector Complementarity and Minimal Element Problems *J. Optimiz. Theory Appl.*77:483-495.

[32] Yang XQ and Chen GY(1992) A Class of Nonconvex Functions and Pre-variational Inequalities *J.Math. Anal.Appl.*169:359-373.

[33] Yang XQ and Craven BD(1991) Necessary Optimality Conditions with a Modified Subdifferential *Optiminzation*22:387-400.

[34] Yu SY and Yao JC(1996) On Vector Variational Inequalities, *J. Optimiz. Theory Appl.*89:749-769.

Department of Mathematics
Aligarh Muslim University
Aligarh - 202 002 (INDIA)

R S PATHAK

A generalization of the wavelet transform

1. Introduction

The wavelet transform of $f \in L^2(\mathbf{R})$ is defined by

$$(w_\psi f)(b,a) = a^{-1/2} \int_{\mathbf{R}} f(t)\, \overline{\psi}((t-b)/a)dt, \; a>0, \; b \in \mathbf{R}, \tag{1.1}$$

where $\psi \in L^2(\mathbf{R})$ is called the mother wavelet. From (1.1) it is clear that the wavelet transform is obtained by applying translation operator $T_b : L^2(\mathbf{R}) \to L^2(\mathbf{R})$ defined by

$$T_b \psi(x) = \psi(x+b), \quad b \in \mathbf{R} \tag{1.2}$$

and dilation operator $D_a : L^2(\mathbf{R}) \to L^2(\mathbf{R})$ given by

$$D_a \psi(x) = a^{-1/2} \psi(x/a), \qquad a>0. \tag{1.3}$$

If ψ satisfies

$$\int_{\mathbf{R}} |\psi(v)| \; |v|^{-1/2} dv = A < \infty \tag{1.4}$$

then we know that $w : L^2(\mathbf{R}) \to L^2(\mathbf{R} x \mathbf{R}_+)$ is a linear, bounded transformation [9].

Setting $a=1/x$ and $(Wf)(b,a) = a^{1/2}(w_\psi f)(b,a)$ in (1.1) we get

$$(Wf)(b,a) = \int_{\mathbf{R}} (T_b f)(t)\, \overline{\psi}(xt)\, dt, \; x>0. \tag{1.5}$$

From (1.5) it follows that the (modified) wavelet transform W may be regarded as the bilateral Watson transform of $(T_b f)(t)$. De Snoo [10] has given a generalization of the classical Watson transform as a bounded linear operator $V : L^2(\mathbf{R}_+) \to L^2(\mathbf{R}_+)$ satisfying certain dilation equation. His ideas are modified here to define the wavelet transform as a bounded linear operator $W : L^2(\mathbf{R}) \to L^2(\mathbf{R} x \mathbf{R}_+)$ satisfying the relation $(Wf)(b,a) = (V T_b f)(a)$.

In Section 2 of the paper, an account of positive, negative and bilateral Mellin transforms [18] is given. In Section 3 a generator of the wavelet transform is defined by means of positive and negative Mellin transforms and dilation operator D_a. Then a general wavelet transform (GWT) is defined by means of its generator and translation operator T_b. This section is concluded by stating basic properties of the general wavelet transform. In Section 4, integral representations in the form (1.5) of the wavelet transform, are derived using the theory of Mellin convolutions.

2. The bilateral Mellin transform

The Fourier transform of $g \varepsilon L^1(\mathbf{R})$ is defined by

$$(\mathcal{F}g)(v) = (2\pi)^{-1/2} \int_{\mathbf{R}} e^{iuv} g(u)\, du \tag{2.1}$$

and the inversion formula is given by

$$g(u) = (2\pi)^{-1/2} \int_{\mathbf{R}} e^{-iuv} (\mathcal{F}g)(v)\, dv. \tag{2.2}$$

The change of variable $u=\ln x$ $(x>0)$ and the substitution $x^{-1/2}g(\ln x)=f(x)$ transforms (2.1) into the positive Mellin transform

$$(M_+^1 f)(y) = (2\pi)^{-1/2} \int_{\mathbf{R}_+} x^{iy-1/2} f(x)\, dx, \quad y \varepsilon \mathbf{R}, \tag{2.3}$$

the corresponding inversion formula becomes

$$f(x) = (2\pi)^{-1/2} \int_{\mathbf{R}} x^{-iy-1/2} (M_+^1 f)(y)\, dy, \quad x \varepsilon (0,\infty). \tag{2.4}$$

Similarly, the change of variable $u=\ln(-x)$ $(x<0)$ and the substitution $(-x)^{-1/2}g(\ln(-x))=f(x)$ leads to the negative Mellin transform

$$(M_-^1 f)(y) = (2\pi)^{-1/2} \int_{\mathbf{R}} (-x)^{iy-1/2} f(x)\, dx, \quad y \varepsilon \mathbf{R} \tag{2.5}$$

with the inversion formula

$$f(x) = (2\pi)^{-1/2} \int_{\mathbf{R}} (-x)^{-iy-1/2} (M_-^1 f)(y)\, dy, \quad x \varepsilon (-\infty,0). \tag{2.6}$$

Let us define

$$x_+ = \begin{cases} x, & x>0 \\ 0, & x \le 0 \end{cases} \qquad x_- = \begin{cases} 0, & x \ge 0 \\ -x, & x<0. \end{cases} \tag{2.7}$$

Then we can join the positive and negative Mellin transforms and define the bilateral Mellin transform as follows [18, p.345]:

$$(M_b f)_\sigma (y) := f_\sigma^{BM}(y) = (2\pi)^{-1/2} \int_R x_\sigma^{iy-1/2} f(x)\, dx,\ \sigma = \pm \tag{2.8}$$

and

$$(M_B^{-1} f_\sigma^{BM})(x) := f(x) = (2\pi)^{-1/2} \sum_{\sigma=\pm} \int_R x_\sigma^{-iy-1/2}\ f_\sigma^{BM}(y) dy,\ x \varepsilon \mathbf{R}. \tag{2.9}$$

We have the Parseval formula

$$\int_R f(x)\ \bar{g}(x)\, dx = \sum_{\sigma=\pm} \int_R (M_B f)_\sigma (y) \overline{(M_B g)_\sigma} (y) dy, \tag{2.10}$$

so that

$$\|f\|_{L^2}^2 = \sum_{\sigma=\pm} \|(M_B f)_\sigma\|_{L^2}^2 . \tag{2.11}$$

Now, let $F:L^2(\mathbf{R}) \rightarrow L^2(\mathbf{R})$ denote the Fourier-Plancherel transform defined by

$$(Ff)(x) = (2\pi)^{-1/2} (d/dx) \int_R (e^{ixy} - 1)/(iy) f(y)\, dy \text{ a.e., } f \varepsilon L^2(\mathbf{R}) . \tag{2.12}$$

Then the isometric isomorphism $U_+ : L^2(\mathbf{R}_+) \rightarrow L^2(\mathbf{R})$ given by

$$(U_+ f)(x) = e^{x/2} f(e^x),\ \ f \varepsilon L^2(\mathbf{R}_+) \tag{2.13}$$

defines the linear, bounded transform $M_+ : L^2(\mathbf{R}_+) \rightarrow L^2(\mathbf{R})$ by

$$(M_+ f)(y) = (FU_+ f)(y). \tag{2.14}$$

Similarly, the isometric isomorphism $U_- : L^2(\mathbf{R}_-) \rightarrow L^2(\mathbf{R})$ given by

$$(U_- f)(x) = e^{x/2} f(-e^x),\ \ f \varepsilon L^2(\mathbf{R}_-) \tag{2.15}$$

defines the negative Mellin transform $M_- : L^2(\mathbf{R}_-) \rightarrow L^2(\mathbf{R})$ by

$$(M_- f)(y) = (FU_- f)(y). \tag{2.16}$$

3. The general wavelet transform

To define the general wavelet transform $W:L^2(\mathbf{R}) \rightarrow L^2(\mathbf{R} x \mathbf{R}_+)$ we shall make use for the following well-known result on Fourier transforms [1].

Theorem 3.1. A linear, bounded transform $B:L^2(\mathbf{R})\to L^2(\mathbf{R})$ commutes with all translations $T_b(b\varepsilon\mathbf{R})$ if and only if there exists a function $K\varepsilon L^\infty(\mathbf{R})$ such that

$$B = F^{-1}\, m[K]\, F \tag{3.1}$$

where $m[K] : L^2(\mathbf{R})\to L^2(\mathbf{R})$ is the linear bounded transform defined by

$$(m[K]f)\,(x) = K(x)\, f(x),\ f\varepsilon L^2(\mathbf{R}). \tag{3.2}$$

Taking $T_{\ln a} = a\, U_+\, D_a\, U_+^{-1}$ and using the relation $M_+U_+^{-1} = F$, we derive from Theorem 3.1:

Theorem 3.2(a). A linear, bounded transform $V_+:L^2(\mathbf{R}_+)\to L^2(\mathbf{R}_+)$ commutes with all dilations $D_a(a\varepsilon\mathbf{R}_+)$ if and only if there exists $K\varepsilon L^\infty(\mathbf{R})$ such that

$$V_+ = M_+^{-1}\, m[K]\, M_+\, . \tag{3.3}$$

Similarly, isomorphism V_- yields

Theorem 3.2(b). A linear, bounded transform $V_- : L^2(\mathbf{R}_-)\to L^2(\mathbf{R}_-)$ commutes with all dilations $D_a\ (a\varepsilon\mathbf{R}_+)$ if and only if there exists $H\varepsilon L^2\ (\mathbf{R})$ such that

$$V_- = M_-^{-1}\, m[H]\, M_-\, . \tag{3.4}$$

Define

$$\begin{aligned} V &= V_+ \ \text{ on } \mathbf{R}_+ \\ &= V_- \ \text{ on } \mathbf{R}_-\, . \end{aligned} \tag{3.5}$$

Then using direct sum notation: $V = V_+ \oplus V_-$ we get

Theorem 3.3. A linear, bounded transform $V:L^2(\mathbf{R})\to L^2(\mathbf{R})$ commutes with all dilations $D_a\ (a\varepsilon\mathbf{R}_+)$ if and only if there exists $K,H\ \varepsilon L^2(\mathbf{R})$ such that

$$V = M_+^{-1}\, m[K]\, M_+ \oplus M_-^{-1}\, m[H]\, M_-\, . \tag{3.6}$$

Now, let the linear, bounded transform $P:L^2(\mathbf{R}\text{-}\{0\})\to L^2(\mathbf{R})$ be defined by

$$(Pf)\,(x) = x^{-1}\, f(x^{-1}),\ \ f\varepsilon L^2(\mathbf{R}). \tag{3.7}$$

Then the following theorem characterises all linear, bounded transforms $V: L^2(\mathbf{R}) \to L^2(\mathbf{R})$ which satisfy

$$D_{1/a} V = V D_a \quad (a>0). \tag{3.8}$$

Theorem 3.4. A linear, bounded transform $V: L^2(\mathbf{R}) \to L^2(\mathbf{R})$ satisfies (3.8) if and only if there exists K, H $\varepsilon L^\infty(\mathbf{R})$ such that

$$V = M_+^{-1} m[K] M_+ P \oplus M_-^{-1} m[H] M_- P. \tag{3.9}$$

Proof. If $K \varepsilon L^\infty(\mathbf{R})$ and V_+ is given by

$$V_+ = M_+^{-1} m[K] M_+ P \tag{3.10}$$

then V_+ is linear and bounded. Since $P^2 = \mathbf{I}$ it follows from Theorem 3.2(a) that V_+P commutes with all D_a, i.e.

$$D_{1/a} V_+ P = V_+ P D_{1/a} \quad \forall a>0 .$$

Since $$PD_a = D_{1/a} P \tag{3.11}$$

we have

$$D_{1/a} V_+ P = V_+ D_a P$$

which gives

$$D_{1/a} V_+ = V_+ D_a .$$

Conversely, if V_+ is linear and bounded on $\mathbf{R}_+$ and satisfies

$$D_{1/a} V_+ = V_+ D_a ,$$

then

$$D_{1/a} V_+ P = V_+ D_a P.$$

Hence by (3.11)

$$D_a V_+ P = V_+ D_{1/a} P = V_+ P D_a .$$

Thus, V_+P commutes with all D_a, $a>0$. Hence by Theorem 3.2(a) there exists $K\varepsilon L^\infty(\mathbf{R})$ such that $V_+P = M_+^{-1}m[K]\,M_+$, which yields (3.10).

Similarly, if $V_- : L^2(\mathbf{R}_-)\to L^2(\mathbf{R}_-)$ satisfies

$$D_{1/a}\,V_- = V_-\,D_a \quad (a>0) \tag{3.12}$$

then Theorem 3.2(b) gives the representation

$$V_- = M_-^{-1}\,m[H]\,M_-P. \tag{3.13}$$

Finally, the direct sum $V = V_+ \oplus V_-$ gives (3.9).

Definition 3.5. The generator of the wavelet transform, $V:L^2(\mathbf{R})\to L^2(\mathbf{R})$, is a linear, bounded transform, which satisfies (3.8).

Definition 3.6. The general wavelet transform

$$W(.,.) : L^2(\mathbf{R}) \to L^2(\mathbf{R} \times \mathbf{R}_+)$$

is a linear, bounded transformation given by

$$(Wf)\,(b,a) : = (V\,T_b)\,(f)\,(a),\quad b\varepsilon\mathbf{R},\ a\varepsilon\mathbf{R}_+ \tag{3.14}$$

$$= (M_+^{-1}\,m[K]\,M_+ \oplus M_-^{-1}\,m[H]\,M_-)\,(P\,T_bf)\,(a) \tag{3.15}$$

where K, $L\varepsilon L^\infty(\mathbf{R})$ and $f\varepsilon L^2(\mathbf{R})$.

In what follows (3.15) will be called wavelet transform and denoted by W.

From Definitions 3.5, 3.6 and Theorems 3.3, 3.4 and the fact that $P^2=I$ we deduce the following properties of generator and wavelet transforms:

(I). The product of an odd number of generators is a generator.

(II). The product of an even number of generators is a transform of type (3.6).

(III). Generators V_1, V_2, V_3 satisfy

$$V_1\,V_2\,V_3 = V_3\,V_2\,V_1 = 1$$

(IV). The transform $P.T_b$ is the "elementary" wavelet transform (when $K=H=1$).

(V). Every wavelet transform can be factorised into the product of the "elementary" wavelet transform and a transform of type (3.6).

(VI). V_+P is the Watson transform studied in [10].

(VII). $W(0,a)$ may be regarded as a bilateral Watson transform.

4. Integral representations

In this section we derive the wavelet transform (1.5) as a special case of the general wavelet transform (3.15). Since M_+ and M_- are isometric isomorphisms from $L^2(\mathbf{R}_+)$ onto $L^2(\mathbf{R})$ and from $L^2(\mathbf{R}_-)$ onto $L^2(\mathbf{R})$ respectively, we have

$$(f,g)_{L^2(\mathbf{R}_+)} = (M_+f, M_+g)_{L^2(\mathbf{R})}, \ f, g \,\varepsilon\, L^2(\mathbf{R}_+) \tag{4.1}$$

and

$$(f,g)_{L^2(\mathbf{R}_-)} = (M_-f, M_-g)_{L^2(\mathbf{R})}, \ f, g \,\varepsilon\, L^2(\mathbf{R}_-) \tag{4.2}$$

Adding these results, for f, $g \varepsilon L^2(\mathbf{R})$, we get

$$(f,g)_{L^2(\mathbf{R})} = \sum_\sigma (M_\sigma f, M_\sigma g), \quad \sigma = \pm\,. \tag{4.3}$$

Let J denote the compex conjugation operator $Jg = \bar{g}$ and Q denote the operator $Qg=g(-x)$. Then we have

$$M_\sigma J = JQ\, M_\sigma \tag{4.4}$$

and

$$M_\sigma P = Q\, M_\sigma\,. \tag{4.5}$$

From (4.1), (4.2), (4.4) and (4.5) we obtain for $f, g \varepsilon L^2(\mathbf{R}_\sigma)$,

$$\int_{\mathbf{R}_\sigma} f(t)\, g(t) dt = \int_{\mathbf{R}} (M_\sigma f)(x)\, (M_\sigma g)(-x)\, dx \tag{4.6}$$

and

$$\int_{\mathbf{R}_\sigma} f(t)\, g(t^{-1})\, t^{-1}\, dt = \int_{\mathbf{R}} (M_\sigma f)(x)\, (M_\sigma g)(x)\, dx \tag{4.7}$$

For $\alpha \varepsilon \mathbf{R}_+$ define $S_\alpha : L^2(\mathbf{R}_\sigma) \to L^2(\mathbf{R}_\sigma)$ by

$$f_\alpha(x) := (S_\alpha f)(x) = f(\alpha x), \ f \varepsilon L^2(\mathbf{R}_\sigma). \tag{4.8}$$

Then we have

$$(M_\sigma f_\alpha)(x) = \alpha^{-\frac{1}{2}-ix}\, (M_\sigma f)(x), \ f \varepsilon L^2(\mathbf{R}_\sigma). \tag{4.9}$$

From (4.6), (4.7) and (4.8) we obtain for f, $g \varepsilon L^2(\mathbf{R}_\sigma)$,

$$(f\circ g)(x) := \int_{R_\sigma} f(xt)\, g(t^{-1})\, t^{-1}\, dt = \int_{R} x^{-\frac{1}{2}-iy} (M_\sigma f)(y)\, (M_\sigma g)(y)\, dy \qquad (4.10)$$

and

$$(f\circ Pg)(x) := \int_{R_\sigma} f(xt)\, g(t)\, dt = \int_{R} x^{-\frac{1}{2}-iy} (M_\sigma f)(y)\, (M_\sigma g)(-y)\, dy\,. \qquad (4.11)$$

Theorem 4.1. (i) Let $K, H \varepsilon L^\infty(\mathbf{R})$, $\phi\ \varepsilon\ L^2(\mathbf{R})$ and $k=(M_+^{-1}\, m[K]\, M_+ \oplus M_-^{-1}\, m[H] M_-)\phi$. Assume further that $g = Wf$ for $f \varepsilon L^2(\mathbf{R})$. Then

$$\int_{R} \phi(xt)\, g(t^{-1})\, t^{-1}\, dt = \int_{R} k(xt)\, (P\, T_b f)(t) dt. \qquad (4.12)$$

If in addition W is unitary then $|M_\sigma k| = |M_\sigma \phi|$.
(ii) Let $k, \phi \varepsilon L^2(\mathbf{R})$, $M_\sigma \phi \neq 0$ a.e.. If for every $f \varepsilon L^2(\mathbf{R})$ there exists $g \varepsilon L^2(\mathbf{R})$ such that

$$\int_{R} \phi(xt)\, g(t^{-1})\, t^{-1}\, dt = \int_{R_\sigma} k(xt)\, (P\, T_b f)(t) dt, \quad \sigma = \pm \qquad (4.13)$$

then

$$K = (M_+ k)/(M_+ \phi)\ \varepsilon\ L^\infty(\mathbf{R}),$$

$$H = (M_- K)/(M_- \phi)\ \varepsilon\ L^\infty(\mathbf{R})$$

and

$$g = Wf. \qquad (4.14)$$

If in addition $|M_\sigma k| = |M_\sigma \phi|$ then the map $f \to g$ is unitary.

Proof. (i) Since $g = Wf$ for $f \varepsilon L^2(\mathbf{R})$, we have

$$(M_+ \oplus M_-)\, g\, (y) = (K(y) M_+ \oplus H(y) M_-)\, P\, T_b f(y),$$

so that

$$(M_+\, g)(y) = K(y)\, M_+\, P\, T_b\, f(y) = K(y)\, M_+\, T_b f(-y)$$

and

$$(M_-\, g)(y) = H(y)\, M_-\, T_b\, f(-y).$$

Therefore for all $x\varepsilon\mathbf{R}$,

$$\int_{\mathbf{R}} x^{-\frac{1}{2}-iy} (M_+\phi)(y)(M_+ g)(y)\,dy = \int_{\mathbf{R}} x^{-\frac{1}{2}-iy} (M_+ \phi)(y) K(y) M_+T_bf(-y)\,dy$$

$$= \int_{\mathbf{R}} x^{-\frac{1}{2}-iy} M_+k(y) M_+T_bf(-y)\,dy \qquad (4.15)$$

and

$$\int_{\mathbf{R}} x^{-\frac{1}{2}-iy} (M_-\phi)(y)(M_- g)(y)\,dy = \int_{\mathbf{R}} x^{-\frac{1}{2}-iy} (M_- \phi)(y) H(y) M_-T_bf(-y)\,dy$$

$$= \int_{\mathbf{R}} x^{-\frac{1}{2}-iy} M_-k(y) M_-T_bf(-y)\,dy. \qquad (4.16)$$

Then from (4.10) and (4.11) we obtain

$$\int_{\mathbf{R}_\sigma} \phi(xt)\, g(t^{-1})\, t^{-1}\,dt = \int_{\mathbf{R}_\sigma} k(xt)\, T_b\, f(t)dt. \qquad (4.17)$$

Since $\mathbf{R} = \Sigma_\sigma \mathbf{R}_\sigma$, (4.14) yields (4.12).

To prove the last part we note that W is unitary if $K \equiv 1$, $H \equiv 1$ and then the result follows from the relations $M_+k = K M_+\phi$ and $M_-k = H M_- \phi$.

(ii) If (4.13) holds for all $x\varepsilon\mathbf{R}$, then (4.15) and (4.16) also hold for all $x\varepsilon\mathbf{R}$. Therefore by uniqueness of the Fourier transform we obtain

$$(M_+ \phi)(y)(M_\sigma g)(y) = (M_\sigma k)(y)(M_\sigma T_b f)(-y). \qquad (4.18)$$

Let us set

$$K := (M_+k)/(M_+\phi)$$

and

$$H := (M_-k)/(M_-\phi).$$

Since by assumption $M_\sigma g\varepsilon L^2(\mathbf{R})$ and $M_\sigma T_b f\varepsilon L^2(\mathbf{R})$ for every $f\varepsilon L^2(\mathbf{R})$ it follows from (4.18) that $K, H\varepsilon L^\infty(\mathbf{R})$. Relation (4.18) yields that

$$(M_+ g)(y) = K(y)(M_+T_bf)(-y)$$

and

$$(M_- g)(y) = H(y)\,(M_- T_b f)(-y),$$

so that

$$g(x) = (M_+^{-1}\, m[K]\, M_+ T_b f(-y))(x) \qquad x > 0$$

$$= (M_-^{-1}\, m[H]\, M_- T_b f(-y))(x) \qquad x < 0.$$

Thus $g = Wf$.

The last part of the theorem is obvious.

The following theorem provides the representation of the wavelet transform in the form (1.1).

Theorem 4.2. If $K, H \varepsilon L^\infty(\mathbf{R}) \cap L^2(\mathbf{R})$ and $\bar{k} = (2\pi)^{-1/2}(M_+^{-1} K \oplus M_-^{-1} H)$, then

$$Wf = \bar{k}\, O\, P\, T_b f \qquad \forall\, f \varepsilon L^2(\mathbf{R}). \tag{4.19}$$

Proof. Let $g = Wf$; and let

$$g_+ = M_+^{-1}\, m[K]\, M_+ (T_b f(-y))$$

$$g_- = M_-^{-1}\, m[H]\, M_- (T_b f(-y)),$$

so that $g = g_+ \oplus g_-$. Then for $A>0$ we have

$$(2\pi)^{-1/2} \int_{-A}^{A} x^{-1/2-iy} (M_+ g)(y)\, dy = (2\pi)^{-1/2} \int_{-A}^{A} x^{-1/2-iy} K(y)\, M_+ T_b f(-y) dy$$

and

$$(2\pi)^{-1/2} \int_{-A}^{A} (-x)^{-1/2-iy} (M_- g)(y)\, dy = (2\pi)^{-1/2} \int_{-A}^{A} (-x)^{-1/2-iy} H(y)\, M_- T_b f(-y) dy.$$

Taking 1.i.m. $A \to \infty$ of both sides we get

$$g_+(x) = \int_{\mathbf{R}} x^{-1/2-iy} M_+ \bar{k}(y)\, M_+ T_b f(-y) dy$$

$$= \int_{\mathbf{R}} \bar{k}(x\text{-}t)\,(T_b f)(t) dt$$

$$= \int_{\mathbf{R}} \bar{k}(x(u-b))\, f(u) du = \int_{\mathbf{R}} \bar{k}(xu)\,(T_b f)(u) du,\ x > 0.$$

Similarly,

$$g_-(x) = \int_{\mathbf{R}} \bar{k}(xu)\,(T_b f)(u)\,du, \qquad x < 0 .$$

Thus

$$g = g_+ \oplus g_- = \bar{k}\,O\,P\,T_b\,f.$$

In the following we obtain conditions for the wavelet transform to be unitary.

Lemma 4.3. Let $K_i, H_i \,\varepsilon L^\infty(\mathbf{R})$ for $i = 1,2$ and let

$$k = (M_+^{-1}\ m[K_1]\,M_+ \oplus M_-^{-1}\ m[H_1]\,M_-)\,\phi$$

and

$$h = (M_+^{-1}\ m[K_2]\,M_+ \oplus M_-^{-1}\ m[H_2]\,M_-)\,\psi.$$

(i) If

$$K_1\ \bar{K}_2 = 1, \quad H_1\ \bar{H}_2 = 1 \tag{4.20}$$

then

$$(k_a, h_b)_{L^2(\mathbf{R})} = (\phi_a, \psi_b)_{L^2(\mathbf{R})}\ \ \forall\ a, b > 0, \tag{4.21}$$

where, as in (4.8), $f_a(x) = f(ax)$.

(ii) Let $M_+\,\phi \neq 0,\ M_+\,\psi \neq 0$ a.e.. If (4.21) holds, then (4.20) holds.

Proof. (i) We have

$$M_+\,k_a = K_1\,M_+\,\phi_a\,,\ \ M_-\,k_a = H_1\,M_-\,\phi_a$$

$$M_+\,h_b = K_2\,M_+\,\psi_b,\ \ M_-\,h_b = H_2\,M_-\,\psi_b$$

$$(k_a, h_b)_{L^2(\mathbf{R}_+)} = (K_1\,M_+\,\phi_a, H_2\,M_+\,\psi_b) = (M_+\,\phi_a, M_+\,\psi_b) = (\phi_a, \psi_b)_{L^2(\mathbf{R}_+)}$$

and

$$(k_a, h_b)_{L^2(\mathbf{R}_-)} = (\phi_a, \psi_b)_{L^2(\mathbf{R}_-)},$$

so that

$$(k_a, h_b)_{L^2(\mathbf{R})} = (\phi_a, \psi_b)_{L^2(\mathbf{R})}.$$

(ii) By (4.1) we have

$$(k_a, h_b)_{L^2(\mathbf{R}_+)} - (\phi_a, \psi_b)_{L^2(\mathbf{R}_+)}$$

$$= (M_+ k_a, M_+ h_b)_{L^2(\mathbf{R})} - (M_+ \phi_a, M_+ \psi_b)_{L^2(\mathbf{R})}$$

$$= (K_1 M_+ \phi_a, K_2 M_+ \psi_b) - (M_+ \phi_a, M_+ \psi_b)$$

$$= (M_+ \phi_a (K_1 \overline{K}_2 - 1), M_+ \psi_b)$$

$$= \int_{\mathbf{R}} a^{-1/2 - iu} M_+ \phi (K_1 \overline{K}_2 - 1) b^{-1/2 + iu} \overline{M_+ \psi(u)} \, du$$

$$= 0.$$

This implies that

$$K_1 \overline{K}_2 - 1 = 0 \text{ a.e.}$$

Theorem 4.4. (i) Let $K, H \varepsilon L^\infty(\mathbf{R})$, $\phi \varepsilon L^2(\mathbf{R})$ and $k = (M_+^{-1} \ m[K] M_+ \oplus M_-^{-1} \ m[H] M_-) \phi$. If W is unitary, then

$$(k_a, k_b) = (\phi_a, \phi_b) \qquad \forall\, a, b > 0 \tag{4.22}$$

(ii) Let $k, \phi \varepsilon L^2(\mathbf{R})$, $M_+ \phi \neq 0$ a.e. and

$$K = (M_+ k)/(M_+ \phi).$$

If (4.22) holds, then W is unitary.

Proof. (i) If W is unitary, then $|K| = 1$, $|H| = 1$ and hence (4.22) follows from (4.21) on setting $K_2 = K_1 = K$, $H_2 = H_1 = H$ and $\phi = \psi$. (ii) If (4.22) holds then from Lemma 4.3 (ii), $|K| = 1$, hence W is unitary.

4.1. A special case

Let us define a function ϕ on $\mathbf{R}$ by

$$\phi(t) = \begin{cases} 0 & |t| < 1 \\ |t|^{-\lambda-1/2} & |t| > 1, \end{cases}$$

where $\lambda > 0$. Then $\phi \varepsilon L^2(\mathbf{R})$.

Now assume that K, $H \varepsilon L^{\infty}(\mathbf{R})$ and $k = (M_+^{-1} m[K] M_+ \oplus M_-^{-1} m[H] M_-)\phi$. Then (4.12) assumes the form

$$\int_{-x}^{x} t^{\lambda-1/2} g(t)dt = |x|^{\lambda+1/2} \int_{-\infty}^{\infty} f(t)\, k(x(t-b))dt \qquad (4.23)$$

which is an analogue of Watson transform [12].

If W is unitary, i.e. $|K| \equiv 1$, $|H| \equiv 1$ then the relation (4.22) gives

$$\int_{-\infty}^{\infty} k(at)\, \overline{k(bt)}dt = (4/\lambda)\,(ab)^{-1/2}\,(a/b)^{\lambda} \quad a < b$$

$$= (4/\lambda)\,(ab)^{-1/2}\,(b/a)^{\lambda} \quad a \geq b.$$

4.2 The space A(R)

Now, we obtain integral representation for the wavelet transform W when k belongs to the space $A(\mathbf{R})$ defined below.

Let $A(\mathbf{R}_\sigma)$ denote the set of functions f defined on $\mathbf{R}_\sigma$ such that

$$||f||_{A(\mathbf{R}_\sigma)} = \int_{\mathbf{R}_\sigma} |t|^{-1/2}\, |f(t)|\, dt < \infty\,. \qquad (4.24)$$

Then the maps U_+ and U_- defined by

$$(U_+f)(x) = e^{x/2} f(e^x) \quad f \,\varepsilon\, A(\mathbf{R}_+)$$

$$(U_-f)(x) = e^{x/2} f(-e^x) \quad f \,\varepsilon\, A(\mathbf{R}_-)$$

are isomeric isomorphisms of $A(\mathbf{R}_+)$ (resp. $A(\mathbf{R}_-)$)onto $L^1(\mathbf{R})$.

We define the positive Mellin transform M_+^1 on $A(\mathbf{R}_+)$ by

$$(M_+^1 f)(x) = (2\pi)^{-1/2} \int_{\mathbf{R}_+} t^{ix-1/2} f(t)dt, \qquad f \varepsilon A(\mathbf{R}_+) \qquad (4.25)$$

and negative Mellin transform M_{-}^{1} on $A(\mathbf{R}_{-})$ by

$$(M^{1} f)(x) = (2\pi)^{-1/2} \int_{\mathbf{R}_{-}} (-t)^{ix-1/2} f(t)dt, \quad f \varepsilon A(\mathbf{R}_{-}) \tag{4.26}$$

For functions f and g for which fog is defined, we obtain the relation

$$U_{\sigma} (f \circ g) = U_{\sigma} f * U_{\sigma} g, \qquad \sigma = \pm \tag{4.27}$$

where * denotes the convolution on $\mathbf{R}$. If $f\varepsilon A(\mathbf{R}_{+})$ and $g\varepsilon L^{2}(\mathbf{R}_{+})$ then $f \circ g \varepsilon L^{2}(\mathbf{R}_{+})$ and

$$M_{+}^{1} (f \circ g) = (2\pi)^{1/2} (M_{+}^{1} f)(M_{+} g). \tag{4.28}$$

Also

$$M_{-}^{1} (f \circ g) = (2\pi)^{1/2} (M_{-}^{1} f)(M_{+} g), \tag{4.29}$$

for $\qquad f\varepsilon A(\mathbf{R}_{-})$ and $g\varepsilon L^{2}(\mathbf{R}_{-})$.

Theorem 4.5. Let $K = (2\pi)^{1/2} M_{+}^{1} \bar{k}$ and $H = (2\pi)^{1/2} M_{-}^{1} \bar{k}$, where $k\varepsilon A(\mathbf{R})$. Then

$$Wf = \bar{k} o P T_{b} f \qquad \forall f\varepsilon L^{2}(\mathbf{R}). \tag{4.30}$$

Proof. Let $f\varepsilon L^{2}(\mathbf{R})$, then by (4.28) and (4.29) we obtain

$$K M_{+} P T_{b} f = (2\pi)^{1/2} (M_{+}^{1} \bar{k})(M_{+} P T_{b} f) = M_{+} (\bar{k} o P T_{b} f)$$

and

$$K M_{-} P T_{b} f = M_{-} (\bar{k} \circ P T_{b} f).$$

Therefore

$$(M_{+}^{-1} m[K] M_{+} \oplus M_{-}^{-1} m[H] M_{-}) P T_{b} f = \bar{k} \circ P T_{b} f.$$

Acknowledgement

The work is supported by NBHM (DAE) Grant No.48/1/94 R&D II.

References

1. Bochner, S and Chandrasekharan, K (1949) *Fourier Transforms*, Annals of Mathematical Studies vol.19, Princeton University Press, Princepton.
2. Chui, C K (1992) *An Introduction to Wavelets*, Academic Press, New York.
3. Diksma, A and de Snoo H S V (1974) *Distributional Watson transforms*, SIAM J. Math. Anal. 5(6), 888-892.
4. Daubechies, I (1992) *Ten Lectures on Wavelets*, SIAM, Philadelphia, PA.
5. Goldberg, R R (1960) *Convolutions and general transforms on L^p*, Duke Math. J. 27, 251-260.
6. Heil, C E and Walnut D F (1989) *Continuous and discrete wavelet transforms*, SIAM Review 31(4), 628-666.
7. Holschneider, M (1995) *Wavelets: An Analysis Tool*, Claredon Press, Oxford.
8. Koornwinder, T H (1993) *The Continuous wavelet transform" in Wavelets: An Elementary Treatment of Theory and Applications*, World Scientific, Singapore, 27-48.
9. Pathak, R S (1995) *Continuity and inversion of the wavelet transform*, to appear in Integral Transform and Special Functions (1997).
10. de Snoo, H S V (1973) *A Note on Watson transforms*, Proc. Camb. Phil. Soc. 33, 83-85.
11. de Snoo, H S V (1973) *On invariant linear manifolds of Watson transforms*, Quart. J. Math. 24, 217-221.
12. de Snoo, H S V (1973) *Integral representations of Watson transforms*, Niew Arch. Wisk, 21, 48-58.
13. de Snoo, H S V (1974) *On the spectrum of Watson Transforms*, J. London Math. Soc. (2)8, 297-305.
14. Taylor, A E (1958) *Introduction to Functional Analysis*, John Wiley, New York.
15. Titchmarsh E C (1948) *Introduction to the Theory of Fourier Integrals*, Clarendon Press, Oxford, 2nd ed.
16. Watson, G N (1933) *General transforms*, Proc. London Math. Soc. 35, 156-199.
17. Widder, D V (1941) *The Laplace Transform*, Princeton University Press, Princeton, N.J.
18. Wolf, K B (1979) *Integral Transforms in Science and Engineering*, Plenum Press, New York.

R.S. Pathak
Department of Mathematics
Banaras Hindu University
Varanasi-221 005, India

A P SINGH

Iterates of meromorphic functions: an introduction

1. Introduction

Given a map $f : S \rightarrow S$ denote the n-th iterate of f by f^n, $n \in \mathbb{N} \cup \{o\}$. Thus $f^o(z) = z$, $f^n(z) = f^{n-1}(f(z))$ for $n \geq 1$. The study of such situation is quite old. One of the earlier instances which is still a subject of interest is the Newton's iterative technique for finding the zeros of f, which is obtained by iterating the function F defined by

$$F(z) = z - \frac{f(z)}{f'(z)}$$

The iterative techniques have also been used by Euler [21] and Abel [1]. However a systematic study on iteration, more precisely the iteration theory for rational functions originated with the works of Fatou and Julia who independently published a number of Compte Rendu notes and later both wrote long memoires [22, 25, 26, 27]. Fatou [24] extended some of the results to transcendental entire functions. Some work on the iterative theory [29, 36, 38, 41, 42] continued, but since the 70's there has been a renewed interest in this field with several mathematicians making significant contributions. The renewed interest in this field is partially due to the beautiful computer graphics related to it (see for instance [17, 32]) and also due to new and analytical techniques involving quasi-conformal maps, topological notions and ideas from dynamical system theory and ergodic theory (see for instance [15, 18, 19, 30, 33, 34, 37]). This interesting field of iteration is still expanding with no end in sight.

The aim of this brief introduction is to acquaint and motivate the reader in this fast growing interesting field of iterations. For a more exhaustive surveys one may refer to [13] for meromorphic functions, [4] for entire functions, [17, 31, 39] for rational functions and also to their references.]

2. Fatou and Julia Sets

One of the basic concepts in the theory of iterates is the notion of normal family.

DEFINITION : A family £ of meromorphic functions in a domain $D \subset \bar{\mathbb{C}}$ is said to be a normal family if every sequence $\{f_n\}$ in £ contains a subsequence that converges uniformly in the spherical metric on compact subsets of D.

By Arzela-Ascoli theorem, the above definition is equivalent to saying that the family is equicontinuous (with respect to the spherical metric) on every compact subset of D. Note that we allow f_n to converage to ∞. Also we shall assume that f is not of the form

$$\frac{az+b}{cz+d} \quad \text{(a, b, c, d constants).}$$

Let $f : \mathbb{C} \rightarrow \bar{\mathbb{C}}$ be a meromorphic function where $\mathbb{C}$ is the complex plane and $\bar{\mathbb{C}} = \mathbb{C} \cup \{\infty\}$. As mentioned earlier f^n will denote the n-th iterate of f. For f entire f^n is defined for all $z \in \mathbb{C}$. If f is rational, then f has a meromorphic extension to $\bar{\mathbb{C}}$. Denoting this extension also by f we see that f^n is defined and meromorphic in $\bar{\mathbb{C}}$. However if f is transcendental meromorphic then there is no reasonable way of defining $f(\infty)$, and thus the iterations in such cases would have to be defined so as to avoid this situation.

Let $F = \{z \in \bar{\mathbb{C}} : \{f^n\}_{n \in \mathbb{N}}$ is defined and normal in some neighbourhood of z$\}$ and $J = \bar{\mathbb{C}} - F$ then F is called Fatou set (or set of normality) and J is called Julia set.

If f is a transcendental entire function then $\{f^n\}_{n \in N}$ is defined for all $z \in \mathbb{C}$ and in this case $\infty \in J$ always. If f is rational then also as mentioned earlier $\{f^n\}_{n \in N}$ is defined for all $z \in \bar{\mathbb{C}}$. If however f is transcendental meromorphic with exactly one pole say z_0 and if this pole is an omitted value (i.e. $f(z) \neq z_0$ for all $z \in \mathbb{C}$) then clearly $\{f^n\}_{n \in \mathbb{N}}$ is defined for all $z \in \bar{\mathbb{C}} - \{z_0, \infty\}$ and in this case $\{z_0, \infty\} \subset J$. Finally if f is transcendental meromorphic with either atleast two poles or only one pole which is not an omitted value, there is an interesting alternate way of obtaining the Fatou set and the Julia sets on using Montels theorem [16]. For a family £ of meromorphic functions defined on a domain D, if there are three fixed values that are omitted by every $f \in$ £ then £ is a normal family.

For $z_0 \in \bar{\mathbb{C}}$, set $f^{-n}(z_0) = \{z : f^n(z) = z_0\}$. The backward orbit of z_0 denoted by $O^-(z_0)$ is then defined by

$$O^-(z_0) = \bigcup_{n=1}^{\infty} f^{-n}(z_0)$$

Clearly $\bar{\mathbb{C}} - \overline{O^-(\infty)}$ is the largest open set where all the iterates of f are defined. Also $f(\bar{\mathbb{C}} - \overline{O^-(\infty)}) \subset \bar{\mathbb{C}} - \overline{O^-(\infty)}$ and clearly from Picards theorem $O^-(\infty)$ is an infinite set, and so by Montel's theorem $\{f^n\}_{n \in N}$ is normal in $\bar{\mathbb{C}} - \overline{O^-(\infty)}$. Thus $F = \bar{\mathbb{C}} - \overline{O^-(\infty)}$ and $J = \overline{O^-(\infty)}$.

In view of the above observations we may classify the set of all meromorphic functions into the following subclasses :

$R = \{f : f$ is rational$\}$

$E = \{f : f$ is transcendental entire$\}$

P = {f : f is transcendental meromorphic, has exactly one pole and this pole is an omitted value}

M = {f : f is transcendental meromorphic and has either atleast two poles or exactly one pole which is not an omitted value}.

In R we shall only be interested with rationals of degree atleast 2.

The Fatou and Julia sets of the above classes have several distinct properties and also several properties may be common to all classes, though the proof need not be the same. We begin by giving some properties of Fatou and Julia sets.

3. Properties of Fatou and Julia Sets

J 1 : By definition, F is open, J is closed and non-empty, infact J (f) contains infinitely many points.

J 2 : For $f \in R, E, P$, $J(f) = J(f^n)$ and $F(f) = F(f^n)$ for all $n \geq 2$. For $f \in M$, f^n is not meromorphic and so $F(f^n)$ and $J(f^n)$ are not defined.

J 3 : F and J are completely invariant. That is $z \in F$ if and only if $f(z) \in F$ and $z \in J$ if and only if $f(z) \in J$.

J 4 : $J(f) = \bar{\mathbb{C}}$ or J has empty interior.

J 5 : A point z_0 is called exceptional if $O^-(z_0)$ is finite. By Picard's theorem a transcenental meromorphic function can have atmost two exceptional points. Thus if $z_0 \in J(f)$ is not exceptional then $\overline{O^-(zo)} = J(f)$.

J 6 : J (f) is perfect (i.e. closed, non-empty and does not contain isolated points).

An important role in the theory of iteration is played by the periodic points. A point z_0 is called a periodic point of f with period n if $f^n(z_0) = z_0$ for some $n \geq 1$ and $f^j(z_0) \neq z_0$ for $1 \leq j < n$. A periodic point of period 1 is called fixed point. For a periodic point z_0 of period n, let $\lambda = (f^n)'(z_0)$. Then z_0 is called attracting, repelling or indifferent according as $|\lambda|$ is less than, greater than or equal to 1. Also $(f^n)'(z_0)$ is called multiplier of z_0. If z_0 is an indifferent periodic point then the multiplier is of the form $e^{2\Pi i\alpha}$ where $0 \leq \alpha < 1$ and in this case z_0 is said to be rationally or irrationally indifferent according as α is rational or irrational.

J 7 : Attracting periodic points lie in F. Repelling and rationally indifferent periodic points lie in J. Irrationally periodic points may lie either in F or in J.

J 8 : J (f) is the closure of the set of repelling periodic points of f.

4. Components of the Fatou Set

Let U_0 be a (connected) component of F and let $n \in \mathbb{N}$. Since F is completely invariant, it follows that there exists a component U_n say of F such that $f^n(U_0) \subset U_n$. If all U_n are different we say that U_0 is a wandering component of F, else U_0 is called pre-periodic. If further $f^p(U_0) \subset U_0$ for some p then U_0 is called periodic and if p = 1, then U_0 is called invariant component.

The behaviour of f^n in the periodic component is fairly well understood.

THEOREM : Let U be a periodic component of period p. Then we have one of the following possibilities.

(i) U is an immediate attractive basin of z_0 : U contains an attracting periodic point z_0 of period p and $f^{np}(z) \to z_0$ as $n \to \infty$ for every $z \in U$.

(ii) U is a Leau domain : ∂U contains a periodic point z_0 of period p and $f^{np}(z) \to z_0$ as $n \to \infty$ for all $z \in U$. Further $(f^p)'(z_0) = 1$ if $z_0 \in \mathbb{C}$ and if $z_0 = \infty$ then $(g^p)'(0) = 1$ where $g(z) = \dfrac{1}{f(1/z)}$.

(iii) U is a Siegel disc : There exists an analytic homeomorphism $\varnothing : U \to D$ where D is the unit disc such that $\varnothing\{f^p[\varnothing^{-1}(z)]\} = e^{2\Pi i\alpha} z$ for some $\alpha \in \mathbb{R} - Q$.

(iv) U is a Herman ring : There exists an analytic homeomorphism $\varnothing : U \to A$ where A is the annulus $A = \{z : 1 < |z| < r\}$, $r > 1$ such that $\varnothing\{f^p[\varnothing^{-1}(z)]\} = e^{2\Pi i\alpha} z$ for some $\alpha \in \mathbb{R} - Q$.

(v) U is a Baker domain : There exists $z_0 \in \partial U$, such that $f^{np}(z) \to z_0$ as $n \to \infty$ for all $z \in U$, but $f^p(z_0)$ is not defined.

For rational f, Baker domain do not exist. If $f \in E$, Baker domain are possible only for $z_0 = \infty$. If $f \in P$ with pole 0 then Baker domain are possible only for $z_0 \in \{0, \infty\}$. For $f \in E$, f does not have Herman ring. For $f \in P$, f may have atmost one Herman ring, but whether this is possible is still not known. For a more detailed account one may refer to Bergweiler [13]. There are several other differences in the properties associated with different classes of meromorphic functions. We mention a few of these differences.

For $f \in E$, M or P, J (f) is unbounded. Hower for $f \in R$, J (f) may be bounded or unbounded.

For instance $J(f) = \{z : |z| = 1\}$ for $f(z) = z^2$

and $J(f) = \bar{\mathbb{C}}$ for $f(z) = \dfrac{(z^2+1)^2}{4z(z^2+1)}$. Infact [25]

$J(f) = \bar{\mathbb{C}}$ if and only if there is some z whose forward orbit $\{f^n(z) : n \geq 1\}$ is dense in $\bar{\mathbb{C}}$.

One of the most important result in the iteration theory of rationals is the Sullivan's theorem [40] : If f is rational function then f has no wandering domain. The proof introduces the subject of quasi-conformal mapping into iteration theory. However for transcendental meromorphic functions, wandering domains are possible. The first result in this direction was given by Baker who constructed a transcendental entire function having wandering domain [2]. Later several other examples were formed with different set of conditions. For instance a wandering domain having $f^n \to \infty$ arbitrarily fast [6], having infinitely many different constant limit functions [20] and existence of composite entire functions having wandering domains [11], have been given.

Sullivan's theorem has been extended to several classes of transcendental functions. For instance [13] if $f \in S, T, N, R$ then f has no wandering domain, where

$S = \{f :$ f has only finitely many critical and asymptotic values$\}$.

$T = \{f :$ f has a representation of the form $f(z) = z + r(z)\, e^{p(z)}$ where r is rational and p is a polynomial$\}$.

$N = \{f :$ f has finite order and $f'(z) = r(z)\, e^{p(z)} [f(z) - z]$ where r is rational and p is a polynomial$\}$

$R = \{f : f'(z) = r(z) [f(z)-z]^2$ or $f'(z) = r(z) [f(z)-z] [f(z)-a]$ where r is rational and $a \in \mathbb{C}\}$.

Even in transcendental functions, functions belonging to

different classes may have different characteristics. A wandering domain of transcendental entire function may be simply connected or multiply connected, and a multiply connected wandering domain must be bounded. Also examples of infinite connectivity can occur. However it is still not known whether the connectivity of wandering domain in E may be finite but different from 1. On the other hand for a meromorphic function in the class P Baker et all [7] have shown that the Fatou set F can have

(i) K-connected bounded wandering domain.

(ii) K-connected unbounded wandering domain.

(iii) bounded wandering domain of infinite connectivity.

(iv) unbounded wandering domain of infinite connectivity.

5. CONCLUSION : As mentioned in the begining, this brief exposition is just to acquaint the reader with some fundamental aspects of iteration theory. This is in no way a survey. As a matter of fact we have not mentioned many of the important and interesting facets of this field. The study of Mandelbrot set, ergodic problems, area and Husdorff dimensions of Julia sets, the Cantor bouquet, the completely invariant domains. Even in the topics touched upon, we have not mentioned about several other interesting topological properties of Julia sets, relation between wandering domain and weakly repelling fixed points, meromorphic functions satisfying certain differential equations and the corresponding existence (or non-existence) of wandering domains. The limiting behaviour of wandering domain is still not completely understood. In short, this field of iteration is still expanding at a rate which one could never have thought of in the 1930's and the end is nowhere in sight.

References :

1. Abel, N.H., Détermination d'une function au moyen d'une fonction qui ne continent qu'une seule variable, Oeuvres complètes II, Christiania, 1881, 36-39.

2. Baker, I.N., An entire function which has wandering domain, J. Australian Math. Soc. (Ser A) 22 (1976) 173-176.

3. Baker, I.N., Wandering domains in the iteration of entire functions, Proc. London Math. Soc. (3) 49 (1984) 563-576.

4. Baker, I.N., Iteration of entire functions : an introductory survey, Lectures on complex analysis, World Scientific, Singapore, New Jersey, London, Hongkong 1987, 1-17.

5. Baker, I.N., Some entire functions with multiply connected wandering domains, Ergod. Theory Dynam. System 5 (1985), 163-169.

6. Baker, I.N., Infinite limits in the iteration of entire functions, Ergod. Th. and Dynam. System (1988) 503-507.

7. Baker, I.N., Kotus, J. and Lü, y., Iterates of meromorphic functions I, Ergod. Th. and Dynam. System. 11 (1991) 241-248.

8. Baker, I.N., Kotus, J and Lü, y., Iterates of meromorphic functions II : Examples of wandering domains, J. London Math. Soc. (2) 42 (1990) 267-278.

9. Baker, I.N., Kotus, J. and Lü, y., Iterates of meromorphic function III : Preperiodic domains, Ergod. Th. and Dynam. System 11 (1991), 603-618.

10. Baker, I.N., Kotus, J. and Lü, y., iterates of meromorphic function IV. Critically finite functions, Results Math. 22 (1992), 651-658.

11. Baker, I.N., and Singh, A.P., Wandering domains in the iteration of composition of entire functions, Annal. Acad. Sci. Fenn. Sr. A. I, Math 20 (1995), 149-153.

12. Beardon, A.F., Iteration of rational functions, Springer, New York, Berlin, Heidelberg 1991.

13. Bergweiler, W., Iteration of meromorphic functions, Bull. Amer Math. Soc. (2) 29 (1993), 151-188.

14. Blanchard, P., Complex analytic dynamics on the Riemann sphere, Bull. Amer. Math. Soc. 11 (1984) 85-141.

15. Brolin, H., Invariant sets under iteration of rational functions, Ark. Mat. 6 (1965), 103-144.

16. Carleson, L. and Gamelin, T.W., Complex Dynamics, Universitext, Springer Verlag 1993.

17. Devane, R.L. and Keen, L. (ed), Chaos and fractals : The mathematics behind computer graphics. Amer. Math. Soc. Providence R.I. 1989.

18. Douady, A. and Hubbard, J.H., Etude dynamique des polynomes complexes, Pub. Mat. Orsay I, 1984 and II, 1985.

19. Douady, A. and Hubbard, J.H., On the dynamics of polynomial like mappings Ann. Sci. Ecole. Normale Sup. 18 (1985) 287-343.

20. Eremenko, A.E. and Lyubich, M.Yu., Examples of entire functions with pathological dynamics, J. London, Math. Soc. (2) 36 (1987) 458-468.

21. Euler, L., De formulis exponentialibus replicatius, Acta Petropolitanae, 1 (1777) 38-60.

22. Fatou, P., Sur les équations fonctionelles, Bull. Soc. Math. France 47 (1919), 161-271.

23. Fatou, P., Sur les equations fonctionelles, Bull. Soc. Math. France 48 (1920), 33-94, 208-314.

24. Fatou, P., Sur líteration des fonctions transcendentes entires, Acta. Math. 47 (1926), 337-360.

25. Herman, M.R., Exemples de fractions rationelles ayant une orbite dense sur la sphere de Riemann, Bull. Soc. Math. France 112 (1984) 93-142.

26. Julia, G., Sur líteration des fonctions rationelles. J. Math. Pure Appl. (7) 4 (1918) 47-245.

27. Julia, G., Sur des problèmes concertant l'iteration des fonctions rationelles, C.R. Acad. Sci. Paris 166 (1918) 153-156.

28. Julia, G., Mémoire sur la permutabilité des fractions rationalles, Ann. Sci. Ecole Norm. Sup. (3) 39 (1922), 131-215.

29. Kneser, H., Reelle analytische LÖsungen der Gleichung $\phi[\phi(x)] = e^x$ und Verwandter Funktional gleichungen, J.f.d. reine u. angew. Math. 187, (1949), 56-67.

30. Lyubich, M.Yu., The dynamics of rational transforms : the topological picture, Russian Math. Surveys (4) 41 (1986) 43-117.

31. Milnor, J., Dynamics in one complex variables : introductory lectures, Stony Brooke Institute for Mathematical Sciences Preprint.

32. Peitgen, H.O. and Richter, P.H., The beauty of fractals, Springer Berlin, Heidelberg, New York, Tokyo 1986.

33. Rees, M., Ergodic rational maps with dense critical point forward orbit, Univ. of Minnesota Report 82-140.

34. Rees, M., Positive measure sets of ergodic rational maps Univ. of Minnesota Report 142-169.

35. Ritt, J.F., On iteration of rational functions, Trans. Amer. Math. Soc. 21 (1920) 348-356.

36. Rosenbloom, P., L'iteration des fonctions entires, C.R. Acad. Sci. Paris 9 (1948), 382-383.

37. Ruelle, D., Repellers for real analytic maps, Ergodic theory and Dynamical System 2 (1982) 99-108.

38. Siegel, C.L., Iteration of analytic functions, Ann. of Math. (2) 43 (1942) 607-612.

39. Steinmetz, N., Rational iteration, Walter de Gruyter, 1993.

40. Sullivan, D., Quasiconformal homeomorphism and dynamics I : Solution of the Fatou-Julia problem on wandering domain, Ann. of Math. 122 (1985) 401-418.

41. Töpfer, H., über die Iteration der ganzen tranzendenten Funktionen, insbesondere von sin z and cos z, Math Ann. 117 (1939) 65-84.

42. Töpfer, H., Komplexe iterationsindizes ganzer und rationaler Funktionen Math. Ann. 121 (1949), 191-222.

Department of Mathematics,
University of Jammu,
Jammu - 180 004
INDIA

Part C

Applications in Science and Industry

H NEUNZERT AND A H SIDDIQI

Industrial mathematics - ideas and examples

Industrial Mathematics - a new subject or just a new name, the „emperor's new clothes"? What should be the difference between applied and industrial mathematics? But what is the difference between applied and pure mathematics, if any at all? „Applied Mathematics is Bad Mathematics" writes Halmos [2] in his article with just that name, „it isn't really, but it is different". And he compares the picture of the real world provided by applied mathematics with the police photograph of a wanted criminal, meanwhile that of pure mathematics with a portrait by Picasso. And now, worse, industrial mathematics - even more useful, but even less beautiful?

Halmos also claims that applied mathematicians do not accept a difference between pure and applied, while pure mathematicians do. This, maybe, is the first difference between applied and „industrial" mathematicians: We do see a difference.

Most applied is „applicable" mathematics - mathematics, which hopes or expects to be applied, but still is not. And this is the second difference: Industrial mathematics has its industrial applications; it treats problems posed by industry, questions of technical, organisatorial or economic nature, which are posed by companies.

Let us finish with looking for differences, defining the subject just by saying how it differs from other subjects; let us say, what it is: Industrial Mathematics means, to transform these technical, organisatorial and economic problems originally posed in a nonmathematical language into mathematical problems, to "solve" these problems in general by approximative methods of analytical and/or numerical nature and reinterprete the results in terms of the original problems.

In short: Industrial Mathematics is Modelling and Scientific Computing of industrial problems. Of course, modelling and asymptotic analysis appear in science and in „classical" applied mathematics as do numerical methods. And, of course, industrial mathematicians need a wide and broad knowledge on mathematical ideas, algorithms and computer. But they need more: They need an understanding of the problems in industry, they have to understand and speak the language spoken in industry.

Industrial Mathematicians are bridge builders: They build bridges from the field of mathematics to the practical world; for that they have to know about both sides, to know problems from some companies and ideas and methods from mathematics.

They have to be generalists, not specialists. If you enter the world of industry, you never know, which kind of problems you will find, which kind of mathematical methods you will need: Problems, where signals or images have to be processed and methods of Bayessian statistics, time series analysis, wavelets, energy functionals etc. are used; problems of optimizing the transport or the storage of industrial goods with all these new methods of combinatorial optimization; problems of predicting the outcome of complicated systems which do not allow to understand the mechanisms behind and methods of adaptive system theory, of learning, of neural networks etc.; problems, where complicated multiphase flow happens in porous media, forming fronts and drops and the methods of homogenization, level set methods, front tracking algorithms etc.; problems of the interaction of flexible structures (fibres, papersheets) with a flow around and multigrid methods, particle methods etc.; problems of optimizing the properties (elasticity, heat conductivity etc.) of compound materials and again methods of homogenization, regularization of inverse problems, optimal shape design.

We may go on and on - the selection here is based on our personal experiences. But we hope, it is obvious: Industrial mathematicians have to be generalists, they have to know a large variety of mathematical concepts and methods; they have to know modern mathematics, should know more and more advanced methods than scientists and engineers: Industrial mathematics is not engineering mathematics. But it is not a new subject in itself like Statistics, Optimization, Computational Mechanics or even Finance Mathematics; it is another kind of getting problems, looking at problems, interpreting and using solutions. It is a new discipline. Is it less beautiful than pure mathematics? This is, of course, a question of taste. But inspite of the beauty of a Picasso painting, which we see as Halmos did, we may have as much fun in developing algorithms for the identification of criminals as in searching for a „grammar“ of Picasso’s paintings. From the point of view of human creativity, industrial mathematics is one of the most exciting, adventurous, joyful activities a mathematician could find.

And, inspite of being different from pure and even applied mathematics, it needs both: Industrial mathematics stands on the shoulders of mathematics at all; industrial mathematicians use ideas of others and, if they do not find appropriate concepts and algorithms, invent new ones. Vice versa, mathematics profits from industrial mathematics, mainly by offering new challenging problems, better chances for young mathematicians, higher prestige outside the mathematical community, financial support.

Very often, Murphys principle holds, if one tries to apply a standard theory to a practical problem: What can go wrong, goes wrong. The right hand side of the differential equation is of course not Lipschitz continuous, the noise is not white, the operator not compact, the system too stiff, the material parameters too much oscillatory, the problem ill-posed etc. etc. Almost each practical problem asks meaningful questions, creates new concepts.

Why this new uprise of industrial mathematics worldwide (there is the European Consortium for Mathematics in Industry = ECMI, which gathers many activities of this kind in Europe, there is the Division of Mathematics and Statistics at CSIRO in Australia, an Institute for Industrial Mathematics in Israel, the IMA in Minneapolis, the Claremont Clinic in Pasadena and some more in the US)?

Modelling has several thousand years of history! But the tool to evaluate the models is new and becomes more powerful every year: Computer! Therefore, industrial mathematics is, to large extent, also numerical mathematics: not numerical analysis, but scientific computing, algorithms and their implementation. In this respect the computer scientist is our partner; but he is not, as some believe, the „owner" of this field. Developing algorithms - to optimize, to solve pde's, to process images, to predict stocks - is a genuinely mathematical task, which mathematicians should claim.

The basis of our statements and the source of our examples is mainly based in Europe. One of us (H.N.) has introduced a new educational master programme „Technomathematics" 18 years ago, which by now was spread out to 15 German and ca. 20 other universities all over the world; he was a co-founder of the above mentioned

ECMI, which organizes eduction and research in industrial mathematics in Europe; and, finally, 1995 he founded a new „Institute of Industrial Mathematics“ (ITWM) in Kaiserslautern, which earns today ca. 65% of the total budget, i.e. ca. 5 million DM, from industrial projects. The other one (A.H.S.) tries since years to implement the discipline in India, quite often facing hurdles. But we are convinced that mathematics will gain new momentum, new prestige, better students by integrating this new discipline in countries like India. Several years ago the Nobel Prize Winner from Pakistan, Prof. A. Salam, who founded the ICTP in Trieste, wrote that it is a kind of Machiavellism to export technology to the third world without developing science there. We would like to add that it seems to be Machiavellism too, to cultivate „pure“ science without explaining how it can contribute to the industrial development. Teaching and practising industrial mathematics could be step away from this Machiavellism.

The paper will give a survey

- on the teaching activities in the field of industrial mathematics in Europe,
- on the research activities mainly carried out at the „Institute for Industrial Mathematics“ in Kaiserslautern,
- on our experiences with implementing the field in 3rd world countries.

I. Teaching Industrial Mathematics: Experiences from the ECMI Programmes

ECMI, the European Consortium for Mathematics in Industry was founded in 1986 by mathematicians from several European Universities, f.e. by Eindhoven (Netherlands), Firenze (Italy), Glasgow (Scottland), Kaiserslautern (Germany), Linz (Austria), Oxford (England). It had three main goals, namely

to promote the use of mathematics in industry,

to educate industrial mathematicians,

to cooperate on a European scale.

During the last decade ECMI was quite successful: It has now centres in more than 10 European countries, where a Postgraduate Programme „Industrial Mathematics" is organized. This programme has two branches: Technomathematics and Economathematics. Technomathematics deals with applications of mathematics in technology and focusses on differential equations and its applications in mechanics, electrodynamics, thermodynamics etc., on signal analysis, image processing and inverse problems and mainly on numerical methods too; economathematics treats organisatorial problems by mathematical methods, is therefore concentrated mainly on discrete mathematics, f.e. combinatorial optimization and graph theory and on statistics.
Both programmes differ from classical applied mathematics programmes by using quite a big part for modelling and scientific computing. Of course, there are general aims. After completing the programme the student should have the following capabilities:

1. Proficiency in the formulation of real world situations into qualitative or quantitative mathematical models, and in the use and the analysis of mathematical models;
2. experience of the use of mathematical models in the industry and in the handling of mathematical methods for answering practical questions and for solving problems from the real world;
3. sufficient knowledge of the mathematics that is relevant for 1 and 2 and further reaching knowledge of some special field of interest;
4. knowledge of numerical methods and simulation, and experience in the intelligent use of computers;
5. background knowledge of computer science and pure mathematics;
6. experience in team work and in making a personal contribution to a team project;
7. sufficient knowledge of some scientific field where mathematics is used in practice (for example: physics, mechanics, chemistry, electrotechnics, economics, industrial engineering), and to be able to communicate with experts in this field;
8. to be able to give oral and written presentation of their scientific work in such a way that not only mathematicians but also interested laymen get the main messages.

But what does all that mean practically? Let us try to describe a bit of the important details:

1. Students, who enter the programme, should have a certain preknowledge, provided by a Bachelor, Vordiplom or equivalent degrees awarded in different countries. Of course, a bachelor here is not the same as a bachelor there; a master at a second class university may know less than a bachelor from a first class institution. Even a PhD in pure mathematics from some places should find the programme to be challenging. In short: Titles are not enough to describe the prerequisites.

   So, let us try otherwise:

   a) Calculus is not enough, but we would expect some knowledge in analysis, for example different kinds of convergence (i.e. a bit of topology), an understanding of the problems in changing limit processes, Poincaré's lemma etc.
   b) Linear algebra should include finite dimensional vector spaces, linear operators and their normal forms, spectral theory. But we would like students to be able to compute the Jordan normal form of a matrix, not only knowing everything about structures but nothing about algorithms.
   c) Ordinary Differential Equations should include the main ideas about existence and uniqueness, phase space interpretations of dynamical systems, linear systems of ode's, stability (including Ljapunov functions) and periodicity.
   d) Numerical methods should deal with linear and nonlinear systems (including linear and nonlinear regression), large symmetric systems and eigenvalue problems, interpolation and approximation (including splines and Bézier techniques), numerical integration.
   e) Functional analysis should include the basic concept: Hilbert and Banach spaces, bounded and compact operators and their spectral properties.
   f) In optimization we expect some knowledge in linear programming and basic ideas like gradient methods in nonlinear optimization.
   g) Probability should consist of „elementary probability theory" (conditional probability, independence, Bernoulli schemes, random walks).

Besides these mathematical preknowledges we expect students to be able to handle a computer and a language (f.e. C^+) and to have solid knowledge in physics (for the Techno branch) or in microeconomy (for the Econo branch)

Of course, not every student entering the programme has this knowledge. Therefore the first semester is used as a possibility to fill the gaps. Quite often there are many gaps and this semester is very stressing. Some have good knowledge in pure mathematics - their gaps are mainly in numerics and computing; others are good in physics and computing, but have their difficulties for example with functional analysis. Wherever the gaps are: After one semester they must have disappeared.

2. The core courses, taught in all centres in a similar way, should offer the basis for Techno and Economathematics respectively. In Technomathematics this means a lot of differential equations: Advanced analytical and numerical methods for ode's and pde's, control and system theory; moreover time series analysis, and nonlinear and discrete optimization. In Economathematics it means mainly advanced methods in optimization (network, graphs etc.) and stochastics (stochastic processes, queuing theory). There were attempts to give introductory „cross courses": Technomathematics for Economathematicians, Economathematics for Technomathematicians. But it does not seem to work.

   The central activity from the 2nd semester onwards is the modelling seminar: Problems from the real (mostly industrial) world are presented in a nonmathematical way to groups of 4-5 students; these groups work one semester under a supervisor, who plays the role of an industrial client but not as a mathematical expert. They transform the problem into a mathematical model (using their knowledge and the library), they evaluate the model using the computer facilities, they reinterprete the results in terms of the original problem, they present their results orally and as a written report. It is a lot of work, but is is more rewarding than any lecture or book: Mathematics, especially „Industrial Mathematics", is only learnt by doing! Modelling seminars are the opposite of „learning by heart", it „creates creativity". And this is, what industrial mathematicians need: Creativity to create innovation.

3. The third semester is open for special courses - besides another modelling seminar. Special courses depend on the special knowledge of each department. Here a list of titles we have seen during the last years, at different ECMI centres - possible topics, changing with time and mainly changing with industrial projects: Asymptotic analysis; Mathematical and computational methods in fluid dynamics, elasticity and plasticity, thermo dynamics; Flow in porous media, free boundary problems; Inverse and ill-posed problems, parameter identificaiton; Multigrid methods, domain decomposition; Boundary and/or Finite Element methods; Particle methods, pseudorandom numbers; Advanced system theory, adaptive systems, neuronal networks, fuzzy control; Signal analysis, image processing and pattern recognition, wavelets; Information theory, coding and cryptography.

   In Economathematics topics could be Insurance mathematics, Finance mathematics; Experimental design, reliability and quality control; robust statistics and nonlinear models in statistics; Dynamic optimization, production planning, location theory; Scheduling and routing, location theory; Simulated annealing.
4. The master thesis should deal with a problem, which comes from industry; the student should spend several months in the company, who posed the problem. However, the scientific responsibility remains with the supervisor at the university; he makes the contacts with industry, negotiates the details, prepares the student before he spends his time in the company, supervises the scientific work after the return of the student and evaluates the thesis. Master theses in the ECMI programme take officially 6 months, inofficially it is often 3 months (the holiday months) more.

So much about a European attempt to organize a common programme on „Industrial Mathematics". There are similar attempts worldwide - some just using the new name to get better support for what they have ever done („the emperor's new clothes" - but not his „new mind"). Others try really hard - then their central activity is always similar to our modelling seminar (see for example the „mathematical clinic" in Claremont, US). There are weeks, months, even years dedicated to work on practical „real world" problems, on modelling and computing.

For 3rd world students, the 2 years master programme „Industrial Mathematics" at Kaiserslautern with ca. 12 scholarships per year seems to be one of the larger efforts in this direction. This programme is connected with the ECMI programme and quite similar to it. Graduates from this programme - more than 50 meanwhile - form a Technomathematics family with close contacts with each other (contacts from Trujillo in Peru to Fuzhou in China, from Kathmandu in Nepal to Malawi) and a summerschool family meeting every 3 years.

The programme has become more popular in industry during the last years - certainly correlated with growing popularity of these countries as future markets and source of labour. Why not using this interest?

II. Examples for a scientific cooperation with industry

Of course the literature is full with examples - we refer only to 8 proceedings volumes from ECMI-conferences or to a series of books by Avner Friedman (see for example [1] and [5]).

They contain to a high percentage problems really posed by industry; but even there we may find 2 categories: Problems, whose solution is academically exciting, but of minor industrial interest - and those, who are of „burning" interest for the company, i.e. the company pays for it. In the academic world the value of work is measured by the number of good publications - in industry it is measured by the price paid for it. Industrial mathematicians are squeezed between these two criteria. The „Institute for Industrial Mathematics" (ITWM), one of us (H.N.) is directing, has to earn money and scientific reputation; the results are paid and published.

We shall try to give a short description of the kind of problems at ITWM; quite some of them are „weakly" classified (since they are paid): The mathematics might be published, but not the use of it. These problems would miss the point we try to make - so we have to restrict our selections.

A) There are some problems in the field „flow in porous media" - but not as the flow of oil in reservoirs (this is more important in „oil countries" like Norway), but as a flow of different liquids through woven and nonwoven fabrics: Flow in diapes, blood flow in test stripes, oil flow in filters, resinflow through fibres to produce compound materials, flow of sweat in shirts etc. etc. Mathematically all that means of course Darcy's law, Richards equation, i.e. nonlinear parabolic or elliptic pde's. But there is a lot of modelling: The structure of a fabric is different from the structure of rocks, the scales are different and lead to corrections. The fluid is different, sometimes Non-Newtonian, we need Darcy's law with memory etc. Many of these equations exist already in literature, but reasonable numerics is often missing. Look for example at a filter, whose task is to extract oil drops from an air flow. There are several layers of nonwoven fabrics and if you look closer to what happens in and between these layers, you see a phantastic world: Oil drops coagulate to become a film, bursting into new droplets and so on. Richards equation, free boundaries, mean curvature flow - front tracking, level set methods and more. Details cannot be given here. The message only is: There are plenty of beautiful mathematical problems in that area (the Oxford study group has discovered this fact many years ago). For flow in porous media we may refer to [3], which offers a typical mathematical approach; a vast engineering literature exists too.

B) Another research field is „interaction of a fluid with flexible structures", the fluttering of sheets or fibres in wind. This is an even more sensitive area and we are not able to describe the real applications. But the problem is clear: A surface or curve is immersed in a gas flow, moved by the flow and changing the flow. Modelling is not so complicated, but the numerics! There is a moving boundary (the surface or - even worse - the 1-dimensional curve), which has its own life. Grids run in troubles, the systems get very stiff; our way out is to use gridfree methods, particle methods: Not as accurate as classical methods (multigrid etc.), but able to handle very complex situations with sufficient precision. The well known particle methods like „Particle in Cell" and its modern relatives, others like „Smoothed Particle Hydrodynamics", have to be improved to do the job - new methods like those based on kinetic ideas, lattice Boltzmann gas etc. must be further developed. Further applications of these methods are found in space flight simulation, where generalized Boltzmann equations are ruling the game, in granular flow (transport of grains in silos, of ashes on conveyor belts) and even in the simulation of traffic flows on highways (see for example [5] and [6]). There is much mathematics used: Measure theory, functional analysis and probability, even number theory.

C) Heat flow is a very traditional subject and many commercial packages are able to handle it - if radiation is not involved. But quite often it is: For example, when glass is cooling down and the stresses have to be controlled. This is a very old problem, but in order to organize an optimal control the transport of heat by radiation should be simulated. In 3d nontrivial configurations this is still a tremendous task - no problem, if diffusion approximation holds, but otherwise ray tracing is still too elaborate even for very fast computers. Our way out is domain decomposition and matching: Do approximations wherever it is possible, but use ray tracing where it is necessary. And find appropriate boundary conditions between the domains to match the computations. This is a general strategy, which seems to become more and more important: To combine analytical (i.e. asymptotic) with numerical methods to get results as cheap as possible but as precise as necessary. Ray tracing has some features of particle methods: Again number theory (Quasi random numbers), but also computational geometry, CAGD-methods and more.

D) Image processing is a wide field, normally „conquered“ by computer scientists, but a genuinely mathematical task. Data compression by fractal image analysis or by wavelets, segmentation by Mumford-Shah or similar functionals, filtering by anisotropic diffusion, pattern recognition by Bayesian statistical methods, stochastic geometry: It is an enormous bunch of mathematical ideas (see for example [4]). And the variety of problems is as big: There is never „the“ solution for a class of problems - each individual problem needs its special trick. Our main task is quality control of surfaces - the homogeneity of fabrics has to be measured, very little errors should be detected in surfaces of cars; the errors must be classified - often a job for neural networks. Medicine offers a wide field for image processing - with the advantage that the time needed for the processing is not as limited as for example in on-line quality control.

E) Adaptive systems are - in general - parametrized input-output mappings, which „learn“ the parameters permanently from the experience and use the newly gained knowledge to predict the future. Whether the class of mappings you consider is given by linear control systems, by neural networks or something different, depends on the problem you have (the less one knows the more one tends to use neural networks - are they really as flexible as the human brain?); in the end it always turns out to be a parameter identification problem - very often ill-posed and highly demanding for regularization. Modelling and optimizing - that's, what you also do in describing chemical reactions and even risk assessment for financial products. Finance mathematics, things like option prizing, is a new highlight for mathematicians all over the world (but again: many talk about it, but only a few outside of the banks produce something useful) - and it provides jobs for our graduates.

We could continue, but want to conclude: Mathematics we need is very wide spread, asks for tools to do modelling and computing and, optimally, both things together. If the industrial problem comes first, one is sure that the mathematics used is applied and not only applicable. If the method comes first, the result may be beautiful, but often you have to refer to future applicability: A completely legitimate argument but only for very few, very good pure mathematicians.

III. Industrial Mathematics in 3rd world countries

Of course, the situation of industrial mathematics in third-world countries is different from that in Europe, US and Australia. One of us (A.H.S.) founded an Indian Society for Industrial and Applied Mathematics in 1992 and implemented a programme at his Aligarh Muslim University. The other one (H.N.) has trained more than 50 third-world students in a two-year programme, who went home and tried to apply their new knowledge in their home countries; almost 30 of them came back to Germany September 1996 and reported on their experiences. We believe to have a quite clear picture of the situation there.

i) Managers of companies in 3rd world countries are even less aware of the role of science and especially of mathematics than their „western“ colleagues. This opinion may originate from the fact that some companies have their research labs in Europe or the US, only doing the production here. Moreover, it is a general belief that mathematics is only useful for high tech companies; if their production or products are not high tech, they do not consider mathematics as a valuable tool. But what is „high tech“? There was a kind of definition by David, who wrote the famous report of the NSF in the US about the support for mathematics: „When we entered the era of high technology, we entered the era of mathematical technology.“ Changing a word of David Hilbert we may say: In any technology is as much „high technology“ as is mathematics in it. The consequences of these statements simply are, that one may change every product into a high tech product by investing more science and especially more mathematics in its production. As mentioned before, mathematics for basic technologies as textile, furniture, glass, metal processing, food has become very popular, the same holds for the simulation of environmental events as avalanches, water waves, pollution, breaches in dams etc.

In short: There is much need for mathematics in any technologies, but less demand - a disadvantage for industry and for mathematics as well.

ii) There is a general tendency of managers in 3rd world countries to prefer software products from industrialized countries. It is more expensive but much easier to buy a commercial software tool even if it is not really appropriate for the problem to be solved than thinking and programming himself. But it is the wrong solution especially since the capacity of producing special software is already available: Why else would high tech companies all over the world establish software branches in India, Indonesia, Korea, Malaysia, China, who produce new software tools?

(iii) Only recently the 3rd world began to receive the message of Industrial Mathematics. Therefore, there is an acute shortage of properly educated mathematicians. There are, of course, some excellent pure mathematicians and quite a number of good applied mathematicians in the traditional British sense. But their work is mainly „method driven“ („I do, what I am able to do“), but not problem driven („I do, what should be done"). Academic careers make this attitude necessary: One has to belong to a scientific family (even if it is very small), one has to publish (even if it is a generalization of a generalization) in order to be promoted. Who dares to work on a modelling problem, when he/she does not know, which kind of mathematics is needed, what the result will be, where it should be published, who (besides maybe a company) will be interested in? However: Real career jumps need surprising results, new problems and new ideas - all that could not be expected without taking a risk! Moreover, senior professors do not need a security belt when working on new problems; they could go ahead and give the young scientists a chance and some credit to work on industrial problems. Obviously, the pressure comes from outside: Governments press scientists to contribute to the economic welfare of their countries. Does this really mean a cultural loss? Once more: The very few very best have some right to maintain that - and it is the duty of the many others to protect them, to protect them by proving that their ideas and mathematics as a whole is one of the most useful and valuable technologies.

References

[1] Friedman, A.: Mathematics in Industrial Problems, The IMA Volumes in Mathematics and its Applications, Springer-Verlag, New-York, Berlin, 1988, onwards

[2] Halmos, P.R.: Applied Mathematics is Bad Mathematics, Mathematics Tomorrow, Springer-Verlag, New-York, Berlin, 1981

[3] Hornung, U.: Homogenization and Porous Media, Interdisciplinary Appl. Math. Series, Springer-Verlag, New York, Berlin, 1997

[4] D. Mumford: The statistical description of visual signals. Proceedings of the 3rd ICIAM Congress, Hamburg, 1995

[5] Neunzert, H.; Struckmeier, J.: Particle Methods for the Boltzmann Equation, Acta Numerica, pp. 417-457, 1995

[6] Neunzert, H.; Klar, A.; Struckmeier, J.: Particle Methods - Theory and Applications (ed. K. Kirchgässner, D. Mahrenholtz, R. Mennicken) ICIAM 95: Proceedings of the

Third International Congress on Industrial and Applied Mathematics, Akademie Verlag Berlin, Vol. 87, 1996

Department of Mathematics
University of Kaiserslautern
D-67663 Kaiserslautern, Germany
(neunzert@mathematik.uni-kl.de)

Department of Mathematics
Aligarh Muslim University
Aligarh - 202002, India
(mmto1ahs@amu.nic.in)

M BROKATE

Elastoplastic constitutive laws of nonlinear kinematic hardening type

1. Introduction

The scope of elastoplasticity. Materials like steel and concrete are omnipresent in our societies. In everyday life as well as in specialized professional environments we are surrounded by tools, devices, machines and structures whose proper functioning depends on them. One basic issue is to understand how specimens made of such materials deform in reaction to mechanical forces. For small enough deformations and forces, a linear elastic law like, for example, Hooke's law

$$\boldsymbol{\sigma} = 2\mu\boldsymbol{\varepsilon} + \lambda(\operatorname{tr}\boldsymbol{\varepsilon})I\,, \tag{1.1}$$

often gives a sufficiently accurate description of the constitutive relation between the stress tensor $\boldsymbol{\sigma}$ and the strain tensor $\boldsymbol{\varepsilon}$ at any given point of the specimen and any given time. For larger stresses, various other effects become relevant or even dominant. (Even if outside forces are fixed, changes in shape or size of a specimen may induce large changes in stresses, due to the spatial effects of the law of equilibrium of forces and due to the fact that stresses are forces per area.) These include plastic flow as well as crack propagation and fracture, often combined with viscous effects. We concentrate here on plastic flow – the key phenomenon addressed in elastoplasticity – with zero viscosity, which is adequate for a broad range of applications involving ordinary steels at ordinary temperatures. (Indeed, since viscosity makes the equations more regular from a mathematical standpoint, the mathematical theory for the zero viscosity case is less well developed.) Plastic flow exhibits memory effects; it is not possible to write elastoplastic stress-strain laws in the form

$$\boldsymbol{\sigma}(t) = f(\boldsymbol{\varepsilon}(t)) \tag{1.2}$$

with a function f which relates only the current values of stress and strain. Additional internal variables, for example the so-called plastic strain $\boldsymbol{\varepsilon}^p$, are commonly used to incorporate the memory of the process.

The role of convex analysis. Since the fundamental work by Brézis, Duvaut, Lions, Moreau and others, the notions and methods of convex analysis and the calculus of variations have proved to be of great importance in the mathematical description and treatment of problems in plasticity. For one thing, it provides a natural formalism for a general thermomechanical framework based on the combination of an energy and a dissipation functional. That framework encompasses a large class of constitutive laws, among them the so-called generalized standard materials [19] which include in particular isotropic and linear kinematic (Melan-Prager) hardening; moreover, consistency with the second law of thermodynamics is guaranteed. Secondly, situations with and without viscosity are handled, at least initially, by the same general formalism. See [30] and also [28] on these points. Thirdly, the convex structure makes it possible to use monotonicity methods. These yield a lot of results concerning the boundary value problem – which arises from the combination of the constitutive law with the balance law (equilibrium of forces or Newton's law, respectively) –, from the standpoint of mathematical as well as of numerical analysis; we refrain from any further comments in that direction and just refer to [18, 20, 29].

Dependence on loading conditions. Ratchetting. The actual stress-strain behaviour of materials like steel under arbitrary (non-monotone, non-uniaxial) loading conditions can be rather complex. This is not surprising at all; materials science tells us that the observed macroscopic behaviour is the result of a complex mixture of processes, which take place on the microscopic scale and whose interaction strongly depends on the exact composition of the specimen. As a consequence, although the mathematical formulations of macroscopic stress-strain laws in plasticity generally make sense for arbitrary time-dependent stresses $\boldsymbol{\sigma} = \boldsymbol{\sigma}(t)$ or strains $\boldsymbol{\varepsilon} = \boldsymbol{\varepsilon}(t)$, they are usually developed in relation to a certain range of loading conditions, outside of which they may be quite useless for predicting the actual stress-strain behaviour, both quantitatively as well as qualitatively. Examples of such (tensorial) loadings, either in stress or in strain space, are monotone loading (from zero along a linear path to a prescribed tensorial value), proportional loading (along an arbitrary path within a one-dimensional subspace of tensor space), cyclic loading (oscillation between two prescribed values along the connecting line) and neutral loading (values vary on a fixed sphere in tensor space). In the context of cyclic loading, the phenomena of relaxation and of ratchetting [12, 13, 14, 15, 21, 22, 23, 25, 31, 33, 39] have been of particular interest. The latter refers to the accumulation of plastic strain under cyclic stress loading with nonzero mean stress. Stress-strain laws which account for that behaviour are often developed from the class of so-called nonlinear kinematic hardening models.

Nonlinear kinematic hardening. These models, often referred to under the name of non-associated plasticity, are usually based on the Armstrong-Frederick equation

$$\dot{\boldsymbol{\sigma}}^b = \frac{2}{3} C \dot{\varepsilon}^p - \gamma \boldsymbol{\sigma}^b \dot{p}\,, \quad \dot{p} = \sqrt{\frac{2}{3}} |\dot{\varepsilon}^p|\,, \tag{1.3}$$

which governs the evolution of the backstress $\boldsymbol{\sigma}^b$, to be discussed in detail in Section 3 below. The model of Chaboche is their most well known representative. Not based on (1.3), but closely related are the Mróz-type models which originate from [37]. In addition to the convex structure induced by the maximal dissipation principle (or one of its variants, the normality rule or Drucker's postulate), they include a nonlinear term like the product $\boldsymbol{\sigma}^b \dot{p}$ in (1.3) which seems to prevent their subsumption under an overall convex framework. Until recently, their rigorous mathematical analysis seems to be missing; on the other hand, engineers use them routinely for qualitative and quantitative modeling, including finite element simulations (for example, [24]).

Hysteresis operators. A different line of research, initiated by the group around Krasnosel'skiĭ during the late sixties, focuses on the mathematical properties of operators which are nowadays called hysteresis operators. By definition, a hysteresis operator $\mathcal{W}$ is rate independent, that is,

$$\mathcal{W}(v \circ \varphi) = (\mathcal{W}v) \circ \varphi\,, \tag{1.4}$$

holds for all monotone time transformations $\varphi : I \to I$ of some time interval $I = [t_0, t_f]$ onto itself and all input functions v within some admissible class of scalar, vectorial or tensorial functions. In plasticity, rate independent (often referred to as time independent) models correspond to the zero viscosity case. While there is a substantial overlap between the theory of hysteresis operators and convex analysis – which is best evident from the presentations in [27] and [42] –, because one of the basic hysteresis operators (the so-called stop) is identical with the solution operator of a linear evolution variational inequality, the overall approach in hysteresis operators is not based upon monotonicity. Indeed, there are now several results available concerning nonlinear kinematic hardening laws, see [6, 8, 9, 10]. They refer to their wellposedness as stress-strain laws; the treatment of the boundary value problem seems to be still open. In addition, the emphasis on the memory aspect – inherent in hysteresis operator theory – has made visible, see [7], a close connection to the rainflow counting method, which has become a standard tool of fatigue analysis in industrial research and development, see [38].

Aim and outline of this manuscript. The main aim of this paper is to provide a reasonably complete description of those parts of the theory of hysteresis operators which contribute to the formal treatment of linear and nonlinear kinematic hardening rules in elastoplasticity, and to specify precisely the connections to their formulation in the engineering sciences. To achieve the latter, we follow the terminology of one of the standard references in solid mechanics, the book of Lemaitre and Chaboche [28], to a large extent. Concerning the former, we have collected (and simplified in exposition) in Section 2 some of the relevant material from books [11, 26, 27, 42] and papers on hysteresis; we strictly remain within the variational inequality context, thus emphasizing the connection to convex analysis. Section 3 is devoted to the various rules for kinematic hardening. We have refrained from the temptation to skip lightly over the basic well-known models, and we also fully prove the general results concerning the wellposedness of the interaction of the variational inequality with the nonlinearity. For reasons of space we then could not include detailed proofs of the statements which deal with the application to individual stress-strain laws; instead, we sketch the main line of argument and provide references.

2. The Play and the Stop Operator

Moving convex sets. Let Z be a closed convex subset of some Hilbert space H, assume that $0 \in Z$. If $w : I \to H$, $I = [t_0, t_f]$, is any given function, the formula

$$C(t) = w(t) + Z\,, \quad t \in I\,, \tag{2.1}$$

specifies a – purely translational – motion of Z from the initial position $C(t_0)$ to the final position $C(t_f)$. Let us also consider a single moving point, described by some function $v : I \to H$, which is not allowed to leave the set $C(t)$, that is, we require $v(t) \in C(t)$ for all $t \in I$ or, equivalently,

$$z(t) \in Z\,, \quad t \in I\,, \tag{2.2}$$

where $z : I \to H$ is defined by

$$v(t) = w(t) + z(t)\,. \tag{2.3}$$

We now couple the two movements in the following way. Assume that the motion v of the point is prescribed. From the many possible motions w of the convex set compatible with (2.2) and (2.3) we want to find those (indeed, there will be exactly one by Theorem 2.1 below) which satisfy, in an appropriate sense,

$$\dot{w}(t) \in N_{C(t)}(v(t)) . \tag{2.4}$$

Here, for any convex set $K \subset H$ and any $y \in K$,

$$N_K(y) = \{u : u \in H \, , \, \langle u, k - y \rangle \leq 0 \textit{ for all } k \in K\} \tag{2.5}$$

denotes the normal cone to K in y. Condition (2.4) implies that the convex set does not move as long as $v(t) \in \operatorname{int} C(t)$; if $v(t) \in \partial C(t)$ and $\dot{v}(t)$ points outward in a non-tangential direction, the combined requirements (2.2) and (2.4) force $C(t)$ to move in the direction of an outward normal. See Figure 1.

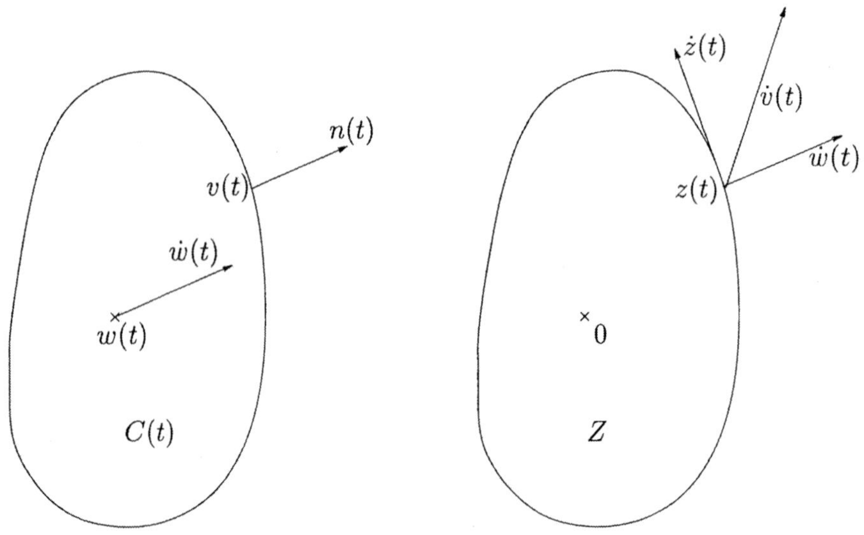

Figure 1: The moving convex set.

With (2.1) and (2.3) we can transform the differential inclusion (2.4) into the evolution variational inequality

$$\langle \dot{w}(t), x - z(t) \rangle \leq 0 \, , \quad \forall \, x \in Z \, , \tag{2.6}$$

with the time-independent convex constraint Z. If we also include an initial condition for the relative position of $C(t_0)$, the system to be solved consists of (2.6) and

$$v(t) = w(t) + z(t)\,, \quad z(t) \in Z\,, \tag{2.7}$$

$$z(t_0) = z^0\,. \tag{2.8}$$

It is well known that this system has unique solutions w and z for given data v and z^0. One can write it in the form of a differential inclusion and use convex analysis or Filippov theory, see e.g. [1], [26], or [42]. One may also eliminate $w = v - z$ and set $f = \dot{v}$ to arrive at the evolution variational inequality

$$z(t) \in Z\,, \quad \langle \dot{z}(t), x - z(t)\rangle \geq \langle f(t), x - z(t)\rangle\,, \ \forall\, x \in Z\,. \tag{2.9}$$

A specific version of the existence and uniqueness result for the system (2.6) – (2.8) in the space $W^{1,1}(I;H)$ of absolutely continuous functions with values in a separable Hilbert space H is given in [27], Theorem 3.1 and Proposition 3.9.

Theorem 2.1 *Let H be a separable Hilbert space, let $v \in W^{1,1}(I;H)$ and $z^0 \in Z$ be given. Then there exist unique functions $w, z \in W^{1,1}(I;H)$ such that (2.8) is satisfied, (2.7) holds for all $t \in I$ and (2.6) holds for almost all $t \in I$.*

According to this theorem,

$$w = \mathcal{P}(v\,; z^0)\,, \quad z = \mathcal{S}(v\,; z^0)\,, \tag{2.10}$$

yields well-defined solution operators

$$\mathcal{P} : W^{1,1}(I;H) \times Z \to W^{1,1}(I;H)\,, \quad \mathcal{S} : W^{1,1}(I;H) \times Z \to W^{1,1}(I;Z)\,. \tag{2.11}$$

Following [26], the operators $\mathcal{P}$ and $\mathcal{S}$ are called the *play operator* and the *stop operator*, respectively. We will write $\mathcal{P}_Z$ respectively $\mathcal{S}_Z$ if we want to specify the set Z explicitly.

The formula

$$\langle \dot{w}(t), \dot{z}(t)\rangle = 0\,, \quad \textit{for a.e. } t \in I \tag{2.12}$$

follows if we set $x = z(t+h)$ respectively $x = z(t-h)$ in (2.6), divide by h and pass to the limit $h \to 0$. Thus, the decomposition

$$\dot{v}(t) = \dot{w}(t) + \dot{z}(t) \tag{2.13}$$

is orthogonal; moreover, since z does not leave the set Z, $\dot{z}(t)$ is an element of the tangent cone $T_Z(z(t))$ to Z in $z(t)$, defined as the dual of the normal cone by

$$T_Z(y) = \{u : u \in H\,,\ \langle u, n\rangle \le 0 \mathit{\ for\ all}\ n \in N_Z(y)\}\,. \tag{2.14}$$

To visualize these relations, we again refer to Figure 1.

We can also couple the two movements the other way round, namely we can let the convex set drive the point. This is what Moreau [34, 35] has termed a *sweeping process*. Here, the unknown function $v = v(t)$ is determined from the known function $C = C(t)$ via the rule

$$-\dot{v}(t) \in N_{C(t)}(v(t))\,. \tag{2.15}$$

Moreau has proved the unique solvability of the initial value problem consisting of (2.15) and an initial condition $v(t_0) = v^0$ for rather general set-valued evolutions $C = C(t)$. In the special case $C(t) = w(t) + Z$ considered here, we can transform (2.15) as above to

$$\langle -\dot{v}(t), x - z(t)\rangle \le 0\,, \quad \forall\, x \in Z\,, \tag{2.16}$$

Since (2.3) can be trivially rewritten as

$$-w(t) = -v(t) + z(t)\,, \tag{2.17}$$

we see that for purely translational motion, the solution operator of the sweeping process (with the initial condition $z(t_0) = z^0$) can be expressed via the play operator as

$$v = -\mathcal{P}(-w\,; z^0)\,, \tag{2.18}$$

or vice versa; in particular, Theorem 2.1 also follows from Moreau's results.

Regularity of the play and stop operator. We first remark that due to the formula

$$v = \mathcal{P}(v\,; z^0) + \mathcal{S}(v\,; z^0)\,, \tag{2.19}$$

many properties immediately carry over from the play to the stop operator and vice versa; we will use that fact without further reference. The basic regularity property results from the standard uniqueness argument for evolution variational inequalities. Let $v_1, v_2 \in W^{1,1}(I;H)$ and $z_1^0, z_2^0 \in Z$ be given, set

$$z_1 = \mathcal{S}(v_1\,;z_1^0)\,, \quad z_2 = \mathcal{S}(v_2\,;z_2^0)\,. \tag{2.20}$$

Using z_2 as a test function in the variational inequality for z_1,

$$\langle \dot{v}_1(t) - \dot{z}_1(t), z_2(t) - z_1(t)\rangle \le 0\,, \quad \textit{for a.e. } t \in I\,, \tag{2.21}$$

and vice versa, we can argue that for almost all $t \in I$ it holds (the norm in H we simply denote by $|\cdot|$)

$$\begin{aligned}|z_1(t) - z_2(t)|\,\frac{\mathrm{d}}{\mathrm{dt}}|z_1(t) - z_2(t)| &= \frac{\mathrm{d}}{\mathrm{dt}}\frac{1}{2}|z_1(t) - z_2(t)|^2\\ &= \langle \dot{z}_1(t) - \dot{z}_2(t), z_1(t) - z_2(t)\rangle \le \langle \dot{v}_1(t) - \dot{v}_2(t), z_1(t) - z_2(t)\rangle\\ &\le |\dot{v}_1(t) - \dot{v}_2(t)|\,|z_1(t) - z_2(t)|\,,\end{aligned} \tag{2.22}$$

so

$$\begin{aligned}|z_1(t) - z_2(t)| &\le |z_1^0 - z_2^0| + \int_{t_0}^{t} \frac{\mathrm{d}}{\mathrm{ds}}|z_1(s) - z_2(s)|\,ds\\ &\le |z_1^0 - z_2^0| + \int_{t_0}^{t} |\dot{v}_1(s) - \dot{v}_2(s)|\,ds\,,\end{aligned} \tag{2.23}$$

for all $t \in I$, and thus in particular

$$\|\, z_1 - z_2 \,\|_\infty \le |z_1^0 - z_2^0| + \mathrm{Var}\,(v_1 - v_2)\,, \tag{2.24}$$

where

$$\|\, v \,\|_\infty = \sup_{t\in I} |v(t)|\,, \quad \mathrm{Var}\,(v) = \int_{t_0}^{t_f} |\dot{v}(t)|\,dt\,, \tag{2.25}$$

denotes the supremum norm and the total variation, respectively. If we equip the space $W^{1,1}(I;H)$ with the norm

$$\|\, v \,\|_{1,1} = |v(t_0)| + \mathrm{Var}\,(v)\,, \tag{2.26}$$

the basic regularity result can be stated as follows.

Theorem 2.2 *Considered as operators with values in the space $C(I;H)$ of continuous H-valued functions, that is,*

$$\mathcal{P},\mathcal{S}: W^{1,1}(I;H)\times Z \to C(I;H)\,, \tag{2.27}$$

the play and the stop operator are Lipschitz continuous.

On one hand, Theorem 2.2 is attractive because it holds for arbitrary closed convex sets Z in arbitrary separable Hilbert spaces H; moreover, with a natural choice of the product metric in $W^{1,1}(I;H)\times Z$, the Lipschitz constant has a value of 1 respectively 2, independent from the form of Z (due to rate independence, it does not depend on the choice of I anyway). On the other hand, going from $W^{1,1}$ to C we suffer a *loss of regularity*. This constitutes a problem in particular when one wants to deal with stress-strain laws where the internal variables are related in a nonlinear (nonconvex) manner. Fortunately, one can improve the estimate (2.23) and thus get different and better results in many special cases. However, they usually require either explicitly or implicitly that $\operatorname{int}(Z)\neq\emptyset$, or equivalently, by translation of Z, that $0\in\operatorname{int}(Z)$. While this is typically satisfied in the finite-dimensional case – which occurs when we look at the stress-strain law in isolation, H then being a space of tensors – it restricts their applicability in infinite-dimensional situations which arise if one wants to solve the elastoplastic boundary value problem; for example we have $\operatorname{int}(Z)=\emptyset$ except in trivial cases if $H=L^2(\Omega;\tilde{H})$ and $Z=\{u : u\in H,\, u(x)\in\tilde{Z}\; a.e.\}$ where $\tilde{Z}$ is a closed convex subset of a finite-dimensional space $\tilde{H}$.

We discuss some of these results, selected from the literature, in particular the monographs [11], [26], [27] and [42]. In order to simplify the exposition, our presentation essentially stays within the calculus of absolutely continuous functions; we refer to the above-mentioned monographs, in particular [27], and to Moreau's contribution in [36] for the extension to continuous functions and to functions of bounded variation.

To begin, we consider the one-dimensional case $H=\mathbb{R}$.

Theorem 2.3 *Let $H=\mathbb{R}$, $Z=[-r,r]$ for some $r>0$. Then we have*

$$\|\mathcal{P}(v_1;z_1^0)-\mathcal{P}(v_2;z_2^0)\|_\infty \le \max\{\|v_1-v_2\|_\infty\,,\,|z_1^0-z_2^0|\} \tag{2.28}$$

for all $v_1,v_2\in W^{1,1}(I)$ and all $z_1^0,z_2^0\in Z$. Thus, the play and the stop operator can be uniquely extended to Lipschitz continuous operators

$$\mathcal{P},\mathcal{S}: C(I)\times Z\to C(I)\,. \tag{2.29}$$

Proof. Let $w_i = \mathcal{P}(v_i\,; z_i^0)$ for $i = 1, 2$. It suffices to show that $|v_1(t) - v_2(t)| \geq |w_1(t) - w_2(t)|$ for all $t \in I$ for which

$$0 < \frac{\mathrm{d}}{\mathrm{d}t}\frac{1}{2}|w_1(t) - w_2(t)|^2 = (\dot{w}_1(t) - \dot{w}_2(t))(w_1(t) - w_2(t))\,. \tag{2.30}$$

Let t be such a point, assume without loss of generality that $\dot{w}_1(t)(w_1(t) - w_2(t) > 0$. Since $\dot{w}_1(t)(z_2(t) - z_1(t)) \leq 0$ from the variational inequality, we have, at the time t, $(w_1 - w_2)(z_2 - z_1) \leq 0$ and therefore

$$|v_1 - v_2| = |(w_1 - w_2) + (z_1 - z_2)| = |w_1 - w_2| + |z_1 - z_2| \geq |w_1 - w_2| \tag{2.31}$$

at this time t. This completes the proof. □

Theorem 2.4 *Let $H = \mathbb{R}$, $Z = [-r, r]$ for some $r > 0$. Then we have*

$$\mathrm{Var}\,(\mathcal{P}(v_1\,; z_1^0) - \mathcal{P}(v_2\,; z_2^0)) \leq |z_1^0 - z_2^0| + \mathrm{Var}\,(v_1 - v_2) \tag{2.32}$$

for all $v_1, v_2 \in W^{1,1}(I)$ and all $z_1^0, z_2^0 \in Z$. Thus, the play and the stop operator are Lipschitz continuous operators

$$\mathcal{P}, \mathcal{S} : W^{1,1}(I) \times Z \to W^{1,1}(I)\,. \tag{2.33}$$

Proof. Let $w_i = \mathcal{P}(v_i\,; z_i^0)$ for $i = 1, 2$. If we can prove that

$$|\dot{w}_1(t) - \dot{w}_2(t)| + \frac{\mathrm{d}}{\mathrm{d}t}|z_1(t) - z_2(t)| = (\dot{v}_1(t) - \dot{v}_2(t))\mathrm{sign}\,(\dot{w}_1(t) - \dot{w}_2(t))\,, \tag{2.34}$$

holds a.e. in I, the estimate (2.32) follows by integration. To prove (2.34) we observe that the variational inequality implies

$$(\dot{w}_1(t) - \dot{w}_2(t))(z_1(t) - z_2(t)) \geq 0 \tag{2.35}$$

a.e. in I. Thus, on the set $I_+ = \{t : t \in I, z_1(t) > z_2(t)\}$, (2.34) reduces to the identity $(\dot{w}_1 - \dot{w}_2) + (\dot{z}_1 - \dot{z}_2) = \dot{v}_1 - \dot{v}_2$ which is true. The same happens for $I_- = \{t : t \in I, z_1(t) < z_2(t)\}$. On $I_0 = \{t : t \in I, z_1(t) = z_2(t)\}$ we have $\dot{v}_1 - \dot{v}_2 = \dot{w}_1 - \dot{w}_2$, so both sides of (2.34) equal $|\dot{w}_1(t) - \dot{w}_2(t)|$. □

If H is multi-dimensional, the analysis has to be more refined. One particular type of results starts from the observation that the difference $|w_1 - w_2|$ with $w_i = \mathcal{P}(v_i\,; z_i^0)$

for $v_i \in W^{1,1}(I;H)$, $i=1,2$, satisfies on any subinterval $[s,t]\subset I$ the estimate

$$\begin{aligned}
&\frac{1}{2}|w_1(t)-w_2(t)|^2-\frac{1}{2}|w_1(s)-w_2(s)|^2= \\
&=\int_s^t \langle \dot{w}_1(\tau)-\dot{w}_2(\tau), w_1(\tau)-w_2(\tau)\rangle\, d\tau \le \int_s^t \langle \dot{w}_1(\tau)-\dot{w}_2(\tau), v_1(\tau)-v_2(\tau)\rangle\, d\tau \\
&\le \| v_1-v_2\|_\infty \left(\int_s^t |\dot{w}_1(\tau)|\, d\tau + \int_s^t |\dot{w}_2(\tau)|\, d\tau\right), \qquad (2.36)
\end{aligned}$$

where we have used (2.21) at the first inequality sign. Setting $s=t_0$ and taking the supremum over t, (2.36) yields the Hölder estimate

$$\| w_1-w_2\|_\infty \le (c(w_1)+c(w_2))\sqrt{\| v_1-v_2\|_\infty}+|z_1^0-z_2^0|+|v_1(t_0)-v_2(t_0)|, \quad (2.37)$$

where

$$c(w)=\sqrt{2\mathrm{Var}\,(w)}\,. \quad (2.38)$$

Consequently, we obtain the 1/2-Hölder continuity of $\mathcal{P}$ and $\mathcal{S}$ on sets of pairs (v,z^0) where the variation of $\mathcal{P}(v\,;z^0)$ is uniformly bounded. Provided that

$$B_r(0)\subset Z\subset B_R(0) \quad (2.39)$$

holds for some $r,R>0$, such a bound can be obtained from the variational inequality on intervals where the oscillation of v is small enough.

Lemma 2.5 *Let (2.39) hold for some $r,R>0$, let $z^0\in Z$ and $v\in W^{1,1}(I;H)$. If $[s,t]\subset I$ and*

$$\sup_{\tau\in[s,t]}|v(\tau)-v(s)|\le r/2\,, \quad (2.40)$$

then $w=\mathcal{P}(v\,;z^0)$ satisfies

$$\int_s^t |\dot{w}(\tau)|\, d\tau \le \frac{R^2}{r}\,. \quad (2.41)$$

More precise estimates can be found in [26] and in [9, Proposition 2.5].

Proof. For an arbitrary $\varphi\in C([s,t];H)$ with $\|\varphi\|_\infty\le 1$ we define the test function $\psi\in C([s,t];H)$ by

$$\psi(\tau)=v(\tau)-v(s)+\frac{r}{2}\varphi(\tau)\,. \quad (2.42)$$

Then $\psi(\tau) \in Z$ for all τ by (2.39) and with $z = \mathcal{S}(v\,; z^0)$ we have

$$\frac{r}{2}\int_s^t \langle \dot{w}(\tau), \varphi(\tau)\rangle \, d\tau = \int_s^t \langle \dot{w}(\tau), v(s) - w(\tau)\rangle \, d\tau + \int_s^t \langle \dot{w}(\tau), \psi(\tau) - z(\tau)\rangle \, d\tau\,. \quad (2.43)$$

Since the rightmost integral in (2.43) is nonpositive and φ is arbitrary, we obtain

$$\begin{aligned} \frac{r}{2}\int_s^t |\dot{w}(\tau)| \, d\tau \;&\le\; \int_s^t \langle \dot{w}(\tau), v(s) - w(\tau)\rangle \, d\tau = -\frac{1}{2}|v(s) - w(\tau)|^2 \Big|_{\tau=s}^{\tau=t} \\ &= \frac{1}{2}|z(s)|^2 - \frac{1}{2}|v(s) - w(t)|^2 \le \frac{R^2}{2}\,. \end{aligned} \quad (2.44)$$

□

Using Lemma 2.5 we can conclude that, for every $v \in W^{1,1}(I;H)$, on the maximum norm neighbourhood $V_\varepsilon = \{\tilde{v} : \tilde{v} \in W^{1,1}(I;H),\ \| \tilde{v} - v \|_\infty \le \varepsilon\}$ it holds

$$\sup_{\tilde{v} \in V_\varepsilon, z^0 \in Z} \operatorname{Var} \mathcal{P}(\tilde{v}\,; z^0) < \infty\,, \quad (2.45)$$

if $\varepsilon > 0$ is sufficiently small and (2.39) holds. We just have to choose $\delta > 0$ small enough such that (2.40) is satisfied for all $|s - t| \le \delta$ and all $\tilde{v} \in V_\varepsilon$ to obtain, by repeated application of Lemma 2.5,

$$\operatorname{Var} \mathcal{P}(\tilde{v}\,; z^0) \le \frac{R^2}{r} \cdot \frac{t_f - t_0}{\delta}\,. \quad (2.46)$$

Indeed, it suffices to choose δ such that v satisfies (2.40) with $r/2$ replaced by $r/4$, and to set $\varepsilon = r/8$. With (2.46), we have obtained on V_ε a uniform bound for the two rightmost integrals in (2.36). This yields the local 1/2-Hölder continuity of the play and the stop and in particular the following continuity result.

Theorem 2.6 *Let Z satisfy $B_r(0) \subset Z \subset B_R(0)$ for some $r, R > 0$. Let $v_n \in W^{1,1}(I;H)$ and $z_n^0 \in Z$ with $z_n^0 \to z^0$ and $v_n \to v \in W^{1,1}(I;H)$ uniformly. Then it holds $\mathcal{P}(v_n\,; z_n^0) \to \mathcal{P}(v\,; z^0)$ uniformly.*

If we only assume $v \in C(I;H)$ in Theorem 2.6, a suitably refined argument shows that the play and the stop operators can be uniquely extended to operators

$$\mathcal{P}, \mathcal{S} : C(I;H) \times Z \to C(I;H)\,. \quad (2.47)$$

which are locally 1/2-Hölder continuous. It does not, however, yield global uniform continuity of $\mathcal{P}$ or $\mathcal{S}$ with respect to the maximum norm, since the number δ in (2.46), related to the modulus of continuity of the function v, can become arbitrarily small. On the other hand, one can say more about the variation of $w = \mathcal{P}(v\,; z^0)$. First, the uniform convergence of $\mathcal{P}(v_n\,; z_n^0)$ to $\mathcal{P}(v\,; z^0)$ and the definitions of variation and limit inferior immediately imply that

$$\operatorname{Var}\mathcal{P}(v\,; z^0) \le \liminf_{n\to\infty} \operatorname{Var}\mathcal{P}(v_n\,; z_n^0)\,. \tag{2.48}$$

Actually, it will turn out that

$$\operatorname{Var}\mathcal{P}(v\,; z^0) = \lim_{n\to\infty} \operatorname{Var}\mathcal{P}(v_n\,; z_n^0)\,, \tag{2.49}$$

if $Z = B_r(0)$. In that case, the normal and the radius vector coincide at every boundary point of Z, therefore $w = \mathcal{P}(v\,; z^0)$ and $z = \mathcal{S}(v\,; z^0)$ satisfy

$$r\dot{w}(t) = |\dot{w}(t)|z(t) \tag{2.50}$$

a.e. in I. Moreover, multiplying (2.50) with $\dot{w}(t)$ we see that

$$r|\dot{w}(t)| = \langle z(t), \dot{w}(t)\rangle \tag{2.51}$$

holds a.e. in I.

Theorem 2.7 *If* $Z = B_r(0)$, *then in the situation of Theorem 2.6 we have in addition that* $\lim_{n\to\infty} \operatorname{Var}\mathcal{P}(v_n\,; z_n^0) = \operatorname{Var}\mathcal{P}(v\,; z^0)$.

This result immediately extends to the case where ∂Z has a unique continuous outer normal, see [27, Proposition 4.11]. On the other hand, simple counterexamples show that, even in the case $H = \mathbb{R}$, it may happen that under the assumptions of Theorem 2.7 the expression $\operatorname{Var}(\mathcal{P}(v_n\,; z_n^0) - \mathcal{P}(v\,; z^0))$ does not converge to zero.

Proof of Theorem 2.7. In view of (2.51), it is sufficient to show that

$$\int_{t_0}^{t_f} \langle z(t), \dot{w}(t)\rangle\, dt = \lim_{n\to\infty} \int_{t_0}^{t_f} \langle z_n(t), \dot{w}_n(t)\rangle\, dt \tag{2.52}$$

holds for $w_n = \mathcal{P}(v_n\,; z_n^0)$, $w = \mathcal{P}(v\,; z^0)$, $z_n = \mathcal{S}(v_n\,; z_n^0)$ and $z = \mathcal{S}(v\,; z^0)$. But this is apparent from the formula

$$\int_{t_0}^{t_f} \langle z_n, \dot{w}_n\rangle\, dt - \int_{t_0}^{t_f} \langle z, \dot{w}\rangle\, dt = \tag{2.53}$$

$$= \int_{t_0}^{t_f} \langle z_n - z, \dot{w}_n\rangle\, dt - \int_{t_0}^{t_f} \langle \dot{z}, w_n - w\rangle\, dt + \langle z(t), w_n(t) - w(t)\rangle \Big|_{t=t_0}^{t=t_f}\,,$$

since $w_n \to w$ uniformly and therefore $z_n \to z$ uniformly by Theorem 2.6. $\square$

We now come to the question of nonlocal uniform, Hölder or even Lipschitz continuity of the play and the stop operator in the multi-dimensional case. As it was already the case in Theorem 2.7, the known results in this direction involve the geometry of the convex set Z in a specific way. One particular type of theorems is related to the curvature of the boundary ∂Z of Z. Given a point $p \in \partial Z$ and a tangent direction $h \in \partial T_Z(p)$, the quantity $\operatorname{dist}(p+h, Z)$ will be zero if and only if the straight line segment L connecting p and $p+h$ belongs to Z; the function

$$\alpha_Z(\rho) = \inf_{p \in \partial Z} \inf_{\substack{h \in \partial T_Z(p) \\ |h|=\rho}} \operatorname{dist}(p+h, Z) \tag{2.54}$$

measures the strict convexity of Z in a uniform manner. Obviously, $\alpha_Z(0) = 0$ and due to the convexity of Z, $\alpha_Z : \mathbb{R}_+ \to \mathbb{R}_+$ is nondecreasing. For example, for the ball $B_r(0)$ of radius r around 0 we have

$$\alpha_{B_r(0)}(\rho) = \sqrt{r^2+\rho^2} - r\,, \quad \alpha_{B_r(0)}^{-1}(a) = \sqrt{2ar + a^2}\,. \tag{2.55}$$

The following theorem constitutes a multi-dimensional generalization of Theorem 2.3.

Theorem 2.8 *Let Z be a closed convex subset of the separable Hilbert space H, assume that $\alpha_Z^{-1} : \mathbb{R}_+ \to \mathbb{R}_+$ exists. Then we have*

$$\| \mathcal{P}(v_1\,; z_1^0) - \mathcal{P}(v_2\,; z_2^0) \|_\infty \le \max\{\alpha_Z^{-1}(\| v_1 - v_2 \|_\infty)\,,\, |z_1^0 - z_2^0|\} \tag{2.56}$$

for all $v_1, v_2 \in W^{1,1}(I; H)$ and all $z_1^0, z_2^0 \in Z$. Thus, the play and the stop operator are uniformly continuous operators

$$\mathcal{P}, \mathcal{S} : C(I;H) \times Z \to C(I;H)\,. \tag{2.57}$$

Before we give the proof of the theorem, we offer some comments. In the case of the ball $Z = B_r(0)$, we see from (2.55) and (2.56) that the play and the stop are globally 1/2-Hölder continuous with respect to the supremum norm. If on the other hand $\operatorname{int}(Z) = \emptyset$, we will have $\alpha_Z(\rho) = 0$ on some finite interval $[0, b]$, so α_Z^{-1} does not exist; moreover, the proof below no longer yields a useful estimate. For polyhedric sets Z in finite dimensions, which are bounded by hyperplanes, the same difficulty arises, but a different approach is possible, see [26] and [27].

Proof of Theorem 2.8. We start with a separate, purely geometric construction which will be needed during the proof. Let $p \in \partial Z$ and $q \in H$ a point outside of Z with $q - p \notin T_Z(p)$, let π_q be the projection of q to Z, see Figure 2.

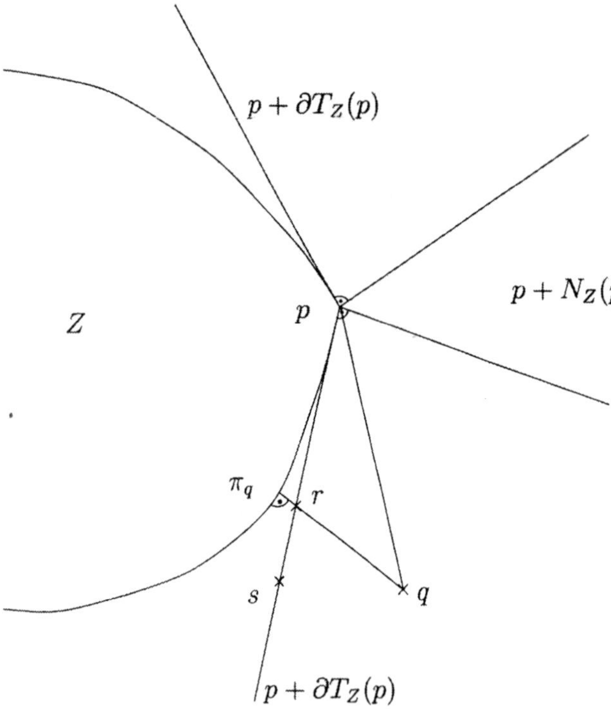

Figure 2: Proof of the uniform continuity.

Let r be the first point of $p + T_Z(p)$ we encounter while moving from q to π_q along the connecting line. Let us now continue the line segment from p to r beyond r to a point s with $|s - p| = |q - p|$. (This makes sense because $|q - p| > |\pi_q - p|$.) We have $s - p \in \partial T_Z(p)$ and $|s - r| \leq |q - r|$, so

$$\alpha_Z(|q - p|) \leq |s - \pi_q| \leq |q - \pi_q| = \text{dist}\,(q, Z). \tag{2.58}$$

Let now $w_i = \mathcal{P}(v_i\,; z_i^0)$ for $i = 1, 2$, assume that at some $t \in I$ we have

$$\frac{\mathrm{d}}{\mathrm{dt}}|w_1(t) - w_2(t)|^2 > 0\,. \tag{2.59}$$

Again, without loss of generality we may assume that $\langle \dot{w}_1(t), w_1(t) - w_2(t)\rangle > 0$. Then we have $0 \neq \dot{w}_1(t) \in N_Z(z_1(t))$, so $z_1(t) \in \partial Z$ and $q := w_1(t) - w_2(t) + z_1(t)$ satisfies

$q - z_1(t) \notin T_Z(z_1(t))$. The construction above and (2.58) show that

$$|v_1(t) - v_2(t)| = |q - z_2(t)| \geq \text{dist}\,(q, Z) \geq \alpha_Z(|w_1(t) - w_2(t)|)\,. \tag{2.60}$$

Therefore, (2.56) holds, since otherwise there must be a point $t \in I$ which satisfies (2.59) but violates (2.60). □

A multi-dimensional analogue of Theorem 2.4 has been obtained recently in [8], but only for the special case $Z = B_r(0)$. In that case, besides the above-mentioned formulas

$$r\dot{w}(t) = |\dot{w}(t)|z(t)\,, \quad r|\dot{w}(t)| = \langle \dot{w}(t), z(t)\rangle\,, \tag{2.61}$$

for almost all $t \in I$, we also have, using (2.12)

$$r^2\dot{w}(t) = \langle \dot{v}(t), z(t)\rangle z(t) \tag{2.62}$$

for almost all $t \in \{t : t \in I, |z(t)| = r\}$.

Theorem 2.9 *Let H be a separable Hilbert space, $Z = B_r(0)$. Then we have*

$$\begin{aligned} &\text{Var}\,(\mathcal{P}(v_1\,; z_1^0) - \mathcal{P}(v_2\,; z_2^0)) \leq \\ &\leq\ |z_1^0 - z_2^0| + \text{Var}\,(v_1 - v_2) + \frac{\sqrt{2}}{r}\int_{t_0}^{t_f} |\dot{v}_1(t)||\mathcal{S}(v_1\,; z_1^0)(t) - \mathcal{S}(v_2\,; z_2^0)(t)|\,dt\,, \end{aligned} \tag{2.63}$$

for all $v_1, v_2 \in W^{1,1}(I; H)$ and all $z_1^0, z_2^0 \in Z$. As a consequence, the operators

$$\mathcal{P}, \mathcal{S} : W^{1,1}(I; H) \times Z \to W^{1,1}(I; H) \tag{2.64}$$

are Lipschitz continuous if the first argument is restricted to a bounded subset of the space $W^{1,1}(I; H)$.

Proof. The assertion concerning the Lipschitz continuity of $\mathcal{P}$ and $\mathcal{S}$ follows from (2.63), if we use (2.24), that is, Theorem 2.2, to estimate the rightmost integral in (2.63). That integral term arises from the geometry of the sphere in the following way. Let $w_i = \mathcal{P}(v_i\,; z_i^0)$ and $z_i = \mathcal{S}(v_i\,; z_i^0)$ for $i = 1, 2$, denote by

$$I_r = \{t : t \in I, |z_1(t)| = |z_2(t)| = r\} \tag{2.65}$$

that part of the time interval where both constraints are active. We claim that (arguments t omitted)

$$|\dot{w}_1 - \dot{w}_2| \leq |\dot{v}_1 - \dot{v}_2| + \frac{\sqrt{2}}{r}|\dot{v}_1||z_1 - z_2|\,. \tag{2.66}$$

holds a.e. on I_r. Indeed, (2.62), applied to both evolutions, implies that

$$r^2|\dot{w}_1 - \dot{w}_2| \leq |\langle \dot{v}_1 - \dot{v}_2, z_2\rangle z_2| + |\langle \dot{v}_1, z_1\rangle z_1 - \langle \dot{v}_1, z_2\rangle z_2| \tag{2.67}$$

holds a.e. on I_r, and using the identity

$$|\langle \dot{v}_1, z_1\rangle z_1 - \langle \dot{v}_1, z_2\rangle z_2|^2 = r^2\langle \dot{v}_1, z_1 - z_2\rangle^2 + \langle \dot{v}_1, z_1\rangle\langle \dot{v}_1, z_2\rangle |z_1 - z_2|^2 , \tag{2.68}$$

we arrive at (2.66) after some computation. We now begin the main argument of the proof. For the first step, we see that (2.61) implies

$$\frac{1}{2}\frac{\mathrm{d}}{\mathrm{dt}}(|z_1|^2 - |z_2|^2) = -r(|\dot{w}_1| - |\dot{w}_2|) + (\langle \dot{v}_1, z_1\rangle - \langle \dot{v}_2, z_2\rangle) \tag{2.69}$$

a.e. on I. The goal of the second step is to prove the estimate which corresponds to (2.34) in the proof of the scalar version, namely

$$r\Big||\dot{w}_1| - |\dot{w}_2|\Big| + \frac{1}{2}\frac{\mathrm{d}}{\mathrm{dt}}\Big||z_1|^2 - |z_2|^2\Big| \leq |\langle \dot{v}_1, z_1\rangle - \langle \dot{v}_2, z_2\rangle| . \tag{2.70}$$

Indeed, on the subset $I_e = \{t : |z_1(t)| = |z_2(t)|\}$ of I, (2.70) holds a.e. because of (2.69), since the left side of (2.69) is equal to zero a.e. on I_e. To see that (2.70) also holds on the complement of I_e in I, we multiply both sides of (2.69) with the sign of $|z_1|^2 - |z_2|^2$ and use the fact that

$$(|\dot{w}_1| - |\dot{w}_2|)(|z_1|^2 - |z_2|^2) \geq 0 , \tag{2.71}$$

which is true a.e. on I since $|\dot{w}_i| \neq 0$ only if $|z_i| = r$. For the third step we claim that, a.e. on I,

$$|\dot{w}_1 - \dot{w}_2| + \frac{1}{2r}\frac{\mathrm{d}}{\mathrm{dt}}\Big||z_1|^2 - |z_2|^2\Big| \leq |\dot{v}_1 - \dot{v}_2| + \frac{\sqrt{2}}{r}|\dot{v}_1||z_1 - z_2| . \tag{2.72}$$

Indeed, (2.72) holds a.e. on I_r because of (2.66). In all points t of I except for a null set, at least one of $\dot{w}_1(t)$ and $\dot{w}_2(t)$ is zero, so $|\dot{w}_1 - \dot{w}_2| = ||\dot{w}_1| - |\dot{w}_2||$, and (2.72) follows from (2.70) and the estimate

$$|\langle \dot{v}_1, z_1\rangle - \langle \dot{v}_2, z_2\rangle| \leq r|\dot{v}_1 - \dot{v}_2| + |\dot{v}_1||z_1 - z_2| . \tag{2.73}$$

The assertion (2.63) is now obtained if we integrate (2.72) over the time interval $I = [t_0, t_f]$. □

3. Linear and Nonlinear Kinematic Hardening

The aim of this section is to discuss constitutive laws of kinematic hardening type, and in particular to provide some mathematical foundation for the more advanced cases of nonlinear kinematic hardening which play an important role in contemporary industrial engineering. We will follow the notation of [28] to some extent. In particular, stress and strain tensors will be denoted by boldface letters, and $\mathbf{a}:\mathbf{b} = \sum_{ij} a_{ij}b_{ij}$ will denote the standard scalar product in the space of second order 3×3-tensors.

Let us fix a point (or volume element) of the material body in question. The constitutive law relates the time histories $\boldsymbol{\sigma} : I \to \mathbf{T}$ and $\boldsymbol{\varepsilon} : I \to \mathbf{T}$, where $I = [t_0, t_f]$ and

$$\mathbf{T} = \{\boldsymbol{\tau} : \boldsymbol{\tau} \in \mathbb{R}^{3,3}\,,\ \boldsymbol{\tau} \text{ symmetric}\}\,, \tag{3.1}$$

of the stress tensor and the strain tensor at that point – we do not consider the temperature dependence. Additional *internal variables* incorporate the memory of the process, the most important being the plastic strain $\boldsymbol{\varepsilon}^p$ defined by the decomposition

$$\boldsymbol{\varepsilon} = \boldsymbol{\varepsilon}^p + \boldsymbol{\varepsilon}^e\,, \tag{3.2}$$

thus $\boldsymbol{\varepsilon}^p$ represents the difference between the total (observed) strain $\boldsymbol{\varepsilon}$ and its elastic part $\boldsymbol{\varepsilon}^e$ given by some linear elastic law (here ':' relates a fourth and a second order tensor, resulting in a second order tensor)

$$\boldsymbol{\varepsilon}^e = \mathbf{A} : \boldsymbol{\sigma}\,, \qquad \boldsymbol{\sigma} = \mathbf{a} : \boldsymbol{\varepsilon}^e\,, \tag{3.3}$$

Hooke's law

$$\mathbf{a} : \boldsymbol{\varepsilon}^e = 2\mu\boldsymbol{\varepsilon}^e + \lambda(\operatorname{tr}\boldsymbol{\varepsilon}^e)\mathbf{1} \tag{3.4}$$

with the Lamé constants $\lambda, \mu > 0$, being its most important special case. The values of the stress are restricted by a condition

$$\boldsymbol{\sigma}(t) \in \boldsymbol{\Sigma}(t)\,, \tag{3.5}$$

where the sets $\boldsymbol{\Sigma}(t)$ are closed and convex subsets of $\mathbf{T}$. The notation '$\boldsymbol{\Sigma}(t)$' is meant to signify that in general the set of admissible stress values changes in time, not that it is an a priori given function; indeed, it will be determined implicitly by the evolution of the internal variables as described below.

For a large class of constitutive laws, the plastic flow is governed by the *normality rule*

$$\dot{\varepsilon}^p(t) \in N_{\Sigma(t)}(\sigma(t)), \tag{3.6}$$

or its equivalent formulation, the so-called *maximal dissipation principle*

$$\dot{\varepsilon}^p(t) : (\tau - \sigma(t)) \leq 0, \tag{3.7}$$

for all $\tau \in \Sigma(t)$. We only consider models where the plastic strain is volume invariant, that is, its trace $\operatorname{tr} \varepsilon^p = \sum_{i=1}^3 \varepsilon_{ii}^p$ satisfies

$$\operatorname{tr} \varepsilon^p = 0. \tag{3.8}$$

It is guaranteed that the evolution governed by (3.6) does not violate (3.8) if $\operatorname{tr} \varepsilon^p(t_0) = 0$ holds initially and if $\Sigma(t)$ has the form

$$\Sigma(t) = \Sigma_d(t) \oplus \mathbf{T}_d^\perp, \tag{3.9}$$

where $\Sigma_d(t)$ is a subset of the *deviatoric space* $\mathbf{T}_d$,

$$\mathbf{T}_d = \{\tau : \tau \in \mathbf{T},\ \operatorname{tr} \tau = 0\}, \tag{3.10}$$

$$\mathbf{T}_d^\perp = \{\tau | \ \tau : \tilde{\tau} = 0 \text{ for all } \tilde{\tau} \in \mathbf{T}_d\} = \{\tau | \ \tau = \alpha \mathbf{1},\ \alpha \in \mathbb{R}\}. \tag{3.11}$$

As usual, by

$$\tau_d = \tau - \frac{1}{3}(\operatorname{tr} \tau)\mathbf{1} \tag{3.12}$$

we denote the *deviatoric part* τ_d of the tensor $\tau \in \mathbf{T}$, which is just the orthogonal projection of τ to the subspace $\mathbf{T}_d$. Indeed, in the case (3.9) the maximal dissipation principle (3.7) is equivalent to

$$\operatorname{tr} \dot{\varepsilon}^p(t) = 0, \quad \dot{\varepsilon}^p(t) : (\tau - \sigma_d(t)) \leq 0, \quad \text{for all } \tau \in \Sigma_d(t). \tag{3.13}$$

The boundary $\partial\Sigma(t)$ of the admissible region is called the *yield surface*, the interior $\operatorname{int} \Sigma(t)$ where necessarily $\dot{\varepsilon}^p = 0$ is called the *elastic region*.

The Prandtl-Reuss model. This model is the basic one in elastoplasticity. Here, $\Sigma(t) \equiv \Sigma$, that is, the yield surface $\partial\Sigma$ does not change during the evolution, and $0 \in \operatorname{int} \Sigma$. Since we can write the decomposition (3.2) as

$$\mathbf{a} : \varepsilon = \mathbf{a} : \varepsilon^p + \mathbf{a} : \varepsilon^e = \mathbf{a} : \varepsilon^p + \sigma, \tag{3.14}$$

and because we can equivalently rewrite the maximal dissipation principle as

$$(\mathbf{A}:\mathbf{a}:\dot{\boldsymbol{\varepsilon}}^p(t)):(\boldsymbol{\tau}-\boldsymbol{\sigma}(t)) \leq 0, \quad \text{for all } \boldsymbol{\tau}\in\boldsymbol{\Sigma}, \tag{3.15}$$

we can express the Prandtl-Reuss model in terms of the stop operator from the previous section as

$$\boldsymbol{\sigma} = \mathcal{S}_Z(\mathbf{a}:\boldsymbol{\varepsilon}\,;\boldsymbol{\sigma}(t_0)), \tag{3.16}$$

if we set $H = \mathbf{T}$, $Z = \boldsymbol{\Sigma}$ and choose the scalar product $\langle\boldsymbol{\sigma},\boldsymbol{\tau}\rangle = (\mathbf{A}:\boldsymbol{\sigma}):\boldsymbol{\tau}$. The initial value is defined by $\boldsymbol{\sigma}(t_0) = \mathbf{a}:(\boldsymbol{\varepsilon}(t_0) - \boldsymbol{\varepsilon}^p(t_0))$, the initial memory is represented by a given value of $\boldsymbol{\varepsilon}^p(t_0)$. Note however that, although it is safe to assume that $\boldsymbol{\Sigma}_d(t)$ is bounded at any fixed time $t \in I$, the set $\boldsymbol{\Sigma}(t)$ is given in (3.9) as the infinite cylinder with the base $\boldsymbol{\Sigma}_d(t)$ and thus is unbounded, so the results presented in the previous section do not apply directly. One way out of this difficulty is to extend the results of the previous section to the case of an unbounded Z, in particular to unbounded cylinders with bounded base; Krejčí has done that in [27]. This is not necessary if the elastic law leaves both $\mathbf{T}_d$ and $\mathbf{T}_d^\perp$ invariant, that is, if

$$(\mathbf{a}:\boldsymbol{\tau})_d = \mathbf{a}:\boldsymbol{\tau}_d \tag{3.17}$$

holds for all $\boldsymbol{\tau}\in\mathbf{T}$, because in that case we can use the decomposition $\mathbf{a}:\boldsymbol{\varepsilon}_d = \boldsymbol{\sigma}_d + \mathbf{a}:\boldsymbol{\varepsilon}^p$ instead of (3.14) and obtain the Prandtl-Reuss model as

$$\boldsymbol{\sigma} = \mathcal{S}_Z(\mathbf{a}:\boldsymbol{\varepsilon}_d\,;\boldsymbol{\sigma}_d(t_0)) + \mathbf{a}:(\boldsymbol{\varepsilon} - \boldsymbol{\varepsilon}_d) \tag{3.18}$$

with $H = \mathbf{T}_d$ and $Z = \boldsymbol{\Sigma}_d$. Hooke's law satisfies (3.17); it actually decomposes over $\mathbf{T}_d \oplus \mathbf{T}_d^\perp$ into (different) scalar multiples of the identity as

$$\mathbf{a}:\boldsymbol{\tau} = 2\mu\boldsymbol{\tau}_d + (2\mu + 3\lambda)(\boldsymbol{\tau} - \boldsymbol{\tau}_d). \tag{3.19}$$

The operator formulation (3.16) respectively (3.18) refers to the strain-controlled form of the Prandtl-Reuss model. It is impossible to invert it to stress-controlled form, since there exist nonconstant strain paths during plastic flow along which the stress does not change – just note that for the stop operator $z = \mathcal{S}(v\,;z^0)$, if $\dot{v} \in N_Z(z)$, it follows that $|\dot{z}|^2 = \langle\dot{v},\dot{z}\rangle \leq 0$, hence $\dot{z} = 0$.

Von Mises criterion. The admissible set $\boldsymbol{\Sigma}$ and the yield surface $\partial\boldsymbol{\Sigma}$ are usually specified as level sets $\{f \leq 0\}$ respectively $\{f = 0\}$ of the so-called *yield function*

f. The most important special case is the *von Mises criterion* where Σ_d is a ball in deviatoric space,

$$f(\tau) = \frac{3}{2}\tau_d : \tau_d - \sigma_s^2 , \tag{3.20}$$

with the *yield stress* $\sigma_s > 0$. The factor 3/2 arises in order that the value of σ_s equals the yield stress in pure tension, that is, it holds $f(\tau) = 0$ if we choose $\tau \in \mathbf{T}$ as $\tau_{11} = \sigma_s$, $\tau_{ij} = 0$ otherwise.

Isotropic hardening. We now discuss models where the yield surface does not remain constant. In a class of models of what is called *isotropic hardening*, the yield surface increases in size in dependence on some internal variable. We obtain such a behaviour in the von Mises case if we generalize (3.20) to

$$f(\tau) = \frac{3}{2}\tau_d : \tau_d - k , \tag{3.21}$$

and assume, for example, that $k = k(p)$ is a nondecreasing function of the *accumulated plastic strain*

$$p(t) = \int_{t_0}^{t} \left(\frac{2}{3}\dot{\varepsilon}^p(s) : \dot{\varepsilon}^p(s)\right)^{\frac{1}{2}} ds \tag{3.22}$$

with $k(0) = \sigma_s^2$.

Kinematic hardening. The Melan-Prager model. In a different class of so-called *kinematic hardening* models the admissible set Σ_d and the yield surface $\partial\Sigma_d$ undergo a translational movement $\Sigma_d(t) = \sigma^b(t) + \Sigma_0$, specified by an additional internal variable $\sigma^b \in \mathbf{T}_d$ usually called the *backstress* (denoted by $\mathbf{X}$ in [28]). In the basic model of *linear kinematic hardening*, which goes back to Melan and Prager, one sets

$$\dot{\sigma}^b = C_0 \dot{\varepsilon}^p , \tag{3.23}$$

with some constant $C_0 > 0$. Due to (3.23) the maximal dissipation principle (3.7) in deviatoric space can be equivalently written for the Melan-Prager model as

$$\dot{\sigma}^b(t) : (\tau - (\sigma_d(t) - \sigma^b(t)) \le 0 , \quad \text{for all } \tau \in \Sigma_0 . \tag{3.24}$$

The obvious decomposition $\sigma_d = \sigma^b + (\sigma_d - \sigma^b)$ yields

$$\sigma^b = \mathcal{P}_Z(\sigma_d ; \sigma^0) , \tag{3.25}$$

with the initial value $\boldsymbol{\sigma}^0 = \boldsymbol{\sigma}_d(t_0) - \boldsymbol{\sigma}^b(t_0)$ and a given initial memory $\boldsymbol{\sigma}^b(t_0) = \boldsymbol{\sigma}_0^b \in \mathbf{T}_d$. Here, the play operator is defined with $H = \mathbf{T}_d$, $Z = \Sigma_0$ and the standard scalar product $\langle \boldsymbol{\sigma}, \boldsymbol{\tau} \rangle = \boldsymbol{\sigma} : \boldsymbol{\tau}$. Thus, in the von Mises case we obtain $Z = B_r(0)$ with $r = \sqrt{2/3}\,\sigma_s$, the corresponding yield function being

$$f(\boldsymbol{\tau}) = \frac{3}{2}(\boldsymbol{\tau}_d - \boldsymbol{\sigma}^b) : (\boldsymbol{\tau}_d - \boldsymbol{\sigma}^b) - \sigma_s^2 . \tag{3.26}$$

The operator form of the stress-controlled Melan-Prager model therefore becomes

$$\boldsymbol{\varepsilon} = \boldsymbol{\varepsilon}^e + \boldsymbol{\varepsilon}^p = \mathbf{a} : \boldsymbol{\sigma} + \frac{1}{C_0}\mathcal{P}(\boldsymbol{\sigma}_d ; \boldsymbol{\sigma}^0) , \tag{3.27}$$

and the results of the foregoing section apply. Its inverse, the strain-controlled Melan-Prager model, can also be derived directly in the case of Hooke's law, where $\boldsymbol{\sigma}_d = 2\mu\boldsymbol{\varepsilon}_d^e$. Let us assume for simplicity that all initial values are zero, then in particular $\boldsymbol{\sigma}^b = C_0\boldsymbol{\varepsilon}^p$ and the decomposition

$$2\mu\boldsymbol{\varepsilon}_d = (C_0 + 2\mu)\boldsymbol{\varepsilon}^p + (\boldsymbol{\sigma}_d - \boldsymbol{\sigma}^b) \tag{3.28}$$

holds. The maximal dissipation principle now yields (we assume zero initial values for $\mathcal{P}$ and $\mathcal{S}$)

$$\boldsymbol{\varepsilon}^p = \frac{1}{C_0 + 2\mu}\mathcal{P}(2\mu\boldsymbol{\varepsilon}_d) , \quad \boldsymbol{\sigma}^b = \frac{C_0}{C_0 + 2\mu}\mathcal{P}(2\mu\boldsymbol{\varepsilon}_d) , \tag{3.29}$$

so

$$\begin{aligned} \boldsymbol{\sigma}_d &= \boldsymbol{\sigma}^b + (\boldsymbol{\sigma}_d - \boldsymbol{\sigma}^b) = \frac{C_0}{C_0 + 2\mu}\mathcal{P}(2\mu\boldsymbol{\varepsilon}_d) + \mathcal{S}(2\mu\boldsymbol{\varepsilon}_d) \\ &= \frac{C_0}{C_0 + 2\mu}2\mu\boldsymbol{\varepsilon}_d + \frac{2\mu}{C_0 + 2\mu}\mathcal{S}(2\mu\boldsymbol{\varepsilon}_d) . \end{aligned} \tag{3.30}$$

Thus,

$$\boldsymbol{\sigma} = \boldsymbol{\sigma}_d + (2\mu + 3\lambda)(\boldsymbol{\varepsilon} - \boldsymbol{\varepsilon}_d) , \tag{3.31}$$

with $\boldsymbol{\sigma}_d$ obtained from (3.30), represents the strain-controlled Melan-Prager model for Hooke's law.

Nonlinear kinematic hardening. While a combination of isotropic and linear kinematic hardening gives quite reasonable results if the loading is nearly monotone, it is usually no longer adequate if the loading is cyclic or even more arbitrary, see [28, p.212]. One important shortcoming (which is corrected somewhat, but in an insufficient manner, by an additional isotropic hardening term) of the Melan-Prager model is the following. In the scalar case $H = \mathbb{R}$, the stop and play operator map time-periodic functions with period T on some interval $[t_0, \infty)$ to functions whose restriction to $[t_0 + T, \infty)$ is also T-periodic. As a consequence, under proportional cyclic loading the Melan-Prager model exhibits closed hysteresis loops from the second cycle on. In contrast to that, one often observes the following behaviour, called *ratchetting*. If we perform unsymmetric proportional cyclic (say, stress-controlled) loading, where $\boldsymbol{\sigma}(t)$ cycles between $\sigma_{min}\tilde{\boldsymbol{\sigma}}$ and $\sigma_{max}\tilde{\boldsymbol{\sigma}}$, with $|\sigma_{min}| \neq |\sigma_{max}|$ and given $\tilde{\boldsymbol{\sigma}} \in \mathbf{T}$, then there will be a non-zero plastic strain increment $\boldsymbol{\varepsilon}^p(t+T) - \boldsymbol{\varepsilon}^p(t)$ in each cycle (assuming, of course, that we do not remain within the elastic region).

We describe some of the modifications and extensions of the Melan-Prager model which have been developed in order to cope, among other things, with the phenomenon of ratchetting. They can be termed summarily as *nonlinear kinematic hardening*; note however, that this name also often is used to refer to the specific models of Armstrong and Frederick [2] and Chaboche [28, p.233f.].

The Armstrong-Frederick model. Here, the equation (3.23) for the backstress is replaced by

$$\dot{\boldsymbol{\sigma}}^b = \frac{2}{3} C \dot{\boldsymbol{\varepsilon}}^p - \gamma \boldsymbol{\sigma}^b \dot{p} , \tag{3.32}$$

where $\dot{p} = \sqrt{2/3}|\dot{\boldsymbol{\varepsilon}}^p|$ denotes the rate of accumulated plastic strain from (3.22), and C and γ are positive constants. As before, it is combined with

$$\boldsymbol{\varepsilon} = \boldsymbol{\varepsilon}^e + \boldsymbol{\varepsilon}^p , \quad \boldsymbol{\sigma} = \mathbf{a} : \boldsymbol{\varepsilon}^e , \tag{3.33}$$

$$\dot{\boldsymbol{\varepsilon}}^p(t) : (\boldsymbol{\tau} - (\boldsymbol{\sigma}_d(t) - \boldsymbol{\sigma}^b(t)) \leq 0 , \quad \text{for all } \boldsymbol{\tau} \in \Sigma_0 . \tag{3.34}$$

To specify the stress-strain behaviour of the model completely, initial values for the memory variables have to be suitably prescribed.

In order to understand the mathematical properties of the Armstrong-Frederick model, we transform it to a system which includes the play and the stop operator. At first, let us discuss the stress-controlled case. We introduce the auxiliary variable $\boldsymbol{\chi}$ in the deviatoric stress space by

$$\boldsymbol{\chi} = \frac{2}{3} C \boldsymbol{\varepsilon}^p + (\boldsymbol{\sigma}_d - \boldsymbol{\sigma}^b) . \tag{3.35}$$

Again, the maximal dissipation principle implies

$$\frac{2}{3}C\boldsymbol{\varepsilon}^p = \mathcal{P}_Z(\boldsymbol{\chi};\boldsymbol{\sigma}^0), \quad \boldsymbol{\sigma}_d - \boldsymbol{\sigma}^b = \mathcal{S}_Z(\boldsymbol{\chi};\boldsymbol{\sigma}^0), \quad \boldsymbol{\sigma}^0 = \boldsymbol{\sigma}_d(t_0) - \boldsymbol{\sigma}^b(t_0), \tag{3.36}$$

where $Z = \boldsymbol{\Sigma}_0 \subset \mathbf{T}_d$ as before. If we form the time derivative of (3.35) and insert (3.32), we have $\dot{\boldsymbol{\chi}} = \dot{\boldsymbol{\sigma}}_d + \gamma\boldsymbol{\sigma}^b\dot{p}$, and from (3.36) we obtain

$$\dot{\boldsymbol{\chi}} = \dot{\boldsymbol{\sigma}}_d + \sqrt{\frac{3}{2}}\frac{\gamma}{C}(\boldsymbol{\sigma}_d - \mathcal{S}(\boldsymbol{\chi};\boldsymbol{\sigma}^0))\left|\frac{\mathrm{d}}{\mathrm{dt}}\mathcal{P}(\boldsymbol{\chi};\boldsymbol{\sigma}^0)\right|, \quad \boldsymbol{\chi}(t_0) = \frac{2}{3}C\boldsymbol{\varepsilon}^p(t_0) + \boldsymbol{\sigma}^0. \tag{3.37}$$

Given $\boldsymbol{\sigma}_d : I \to \mathbf{T}_d$, we can hope to solve (3.37) uniquely for $\boldsymbol{\chi} : I \to \mathbf{T}_d$ if we fix initial conditions

$$\boldsymbol{\sigma}^b(t_0) = \boldsymbol{\sigma}_0^b, \quad \boldsymbol{\varepsilon}^p(t_0) = \boldsymbol{\varepsilon}_0^p, \tag{3.38}$$

for the memory variables. Once we know $\boldsymbol{\chi}$, we obtain $\boldsymbol{\varepsilon}$ from

$$\boldsymbol{\varepsilon} = \boldsymbol{\varepsilon}^e + \boldsymbol{\varepsilon}^p = \mathbf{a}:\boldsymbol{\sigma} + \frac{3}{2C}\mathcal{P}(\boldsymbol{\chi};\boldsymbol{\sigma}^0). \tag{3.39}$$

Setting $\gamma = 0$, we recover the Melan-Prager model for which (3.37) reduces to the trivial equation $\dot{\boldsymbol{\chi}} = \dot{\boldsymbol{\sigma}}_d$; in the Armstrong-Frederick case $\gamma > 0$, the variable $\boldsymbol{\chi}$ becomes a separate mathematical entity to which, however, we are unable to attach a specific mechanical interpretation.

An analogous procedure can be carried out in the strain-controlled case if we assume Hooke's law for the linear elastic part. The auxiliary variable $\boldsymbol{\chi}$ is now chosen as

$$\boldsymbol{\chi} = \left(\frac{2}{3}C + 2\mu\right)\boldsymbol{\varepsilon}^p + (\boldsymbol{\sigma}_d - \boldsymbol{\sigma}^b). \tag{3.40}$$

Using

$$\left(\frac{2}{3}C + 2\mu\right)\boldsymbol{\varepsilon}^p = \mathcal{P}(\boldsymbol{\chi};\boldsymbol{\sigma}^0), \tag{3.41}$$

the differential equation for $\boldsymbol{\chi}$ becomes

$$\dot{\boldsymbol{\chi}} = 2\mu\dot{\boldsymbol{\varepsilon}}_d + \sqrt{\frac{3}{2}}\frac{\gamma}{C + 3\mu}\boldsymbol{\sigma}^b\left|\frac{\mathrm{d}}{\mathrm{dt}}\mathcal{P}(\boldsymbol{\chi};\boldsymbol{\sigma}^0)\right|. \tag{3.42}$$

Here, the elimination of $\boldsymbol{\sigma}^b$ is less straightforward than in the stress-controlled case. One can replace it with an expression involving only $\boldsymbol{\varepsilon}$, $\boldsymbol{\chi}$ and $\mathcal{P}(\boldsymbol{\chi};\boldsymbol{\sigma}^0)$, using (3.41) as well as the identity

$$\boldsymbol{\sigma}^b = \frac{2}{3}C\boldsymbol{\varepsilon}^p + 2\mu\boldsymbol{\varepsilon}_d - \boldsymbol{\chi}, \tag{3.43}$$

One can also proceed as described below (3.70) for the Chaboche model. In any case, we end up – as before in the stress-controlled situation – with an initial value problem for χ whose right hand side includes the term $|(d/dt)\mathcal{P}(\chi\,;\sigma^0)|$. Due to this distinctive nonstandard feature, such initial value problems require a specific analysis. This analysis has been carried out in [8] and [9] for the special case $Z = B_r(0)$ corresponding to the von Mises criterion. We now present it in a somewhat simplified form.

The abstract initial value problem. Within the Hilbert space H, which from now on we assume to be finite-dimensional, we consider the initial value problem

$$\dot{v} = \dot{\theta} + \mathcal{M}(\theta, v; q)|\dot{w}|\,, \quad w = \mathcal{P}(v\,; z^0)\,, \quad v(t_0) = v^0\,, \tag{3.44}$$

where $\mathcal{P} = \mathcal{P}_Z$ denotes the play operator with $Z = B_r(0)$, $r > 0$, and $\mathcal{M}$ is a causal operator. The additional parameter q has values in some metric space Q; its purpose is to represent all the initial memory (of which z^0 is a part). For any given input function $\theta \in W^{1,1}(I; H)$, initial values $z^0 \in Z$ and $v^0 \in H$, and parameter value $q \in Q$, we look for a solution $v \in W^{1,1}(I; H)$ of the associated integral equation

$$v(t) = v^0 + \theta(t) - \theta(t_0) + \int_{t_0}^{t} \mathcal{M}(\theta, v\,; q)(s)|\dot{w}(s)|\, ds\,, \quad w = \mathcal{P}(v\,; z^0)\,. \tag{3.45}$$

One observes that 1 is a critical value for $|\mathcal{M}|$ in (3.44). Indeed, if $\dot{\theta} > 0$ the scalar equation $\dot{v} = \dot{\theta} + |\dot{w}|$ does not possess solutions with $\dot{w} \neq 0$, since that would imply $\dot{v} > |\dot{w}| = |\dot{v}|$. Accordingly, the condition

$$\| \mathcal{M}(\theta, v\,; q) \|_\infty \leq 1 - \kappa \tag{3.46}$$

with some $\kappa > 0$ plays a crucial role in the discussion of wellposedness of (3.44). Condition (3.46) implies the a priori estimate

$$\kappa|\dot{v}(t)| \leq |\dot{\theta}(t)|\,, \quad \text{for a.a. } t \in I\,, \tag{3.47}$$

for every solution of (3.44), recall that $|\dot{w}| \leq |\dot{v}|$. A slightly weaker, nonlocal form of (3.47) is given by

$$\kappa\beta_v \leq \beta_\theta\,, \tag{3.48}$$

that is, $\kappa\beta_v(\delta) \leq \beta_\theta(\delta)$ for all $\delta > 0$, where $\beta_v : [0, \infty) \to [0, \infty)$ denotes the modulus of absolute integrability

$$\beta_v(\delta) = \sup\{\int_F |\dot{v}(s)|\, ds : F \subset I \text{ measurable}\,,\ \text{meas}\,(F) \leq \delta\} \tag{3.49}$$

of the function $v \in W^{1,1}(I;H)$. These considerations motivate the appearance of the set

$$V_{\kappa,\theta,v^0} = \{v : v \in W^{1,1}(I;H),\ \kappa\beta_v \le \beta_\theta,\ v(t_0) = v^0\}\,, \tag{3.50}$$

in the following existence theorem whose proof relies on the fact that V_{κ,θ,v^0} is relatively compact in the weak topology of $W^{1,1}(I;H)$ by the Dunford-Pettis theorem. We assume the operator $\mathcal{M}$ to be causal, that is, for all $t \in I$ it holds $\mathcal{M}(\theta_1, v_1\,; q) = \mathcal{M}(\theta_2, v_2\,; q)$ on $[t_0, t]$ whenever $\theta_1 = \theta_2$ and $v_1 = v_2$ on $[t_0, t]$.

Theorem 3.10 *Let $\theta \in \Theta \subset W^{1,1}(I;H)$, $v^0 \in H$, $z^0 \in Z = B_r(0)$ and $q \in Q$ be given. Assume that for some given $\kappa > 0$, the operator*

$$\mathcal{M} : \Theta \times V_{\kappa,\theta,v^0} \times Q \to W^{1,1}(I;H) \tag{3.51}$$

is causal, satisfies (3.46) for all $v \in V_{\kappa,\theta,v^0}$, and that $\mathcal{M}(\theta, v_n\,; q) \to \mathcal{M}(\theta, v\,; q)$ uniformly if $v_n \to v$ uniformly in V_{κ,θ,v^0}. Then there exists a solution $v_* \in V_{\kappa,\theta,v^0}$ of (3.45).

Proof. We use the method of the retarded argument. For $\rho > 0$ we introduce the shift operator $\tau^\rho : W^{1,1}(I;H) \to W^{1,1}(I;H)$ by the formula

$$(\tau^\rho v)(t) = \begin{cases} v(t-\rho)\,, & t \in [t_0 + \rho, t_f]\,, \\ v(t_0)\,, & t \in [t_0, t_0 + \rho]\,. \end{cases} \tag{3.52}$$

We consider the integral equation for the unknown function v_ρ

$$v_\rho(t) = v^0 + \theta(t) - \theta(t_0) + \int_{t_0}^{t} \mathcal{M}(\theta, \tau^\rho v_\rho\,; q)(s)|\dot{w}_\rho(s)|\,ds\,, \quad w_\rho = \mathcal{P}(\tau^\rho v_\rho\,; z^0)\,. \tag{3.53}$$

We can construct its unique solution $v_\rho \in V_{\kappa,\theta,v^0}$ successively on the intervals $I_k = [t_0, t_0 + k\rho]$, because $\kappa\beta_{v_\rho} \le \beta_\theta$ on $I_k = [t_0, t_0 + k\rho]$ implies by assumption on $\mathcal{M}$ that $|\mathcal{M}(\theta, \tau^\rho v_\rho\,; q)| \le 1 - \kappa$ on I_{k+1}; for every measurable $F \subset I_{k+1}$ we then get

$$\int_F |\dot{v}_\rho(s)|\,ds \le \int_F |\dot{\theta}(s)|\,ds + (1-\kappa)\int_{F-\rho} |\dot{v}_\rho(s)|\,ds \tag{3.54}$$

which in turn implies that $\kappa\beta_{v_\rho} \le \beta_\theta$ holds on I_{k+1}. By virtue of the Dunford-Pettis and the Arzelà-Ascoli theorem, there exists a sequence $\rho_n \to 0$ and a function $v_* \in$

$W^{1,1}(I;H)$ such that $\dot{v}_{\rho_n} \to \dot{v}_*$ weakly in $L^1(I;H)$ and $v_{\rho_n} \to v_*$ uniformly. Moreover, it holds $\tau^{\rho_n} v_{\rho_n} \to v_*$ uniformly. By Theorem 2.7, the sequence $\{W_n\}$ defined by

$$W_n(t) = \int_{t_0}^{t} |\dot{w}_{\rho_n}(s)|\, ds \tag{3.55}$$

converges pointwise (hence uniformly) to $W(t) = \int_{t_0}^{t} |d/ds\mathcal{P}(v_*\,; z^0)(s)|\, ds$. Due to the continuity of $\mathcal{M}$, we can apply a standard convergence theorem for Stieltjes integrals to pass to the limit $\rho_n \to 0$ for this subsequence in (3.53) and to conclude that v_* is a solution of (3.45) satisfying (3.46), therefore $v_* \in V_{\kappa,\theta,v^0}$. □

Theorem 3.11 below shows that the solution of (3.44) is unique and depends Lipschitz continuously in the norm of $W^{1,1}(I;H)$ upon the data, if $\mathcal{M}$ satisfies the Lipschitz condition

$$\| \mathcal{M}(\theta_1, v_1\,; q_1) - \mathcal{M}(\theta_1, v_1\,; q_1) \|_{\infty,t} \le K(\| \theta_1 - \theta_2 \|_{1,1,t} + \| v_1 - v_2 \|_{1,1,t} + |q_1 - q_2|)\,, \tag{3.56}$$

for all $t \in I$, where $\| \cdot \|_{\infty,t}$ and $\| \cdot \|_{1,1,t}$ denote the norms of C and $W^{1,1}$ as before, but restricted to the interval $[t_0, t]$. (Except for unusual circumstances, due to the causality of $\mathcal{M}$ it actually suffices to require (3.56) to hold for $t = t_f$ only.) Note that a loss of time regularity is allowed in (3.56); it will be compensated for in the proof of the following theorem by a judicious use of the estimates for the play and the stop from the preceding section.

Theorem 3.11 *Let $v_1, v_2 \in V_{\kappa,\theta,v^0}$ be solutions of the initial value problem (3.44) to the data $\theta_1, \theta_2 \in \Theta$, $u_1^0, u_2^0 \in H$, $q_1, q_2 \in Q$ and $z_1^0, z_2^0 \in B_r(0)$. If, in addition to the assumptions of Theorem 3.10,*

$$\| \mathcal{M}(\theta_1, v_1\,; q_1) - \mathcal{M}(\theta_1, v_1\,; q_1) \|_{\infty,t} \le K(\| \theta_1 - \theta_2 \|_{1,1,t} + \| v_1 - v_2 \|_{1,1,t} + |q_1 - q_2|)\,, \tag{3.57}$$

holds for all $t \in I$, then there holds

$$\| v_1 - v_2 \|_{1,1} \le L(\| \theta_1 - \theta_2 \|_{1,1} + |u_1^0 - u_2^0| + |z_1^0 - z_2^0| + |q_1 - q_2|)\,, \tag{3.58}$$

and the Lipschitz constant L depends only on K, κ, r and $c = \max\{\| \theta_1 \|_{1,1}, \| \theta_2 \|_{1,1}\}$.

Proof. We subtract the differential equations (3.44) for the two solutions, use the triangle inequality and the estimate $|\dot{w}_1| \le \|\dot{v}_1|$, and integrate over $[t_0, t]$ to obtain the estimate

$$\int_{t_0}^{t} |\dot{v}_1(s) - \dot{v}_2(s)|\,ds \;\le\; \int_{t_0}^{t} |\dot{\theta}_1(s) - \dot{\theta}_2(s)|\,ds + (1-\kappa)\int_{t_0}^{t} |\dot{w}_1(s) - \dot{w}_2(s)|\,ds$$
$$+ \int_{t_0}^{t} |\dot{v}_1(s)||\mathcal{M}(\theta_1, v_1\,; q_1)(s) - \mathcal{M}(\theta_1, v_1\,; q_1)(s)|\,ds\,. \quad (3.59)$$

We estimate the difference $\int |\dot{w}_1 - \dot{w}_2|$ by (2.63) and the difference $|z_1 - z_2| = |\mathcal{S}(v_1) - \mathcal{S}(v_2)|$ appearing there by (2.23) and use the inequality $\kappa|\dot{v}_1| \le |\dot{\theta}_1|$ to arrive at

$$\int_{t_0}^{t} |\dot{w}_1(s) - \dot{w}_2(s)|\,ds \;\le\; \Big(1 + \frac{\sqrt{2}}{r\kappa}\int_{t_0}^{t} |\dot{\theta}_1(s)|\,ds\Big)|z_1^0 - z_2^0| + \int_{t_0}^{t} |\dot{v}_1(s) - \dot{v}_2(s)|\,ds$$
$$+ \frac{\sqrt{2}}{r\kappa}\int_{t_0}^{t} |\dot{\theta}_1(s)| \int_{t_0}^{s} |\dot{v}_1(\tau) - \dot{v}_2(\tau)|\,d\tau\,ds\,. \quad (3.60)$$

We insert (3.60) and (3.57) and obtain, after some rearrangements,

$$\int_{t_0}^{t} |\dot{v}_1(s) - \dot{v}_2(s)|\,ds \le B + c_1\int_{t_0}^{t} |\dot{\theta}_1(s)| \int_{t_0}^{s} |\dot{v}_1(\tau) - \dot{v}_2(\tau)|\,d\tau\,ds\,, \quad (3.61)$$

where $B = c_2(\|\,\theta_1 - \theta_2\,\|_{1,1} + |u_1^0 - u_2^0| + |z_1^0 - z_2^0| + |q_1 - q_2|)$ with a constant c_2 depending only on K, κ, r and c as required. By (3.61), the expression $d(t) = c_1\int_{t_0}^{t} |\dot{\theta}_1| \int_{t0}^{s} |\dot{v}_1 - \dot{v}_2| d\tau ds$ satisfies the differential inequality

$$\dot{d}(t) \le c_1|\dot{\theta}_1(t)|(B + d(t))\,. \quad (3.62)$$

Gronwall's inequality yields the explicit estimate $|d(t)| \le B(\exp\int_{t_0}^{t} c_1|\dot{\theta}_1|ds - 1)$, which we insert into the right hand side of (3.61) to obtain the assertion. □

Application to nonlinear kinematic hardening. The Armstrong-Frederick equation (3.32) implies that its solutions are bounded by

$$|\boldsymbol{\sigma}^b(t)| \le \sqrt{\frac{2}{3}\frac{C}{\gamma}} \quad (3.63)$$

no matter what the specific loading history is. This follows from the estimate

$$\frac{\mathrm{d}}{\mathrm{dt}}\frac{1}{2}|\boldsymbol{\sigma}^b(t)|^2 = \langle \dot{\boldsymbol{\sigma}}^b(t), \boldsymbol{\sigma}^b(t)\rangle \le |\dot{\varepsilon}^p(t)||\boldsymbol{\sigma}^b(t)|\left(\frac{2}{3}C - \sqrt{\frac{2}{3}}\gamma|\boldsymbol{\sigma}^b(t)|\right), \quad (3.64)$$

which shows that $(d/dt)|\boldsymbol{\sigma}(t)|$ would have to be negative if the bound (3.63) would be exceeded. For the special case of monotone loading in pure tension, the scalar problem resulting from the Armstrong-Frederick model (3.32) – (3.34) can be solved exactly, see [28, p.222], to show that the plastic strain ε^p tends to infinity as σ^b approaches the bound (3.63). Since the model possesses spherical symmetry, the same is true for any other monotone proportional loading.

One can derive the wellposedness of the Armstrong-Frederick model from Theorems 3.10 and 3.11 as follows. In the strain-controlled case, we set $v = \chi$, $H = \mathbf{T}_d$, $q = z^0 = \sigma^0$ and, according to (3.42) – (3.41), $\theta = 2\mu\varepsilon_d$ and

$$\mathcal{M}(\theta, v\,; q) = \sqrt{\frac{3}{2}}\frac{\gamma}{C+3\mu}\sigma^b\,, \quad \sigma^b = \frac{C}{C+3\mu}\mathcal{P}(\chi\,; \sigma^0) + 2\mu\varepsilon_d - \chi\,. \tag{3.65}$$

From the bound (3.63) we expect the bound

$$\|\,\mathcal{M}(\theta, v\,; q)\,\|_\infty \leq 1 - \kappa\,, \quad \kappa = \frac{3\mu}{C+3\mu}\,, \tag{3.66}$$

to hold globally. Moreover, the continuity properties of $\mathcal{M}$ required in Theorems 3.10 and 3.11 follow from the corresponding properties of the play and the stop as presented in Section 2. Indeed, a formal argument can be carried out to prove – see Theorem 3.11 in [8] for a more detailed statement, in particular concerning the initial conditions – the wellposedness of the model with respect to the $W^{1,1}$-norm.

Theorem 3.12 *The strain-controlled Armstrong-Frederick model with the von Mises yield criterion is wellposed in the space* $W^{1,1}(I; \mathbf{T})$.

Let us emphasize that this theorem, as well as the ones below, encompasses arbitrary multiaxial loading histories, no matter how complex their form is, the only restriction being the assumption of absolute continuity which is just a matter of mathematical technique (and, moreover, can be relaxed further, see [9]).

In the stress-controlled case one has to account for the singularity $\varepsilon^p \to \infty$ as σ^b approaches the bound (3.63). Indeed, if we proceed as before and set, according to (3.35), $v = \chi$, $H = \mathbf{T}_d$, $q = z^0 = \sigma^0$ and

$$\mathcal{M}(\theta, v\,; q) = \sqrt{\frac{3}{2}}\frac{\gamma}{C}\sigma^b\,, \quad \sigma^b = \sigma_d - \mathcal{S}(\chi\,; \sigma^0)\,, \tag{3.67}$$

the bound (3.63) yields $\|\,\mathcal{M}\,\|_\infty \leq 1$ which is not good enough. On the other hand, since the difference $\sigma_d - \sigma^b$ cannot exceed the yield stress, we can expect wellposedness only for loads which satisfy

$$\|\,\sigma_d\,\|_\infty < \sqrt{\frac{2}{3}}\left(\sigma_s + \frac{C}{\gamma}\right)\,. \tag{3.68}$$

It is shown in [8] how to use (3.68) to derive the necessary a priori estimate $\| \mathcal{M} \|_\infty \leq 1 - \kappa$ locally in time, with the following result.

Theorem 3.13 *The stress-controlled Armstrong-Frederick model with the von Mises yield criterion is wellposed in the space $W^{1,1}(I; \mathbf{T})$ subject to the restriction (3.68).*

The Armstrong-Frederick equation serves as a basis from which several extensions and modifications have been developed. Probably the most widespread and famous of those is the multilayer formulation, the model of Chaboche. Here, the backstress $\boldsymbol{\sigma}^b$ is decomposed as

$$\boldsymbol{\sigma}^b = \sum_{j=1}^{J} \boldsymbol{\sigma}_j^b + \boldsymbol{\sigma}_L^b , \tag{3.69}$$

where the components $\boldsymbol{\sigma}_j^b$ satisfy the Armstrong-Frederick equation

$$\dot{\boldsymbol{\sigma}}_j^b = \frac{2}{3} C_j \dot{\boldsymbol{\varepsilon}}^p - \gamma_j \boldsymbol{\sigma}_j^b \dot{p} , \tag{3.70}$$

with positive constants C_j and γ_j. The component $\boldsymbol{\sigma}_L^b$ represents a linear (Melan-Prager) element

$$\boldsymbol{\sigma}_L^b = C_0 \boldsymbol{\varepsilon}^p . \tag{3.71}$$

Again, the full model results from the combination with the linear elastic part and the maximal dissipation principle

$$\boldsymbol{\varepsilon} = \boldsymbol{\varepsilon}^e + \boldsymbol{\varepsilon}^p , \qquad \boldsymbol{\sigma} = \mathbf{a} : \boldsymbol{\varepsilon}^e , \tag{3.72}$$

$$\dot{\boldsymbol{\varepsilon}}^p(t) : (\boldsymbol{\tau} - (\boldsymbol{\sigma}_d(t) - \boldsymbol{\sigma}^b(t)) \leq 0 , \quad \textit{for all } \boldsymbol{\tau} \in \Sigma_0 . \tag{3.73}$$

The presence respectively absence (if $C_0 = 0$) of the Melan-Prager term $\boldsymbol{\sigma}_L^b$ results in a quite different qualitative behaviour of the model, already to be seen in proportional cyclic loading. In the 'pure' version $C_0 = 0$, and in particular in the original Armstrong-Frederick model, unsymmetric cyclic stress-controlled loading results in progressive ratchetting with asymptotically unbounded plastic strains. In contrast to that, the presence of the Melan-Prager element with $C_0 > 0$ eliminates the singularity of the stress-controlled Armstrong-Frederick model and enforces globally bounded plastic strains for bounded loading. It appears to be an open problem whether, for periodic loading, the periodic limit in the case $C_0 > 0$ is asymptotically stable; numerical simulations suggest this might be the case, see [23].

The wellposedness of the Chaboche model, too, can be derived from the wellposedness of the abstract evolution problem (3.44). For example, in the stress-controlled case we define the auxiliary variable as

$$\chi = \left(C_0 + \frac{2}{3} \sum_{j=1}^{J} C_j \right) \varepsilon^p + (\sigma_d - \sigma^b) . \tag{3.74}$$

The differential equation for χ becomes

$$\dot{\chi} = \dot{\sigma}_d + \sum_{j=1}^{J} \gamma_j \sigma_j^b \dot{p} . \tag{3.75}$$

In contrast to the single Armstrong-Frederick equation, a direct elimination of the backstresses σ_j^b appears no longer possible. Instead, one can use the variation of constants formula to represent the backstresses in terms of ε^p as

$$\sigma_j^b(t) = e^{-\gamma_j p(t)} \left[\sigma_0^b + \int_{t_0}^{t} e^{\gamma_j p(s)} \cdot \frac{2}{3} C_j \dot{\varepsilon}^p(s)\, ds \right] . \tag{3.76}$$

If we moreover substitute the derivative $\dot{\varepsilon}^p(s)$ according to the formula (valid for the von Mises criterion)

$$\dot{\varepsilon}^p = \frac{1}{r} |\dot{\varepsilon}^p| (\sigma_d - \sigma^b) = \frac{1}{r} |\dot{\varepsilon}^p| \mathcal{S}(\chi ; \sigma^0) , \quad r = \sqrt{\frac{2}{3}} \sigma_s , \tag{3.77}$$

we can express the backstresses σ_j^b in terms of χ, $\mathcal{P}(\chi ; \sigma^0)$ and $\mathcal{S}(\chi ; \sigma^0)$ in such a manner that the resulting operator has the continuity properties required of $\mathcal{M}$ in the general existence and uniqueness theorems. If $C_0 = 0$, in the stress controlled case one again has to respect the possible singularity which leads to the restriction

$$\| \sigma_d \|_\infty < \sqrt{\frac{2}{3}} \left(\sigma_s + \sum_{j=1}^{J} \frac{C_j}{\gamma_j} \right) . \tag{3.78}$$

The details are worked out in [10], where the following theorem is proved.

Theorem 3.14 *The Chaboche model with the von Mises yield criterion is wellposed in the space $W^{1,1}(I; \mathbf{T})$ both in the strain-controlled as in the stress-controlled case, the latter subject to the restriction (3.78) in the case $C_0 = 0$.*

One can also refine the Armstrong-Frederick model by modifying its nonlinear term, for example as

$$\dot{\boldsymbol{\sigma}}^b = \frac{2}{3} C \dot{\boldsymbol{\varepsilon}}^p - \gamma(\boldsymbol{\sigma}^b) \boldsymbol{\sigma}^b \dot{p} . \tag{3.79}$$

Here, γ is no longer a constant, but a function of $\boldsymbol{\sigma}^b$. Based on experimental observations and theoretical reasoning, several forms of γ as well as other variants have been discussed, see e.g. [13, 14, 15, 21, 22, 31, 33, 39]. Those which satisfy $0 \leq \gamma \leq 1$ can be interpreted as an interpolation between the Melan-Prager model given by $\gamma \equiv 0$ and the Armstrong-Frederick model which corresponds to $\gamma \equiv 1$. So far, no mathematical results – in the form of theorems and proofs – seem to have been published concerning these models; some work along the lines of this paper is in progress.

Let us note that the mathematical results for the Chaboche model remain true if we replace the sum $\sum_j \boldsymbol{\sigma}_j^b$ in (3.69) by an integral with respect to some measure over a measure space; this includes the case of a continuum of backstress components. For nonlinear kinematic hardening models of Mróz type, we face a different situation. While the original discrete formulation can be subsumed under the abstract evolution problem (3.44), see [8], the continuous 1-parameter version of the model presented in [16, 17] seems to require a completely different approach which is not based on a variational inequality formulation, see [6].

References

[1] J.P. AUBIN and A. CELLINA, *Differential inclusions*, Springer-Verlag, Berlin 1984.

[2] P.J. ARMSTRONG and C.O. FREDERICK, *A mathematical representation of the multiaxial Bauschinger effect*, C.E.G.B., Report RD/B/N 731, 1966.

[3] A.F. BOWER, Cyclic hardening properties of hard-drawn copper and rail steel, *J. Mech. Phys. Solids* **37** (1989), 455 – 470.

[4] H. BREZIS, Problèmes unilateraux, *J. Math. Pures et Appl.* **51** (1972), 1 – 168.

[5] H. BREZIS, *Opérateurs maximaux monotones et semi-groupes de contractions dans les espaces de Hilbert*, North-Holland, Amsterdam 1973.

[6] M. BROKATE, K. DRESSLER and P. KREJČÍ, On the Mróz model, *Eur. J. Appl. Math.* **7** (1996), 473 – 497.

[7] M. BROKATE, K. DRESSLER and P. KREJČÍ, Rainflow counting and energy dissipation for hysteresis models in elastoplasticity, *Eur. J. Mech. A/Solids* **15** (1996), 705 - 737.

[8] M. BROKATE and P. KREJČÍ, Wellposedness of kinematic hardening models in elastoplasticity, *Math. Model. Numer. Anal.*, to appear.

[9] M. BROKATE and P. KREJČÍ, Maximum norm wellposedness of nonlinear kinematic hardening models, submitted.

[10] M. BROKATE and P. KREJČÍ, On the wellposedness of the Chaboche model, to appear in the Proceedings of the 1996 Vorau conference on Control and Estimation of Distributed Parameter Systems, Birkhäuser.

[11] M. BROKATE and J. SPREKELS, *Hysteresis and phase transitions*, Springer-Verlag, Berlin 1996.

[12] O.T. BRUHNS, T. LEHMANN and A. PAPE, On the description of transient cyclic hardening behaviour of mild steel CK 15, *Int. J. Plasticity* **8** (1992), 331 - 359.

[13] J.-L. CHABOCHE, Constitutive equations for cyclic plasticity and cyclic viscoplasticity, *Int. J. Plasticity* **5** (1989), 247 - 302.

[14] J.-L. CHABOCHE, On some modifications of kinematic hardening to improve the description of ratchetting effects, *Int. J. Plasticity* **7** (1991), 661 - 678.

[15] J.-L. CHABOCHE, Modeling of ratchetting: evaluation of various approaches, *Eur. J. Mech., A/Solids* **13** (1994), 501 - 518.

[16] C.C. CHU, A three–dimensional model of anisotropic hardening in metals and its application to the analysis of sheet metal formability, *J. Mech. Phys. Solids* **32** (1984), 197 - 212.

[17] C.C. CHU, The analysis of multiaxial cyclic problems with an anisotropic hardening model, *Int. J. Solids Structures* **23** (1987), 567 - 579.

[18] G. DUVAUT and J.-L. LIONS, *Inequalities in mechanics and physics*, Springer-Verlag, Berlin 1976. French edition: Dunod, Paris 1972.

[19] B. HALPHEN and Q.S. NGUYEN, Sur les matériaux standard généralisés, *J. de Mécanique* **14** (1975), 39 - 63.

[20] W. HAN and B.D. REDDY, Computational plasticity: the variational basis and numerical analysis, *Comp. Mech. Advances* **2** (1995), 283 - 400.

[21] Y. JIANG and H. SEHITOGLU, Cyclic ratchetting of 1070 steel under multiaxial stress states, *Int. J. Plasticity* **10** (1994), 579 - 608.

[22] Y. JIANG and H. SEHITOGLU, Multiaxial cyclic ratchetting under multiple step loading, *Int. J. Plasticity* **10** (1994), 849 - 870.

[23] M. KAMLAH, M. KORZEŃ and CH. TSAKMAKIS, Uniaxial ratchetting in rate-independent plasticity laws, *Acta Mechanica* **120** (1997), 173 - 198.

[24] V.B. KÖTTGEN, R.J. ANTHES and T. SEEGER, Implementation des Werkstoffmodells von Mróz in das Finite Element Programm ABAQUS - Teil 1: Grundlagen, Report FF-7/1991, Fachgebiet Werkstoffmechanik, TH Darmstadt, Darmstadt 1991.

[25] M. KORZEŃ and CH. TSAKMAKIS, Uniaxial ratchetting in linear viscoelasticity, *Acta Mechanica* **120** (1997), 157 - 172.

[26] M.A. KRASNOSEL'SKII and A.V. POKROVSKII, *Systems with hysteresis*, Springer-Verlag, Berlin 1989. Russian edition: Nauka, Moscow 1983.

[27] P. KREJČÍ, *Hysteresis, convexity and dissipation in hyperbolic equations*, Gakkōtosho, Tokyo 1996.

[28] J. LEMAITRE and J.-L. CHABOCHE, *Mechanics of solid materials*, Cambridge University Press, Cambridge 1990. French edition: Dunod, Paris 1985.

[29] P. LE TALLEC, *Numerical analysis of viscoelastic problems*, Springer-Verlag, 1990.

[30] G.A. MAUGIN, *The thermomechanics of plasticity and fracture*, Cambridge University Press, Cambridge 1992.

[31] D.L. McDOWELL, Description of nonproportional cyclic ratchetting behaviour, *Eur. J. Mech., A/Solids* **13** (1994), 593 - 604.

[32] E. MELAN, Zur Plastizität des räumlichen Kontinuums, *Ingenieur-Archiv* **9** (1938), 116 – 126.

[33] J.C. MOOSBRUGGER and D.L. McDOWELL, On a class of kinematic hardening rules for nonproportional cyclic plasticity, *J. Eng. Mat. Techn.* **111** (1989), 87 – 98.

[34] J.-J. MOREAU, Problème d'évolution associé à un convexe mobile d'un espace hilbertien, *C.R. Acad. Sci. Paris Sér. A-B* **273** (1973), A791 – A794.

[35] J.-J. MOREAU, Evolution problem associated with a moving convex set in a Hilbert space, *J. Diff. Eq.* **26** (1977), 347 – 374.

[36] J.-J. MOREAU, P.D. PANAGIOTOPOULOS and G. STRANG (Eds.), *Topics in nonsmooth mechanics*, Birkhäuser, Basel 1988.

[37] Z. MRÓZ, On the description of anisotropic workhardening, *J. Mech. Phys. Solids* **15** (1967), 163 – 175.

[38] Y. MURAKAMI (Ed.), *The rainflow method in fatigue*, Butterworth & Heinemann, Oxford 1992.

[39] N. OHNO and J.-D. WANG, Kinematic hardening rules with critical state of dynamic recovery, Part I: Formulation and basic features for ratchetting behaviour, *Int. J. Plasticity* **9** (1993), 375 – 390.

[40] W. PRAGER, Recent developments in the mathematical theory of plasticity, *J. Appl. Phys.* **20** (1949), 235 – 241.

[41] CH. TSAKMAKIS, *Methoden zur Darstellung inelastischen Materialverhaltens bei kleinen Deformationen*, Habilitation Thesis, Institut für Mechanik, Universität Gesamthochschule Kassel, Kassel 1994.

[42] A. VISINTIN, *Differential models of hysteresis*, Springer-Verlag, Berlin 1994.

Martin Brokate
Mathematisches Seminar II
Universität Kiel
D – 24098 Kiel, Germany
http://www.numerik.uni-kiel.de/

G A EVANS

Singular integrals and their application to finite element methods

1. Introduction and background

There is a general requirement for the evaluation of integrals as illustrated by the demand for books such as Gradshteyn and Ryzhik(1965) which lists a large percentage of analytically known integrals, and more recently marketed packages such as Maple and Mathematica. By natural extension, numerical techniques take up this role once analytic reserves are exhausted. Such numerical methods are as old as antiquity in that the trapezoidal rule was probably used by the Babylonians, and was certainly used by the Greeks. Quadrature methods form the basis of algorithms for the solution of ordinary differential equations and partial differential equations, in particular when these are solved by integral based finite element methods and their variations.

The most simple quadrature rules are normally designated the general purpose methods and have the form:

$$\int_a^b w(x)f(x)\,dx = \sum_{i=1}^{n} a_i f(x_i) \qquad 1)$$

where $w(x)$ is a weight function on which the crucial coefficients of the rule depend, these being the weights a_i and the abscissae x_i. Methods of this type include well-known rules such as Simpson's rule(1743), the Gaussian rules(1866) (commonly employed in finite element integration), the method of Clenshaw-Curtis(1960), and Patterson(1968). These are covered in general works such as Davis and Rabinowitch(1984) and more recently Evans(1993). However, it is very easy to find examples which defeat these methods, or at least render them very ineffective. One such group is the group of integrals with oscillatory kernels which arise in all manner of wave-like applications such as in optics, quantum mechamics, fluid waves and any problem solved by integral transforms. Typical of these problems would be an integral of the form

$$\int_a^b f(x) \sin(1000x)\, dx \qquad 2)$$

whose integrand would be a graph which was virtually filled in, and a general purpose method would sample this integrand at effectively random points, so yielding nonsensical results.

A second such area in which general purpose methods are defeated is that in which the integrand is singular. This case is relevant here, as one of the major applications is in boundary element analysis where integration over an element will be singular for any point within the same element. Hence in the assembly process, the matrix will be made up of a substantial number of elements whose computation requires integration of a singular integrand. Hence there is a requirement to evaluate efficiently these singular integrals. In addition, these integrals arise in a wide range of other applications in their own right. For example, many integral equation formulations such as the Hartree-Fock equations of quantum mechanics have singular kernels, forcing the numerical solution to rely on being able to evaluate singular integrals. The wider scene includes application to any problem in partial differential equations of Green's functions, and to integrals arising in gas dynamics using thin-wing approximations.

The object now is to present a survey of methods for dealing with singular integrals, and hence demonstrate how only the most elementary of these methods are used in applications, in particular in finite element work. To this end, the effectiveness of the methods is demonstarted on a large set of typical test problems analagous to those arising in practice.

2. A review of approaches - the use of accelerators

A somewhat light-hearted but illustrative classification of methods was made by Professor Philip Rabinowitch who coined the terms: 'ignoring the singularity', 'creeping up on the singularity' and 'transforming the singularity'. The first of these approaches is the crudest and involves the use of open rules which do not require the evaluation of the integrand at the singular point. Typically Gaussian rules are used with end-point singularities, and indeed simple sub-divisions will always ensure that singularities can be taken without loss of generality at the end points. The accuracy of the method is degraded by the presence of the singular point and the resulting errors have been analysed by Fox and Hayes(1970). This degrade is quite marked and occurs even with singular derivatives in the integrand. The normal bound based on a high derivative of the integrand no longer exists. An illustrative example is to try to find:

$$\int_0^{\pi/2} \sin^{1/2}(x)\, dx \qquad 3)$$

to high accuracy.

The second classic approach is to absorb the singularity into the weight function of a quadrature rule, leaving only the non-singular part to be derived. Typical of these rules are the Gauss-log rules with the singularity $\ln x$. Inevitably the resulting weights and abscissae are dependent on the weight function, and hence only relatively common singularities can be handled in this way as the generation of a new Gaussian rule is an expensive and time- consuming computational exercise.

A variation on this theme which is more recent, and still needs comparative evaluation as a method, is the generalized Gaussian rules by Ma, Rokhlin and Wandzura(1996). They generalize Gaussian rules from being exact for powers of x to being exact for special combinations of functions typified by 1, $\ln x$, x, $x \ln x$, x^2, $x^2 \ln x \ldots$, so encompassing specific singularities. These rules exhibit all the major characteristics of the classical rules including having positive weights (and therefore being stable) and having rapid convergence properties.

A third approach is creeping up on the singularity. Here the idea is to generate a sequence whose limit is the required integral

$$S = \int_0^1 Q(x)dx = \int_0^1 f(x)g(x)dx \qquad 4)$$

in which $g(x)$ is analytic and $f(x)$ is singular at $x = 0$. There is no loss of generality, as integrals over the range $[a, b]$ with an end-point singularity may be mapped onto $[0, 1]$ by a linear transformation, and if an internal singularity occurs, sub-division at the singularity and subsequent linear transformations yield two integrals of the form 4).

Such a sequence $\{S_n\}$ may be defined by

$$S_n = \int_{x_n}^1 Q(x)dx \qquad 5)$$

where $x_n \to 0$ as $n \to \infty$. The successive terms in the sequence are generated according to

$$S_n = \sum_{i=1}^{n} T_i \qquad 6)$$

where

$$T_i = \int_{x_i}^{x_{i-1}} Q(x)dx \qquad (i \geq 1) \qquad 7)$$

and $x_0 = 1$ and $x_n \to 0$ as $n \to \infty$.

Various possibilities are feasible for the monotonic decreasing sequence $\{x_n\}$ and the choice

$$x_n = \theta^n \qquad (0 < \theta < 1) \qquad 8)$$

is found to be most effective in practice. The parameter θ is at our disposal to enhance the performance of the sequence accelerators which are described next. These are used to obtain the required limit without having to evaluate an unreasonable number of terms of S_n. These accelerators are briefly described here for completeness.

The earliest of the accelerator methods is that of Shanks(1955) who considered the sequence denoted by

$$A = \{A_r\} \qquad r = 0, 1, 2, \ldots \tag{9}$$

The root of the method is to try to fit the elements of the sequence to a pre-defined form which has free constants. These constants are then computed, and the limit of the pre-defined form gives an estimate of the required limit. This process is applied in a local manner, and the sequence of estimated limits itself forms a sequence, and it is this sequence which should converge more rapidly than the original sequence, but to the same limit. The pre-defined form used by Shanks is

$$A_r = B_{kn} + \sum_{i=1}^{k} \alpha_{in} q_{in}^r \quad n - k \leq r \leq n + k \quad \text{and} \quad n \geq k \tag{10}$$

and $q_{in} \neq 1$ or 0. The term B_{kn} is called the local base and for $|q_{in}| \leq 1$ it is clear that B_{kn} will be the required limit as $r \to \infty$. Hence there are $2k + 1$ non-linear equations for the free constants B_{kn}, α_{in} and q_{in}, and the requirement is to compute B_{kn}. The special form of these non-linear equations allows them to be solved in closed form to yield:

$$B_{kn} = \frac{\begin{vmatrix} A_{n-k} & \cdots & A_{n-1} & A_n \\ \triangle A_{n-k} & \cdots & \triangle A_{n-1} & \triangle A_n \\ \triangle A_{n-k+1} & \cdots & \triangle A_n & \triangle A_{n+1} \\ \vdots & & & \vdots \\ \triangle A_{n-1} & \cdots & \triangle A_{n+k-2} & \triangle A_{n+k-1} \end{vmatrix}}{\begin{vmatrix} 1 & \cdots & 1 & \\ \triangle A_{n-k} & \cdots & \triangle A_{n-1} & \triangle A_n \\ \triangle A_{n-k+1} & \cdots & \triangle A_n & \triangle A_{n+1} \\ \vdots & & & \vdots \\ \triangle A_{n-1} & \cdots & \triangle A_{n+k-2} & \triangle A_{n+k-1} \end{vmatrix}} \tag{11}$$

The simplest case arises with determinants of order 2 and yields the Aitken extrapolation rule:

$$B_{1n} = \frac{A_{n-1}A_{n+1} - A_n^2}{A_{n+1} - 2A_n + A_{n-1}} \qquad 12)$$

This formula can be applied to successive sequences of three terms to generate the new sequence $\{B_{1n}, B_{2n}, \ldots\}$ which is usually denoted by the operator e_1 where

$$e_1(A) = \{B_{1n}\} \qquad n \geq 1 \qquad 13)$$

and more generally the operator e_k is defined by

$$e_k(A) = \{B_{kn}\} \qquad n \geq k \qquad 14)$$

A further operator called the diagonal operator e_d is defined by

$$e_d(A) = \{B_{nn}\}. \qquad 15)$$

These operators can be applied to produce a table of new sequences, the first column being the original sequence, the second being e_1, the third e_2, and so on. The sequence from the operator e_d then appears along the top of the display. This table is sometimes called a Shanks' table and can be extended downwards until convergence is obtained. An alternative strategy is the repeated use of the simple transformation by applying e_1 to the second column to get the third, giving columns with the transformations e_1, e_1^2, and so on.

The implementation of the Shanks' accelerators can be made very succinct using the ϵ-algorithm, the link between this algorithm and the determinantal form of the Shanks' sequences being made by Wynn(1956). The algorithm is very simple and uses the recurrence relations

$$\begin{aligned} \epsilon_0^{(-1)} &= 0 \\ \epsilon_n^{(0)} &= A_n \qquad\qquad 16) \\ \text{and} \quad \epsilon_n^{(p)} &= \epsilon_{n+1}^{(p-2)} + [\epsilon_{n+1}^{(p-1)} - \epsilon_n^{(p-1)}]^{-1} \end{aligned}$$

and in terms of these ϵ's the accelerated sequences are given by

$$B_{kn} = \epsilon_{n-k}^{2k} \tag{17)}$$

It is not difficult to generate a sequence which will defeat the Shanks' transformations. For example, $A_r = 1/r^2$, will not have the form which was predetermined in equation 10). To improve the effectiveness of the transformations, Levin(1973) fitted the sequence in a different way. The assumption here is that

$$A_r = T_{kn} + R_r \sum_{i=0}^{k-1} \gamma_{in} f_i(r) \quad n \le r \le n+k \tag{18)}$$

form a set of linear algebraic equations to determine the unknown constants, where R_r has been used for $R_1(r)$. In particular Cramer's rule can be applied to give the following determinantal form for T_{kn}, namely

$$T_{kn} = \frac{\begin{vmatrix} A_n & A_{n+1} & \cdots & A_{n+k} \\ R_n f_0(n) & R_{n+1} f_0(n+1) & \cdots & R_{n+k} f_0(n+k) \\ R_n f_1(n) & R_{n+1} f_1(n+1) & \cdots & R_{n+k} f_1(n+k) \\ \vdots & \vdots & & \vdots \\ R_n f_{k-1}(n) & R_{n+1} f_{k-1}(n+1) & \cdots & R_{n+k} f_{k-1}(n+k) \end{vmatrix}}{\begin{vmatrix} 1 & 1 & \cdots & 1 \\ R_n f_0(n) & R_{n+1} f_0(n+1) & \cdots & R_{n+k} f_0(n+k) \\ R_n f_1(n) & R_{n+1} f_1(n+1) & \cdots & R_{n+k} f_1(n+k) \\ \vdots & \vdots & & \vdots \\ R_n f_{k-1}(n) & R_{n+1} f_{k-1}(n+1) & \cdots & R_{n+k} f_{k-1}(n+k) \end{vmatrix}} \tag{19)}$$

By now taking $f_i(r) = r^{-i}$ and by dividing successive columns by R_n, $R_{n+1}, \ldots$, R_{n+k}, an expansion by the first row gives

$$T_{kn} = \frac{\sum_{j=0}^{k} (-1)^j \binom{k}{j} \left(\frac{n+j}{n+k}\right)^{k-1} \frac{A_{n+j}}{R_{n+j}}}{\sum_{j=0}^{k} (-1)^j \binom{k}{j} \left(\frac{n+j}{n+k}\right)^{k-1} \frac{1}{R_{n+j}}} \tag{20)}$$

The so-called t-transformation is then obtained by taking $R_r = a_r = \Delta A_{r-1}$, and the operator t_k analagous to e_k is defined as

$$t_k(A) = \{T_{kn}\} \tag{21)}$$

with the corresponding diagonal form

$$t_d(A) = \{T_{n1}\} \tag{22}$$

Two modifications of the t-transformation are derived by differing choices of R_r. Hence if $R_r = ra_r$ the u-transformation is produced with the formula

$$U_{kn} = \frac{\sum_{j=0}^{k}(-1)^j \binom{k}{j} \left(\frac{n+j}{n+k}\right)^{k-2} \frac{A_{n+j}}{a_{n+j}}}{\sum_{j=0}^{k}(-1)^j \binom{k}{j} \left(\frac{n+j}{n+k}\right)^{k-2} \frac{1}{a_{n+j}}} \tag{23}$$

The second modification builds in a simple form of the Shanks' transformation to the t-transform. The simple Aitken extrapolation rule in equation 13) can be written as

$$\begin{aligned} e_1(A)_n &= \frac{A_{n-1}a_{n+1} - A_n a_n}{a_{n+1} - a_n} \\ &= A_n + \frac{a_n a_{n+1}}{a_n - a_{n+1}} \end{aligned} \tag{24}$$

and hence the v-transformation is derived using

$$R_r = \frac{a_r a_{r+1}}{a_{r+1} - a_r} \tag{25}$$

which gives the definition

$$V_{kn} = \frac{\sum_{j=0}^{k}(-1)^j \binom{k}{j} \left(\frac{n+j}{n+k}\right)^{k-1} \frac{a_{n+j-1}-a_{n+j}}{a_{n+j}a_{n+j+1}} A_{n+j}}{\sum_{j=0}^{k}(-1)^j \binom{k}{j} \left(\frac{n+j}{n+k}\right)^{k-1} \frac{a_{n+j-1}-a_{n+j}}{a_{n+j}a_{n+j+1}}} \tag{26}$$

3. Examples using accelerators

Some illustrative examples are presented which are listed in Table 1.

Test Integrals
$I_1 = \int_0^1 x^{-\alpha} dx = 10^6 \qquad (\alpha = 1 - 10^6)$
$I_2 = \int_0^1 x^{0.95} e^x dx = 1.020457359$
$I_3 = \int_0^1 (1 + x^2)^{-1} (\ln x)^2 dx = 1.937892293$
$I_4 = \int_0^1 x^{\frac{1}{2}} e^{-x} (1 + x)^{-1} dx = 1.237643927$

Table 1

I_1 demonstrates pure geometric convergence of the sequence, but with an integral which is 'nearly infinite'. I_2 is an example considered by Cohen(1980) which poses some problems for the cautious Romberg strategy. I_3 and I_4 exhibits some common singularities.

The results obtained are presented in Table 2 where the total number of function evaluations required to attain a relative error of $E = 10^{-10}$ are depicted for various values of the parameter θ.

Number of function evaluations required to attain relative error $E = 1.0(-10)$ for Table 1 integrals			
Integral	$\theta = 0.1$	$\theta = 0.2$	$\theta = 0.5$
I_1	45 4.3(-6)	21 4.4(-6)	21 2.3(-6)
I_2	32	27	40
I_3	104	111	91
I_4	105	127	84

Table 2

Note that, for the near pathological integral I_1, the successive terms T_i are very close to one another, the common ratio θ^γ of equation (8) being nearly unity in each case. The approximate value of the ratio is, in fact, $1 + 10^{-6} \ln \theta$. Hence although it is possible to obtain the T_i values accurately using $N = 7$ or 15 points in a Patterson quadrature formula for each cycle, the ϵ-algorithm is subject to severe cancellation when the differences of the T_i values are calculated. Consequently, in a single precision calculation to accuracy $p \simeq 11$ figures, it is possible to obtain I_1 to only about 6 significant figures when extrapolation to the limit is carried out. This fact is reflected by quoting the actual relative errors obtained in Table 2. Nevertheless, it is quite remarkable that such an accurate result is obtained, as predicted, from only the first three terms $\{S_1, S_2, S_3\}$ of the sequence, even for this extreme value of α. All the other methods tried, failed completely on this integral.

The results for the other integrals show that accurate values are obtainable for moderate expenditure in function evaluations, especially for Cohen's integral I_2. The algorithm was implemented completely automatically in that the quadrature results were monitored to investigate convergence to a specified precision, as the number of points was increased.

Examination of Table 2 indicates that the results are not particularly sensitive to the choice of the parameter θ. The 'extreme' values $\theta = 0.1$ and $\theta = 0.5$ may occasionally produce rapid convergence for a particular integral but it is considered safer, in general, to adopt the compromise value of $\theta = 0.2$.

Clearly these methods require each cycle to be integrated, and though very effective, a more efficient procedure might be anticipated if an approach could be found which involved just one quadrature. In effect it would be beneficial to abandon what is in reality a sub-division strategy. The following methods assume this approach.

4. Transformation methods

A simple polynomial transform was employed by Evans, Forbes and Hyslop (1983), and resulted in a group of very effective methods for dealing with integrals with end-point singularities. It is sufficient to consider singular integrals of the form

$$S = \int_0^1 f(x)dx \qquad 27)$$

where $f(x)$ and/or its derivatives are singular at one or both of the end-points, $x = 0$ and $x = 1$, of the normalized range. More general finite range integrals are easily reducible to this form.

In general the transformation

$$x = \psi(t) \qquad 28)$$

is utilized which, for convenience, is assumed to map the interval $(0, 1)$ onto itself. Various simple polynomial forms for $\psi(t)$ are tested for the transformed integral

$$S = \int_0^1 f(\psi(t))\dot{\psi}(t)dt \qquad 29)$$

It is the derivative $\dot{\psi}(t)$ which produces the alleviation of the singularity and it is of interest to investigate the effect of increasing the degree of the polynomial to achieve this end for various types of singularity.

It will be noted that care must be exercised in computing terms like, $f(1-x/2)$, in the integrand as $x \to 0$. This point has been commented on by many authors and the commonly adopted remedy is to express $f(z)$ as $z \to 1$ in the form

$$f(z) = f(1 - \delta) \qquad 30)$$

where the 'residual' δ as well as the variable z must be accurately available in the function routine. In the present context , $\delta = t^n/2$ and even though t may be of order 10^{-7} when implementing the highest order ($N = 511$) Patterson (1968) quadrature formula, underflow may be avoided, even for fairly large values of n. Consequently terms such as $(1-x)^{-\beta}$ may be evaluated without overflow occurring since cancellation is avoided. Note also that, for certain 'composite' singularities such as $\ln \ln x^{-1}$, this device is not successful since $\ln \ln(1-\delta)^{-1}$ cannot be evaluated without overflow under these extreme circumstances, unless arithmetic precision is increased prohibitively. It is necessary in these cases to use analytic expansions such as $\ln(\delta + \delta^2/2 + \ldots)$ in the function routines.

To test and compare these methods, some further integrals are presented in Table 3 which have been taken from the various test sets invoked by other authors.

Test Integrals of Chisholm et al
$J_1 = \int_0^{2\pi} \ln x \sin x dx = -2.437653393$
$J_2 = \int_0^1 x^{\frac{3}{2}} dx = 0.4$
$J_3 = \int_0^1 x^{\frac{1}{2}} \ln x dx = 4/9$
$J_4 = \int_0^1 x^{\frac{3}{4}} \cos x dx = 0.4451649239$
$J_5 = \int_0^1 x^{-\frac{1}{2}} dx = 2$
$J_6 = \int_0^1 (x^{\frac{1}{2}} + x^{\frac{1}{3}})^{-1} dx = 0.8411169166$
$J_7 = \int_0^1 \ln(1 - \cos x) dx = -2.721065445$
$J_8 = \int_0^1 x^{-\frac{1}{2}} \ln x dx = -4$
$J_9 = \int_0^\infty u e^{-u} (1 + u^2)^{-1} du = \int_0^1 u(1 + u^2)^{-1} dx = 0.3433779615$
$J_{10} = \int_0^\infty e^{-u} (1 + u)^{-\frac{1}{2}} du = \int_0^1 (1 + u)^{-\frac{1}{2}} dx = 0.7578721561$
$J_{11} = \int_0^\infty e^{-u} u^{\frac{7}{2}} du = \int_0^1 u^{\frac{7}{2}} dx = 11.63172840$
$J_{12} = \int_0^\infty e^{-u} u^{-\frac{1}{2}} (1 + u)^{-1} du = \int_0^1 u^{-\frac{1}{2}} (1 + u)^{-1} dx = 1.343293422$
$(u = -\ln x)$

Table 3

A further set of test integrals was provided by Harris and Evans (1977) which appear in Table 4.

Test integrals of Harris and Evans
$K_1 = \int_0^1 x^{\frac{1}{2}} dx = 2/3$
$K_2 = \int_0^1 x^{-\frac{1}{3}} dx = 3/2$
$K_3 = \int_0^1 x^{-\frac{2}{3}} dx = 3$
$K_4 = \int_0^1 x^{\frac{7}{2}} dx = 2/9$
$K_5 = \int_0^1 (\ln x)^2 dx = 2$
$K_6 = \int_0^1 (\ln x)^4 dx = 24$
$K_7 = \int_0^1 (1+x^2)^{-1} dx = \pi/4$
$K_8 = \int_0^1 x^{-\frac{1}{2}} \ln x dx = -4$
$K_9 = \int_0^1 (1-\ln x)^{-1}(-\ln x)^{-\frac{1}{2}} dx = 1.343293422$

Table 4

Finally as a very stringent test of the various procedures, the set of integrals shown in Table 5 was considered.

Test integrals with singularities at both end-points
$L_1 = \int_0^1 x^{-\frac{1}{2}}(1-x)^{-\frac{1}{2}}dx = \pi$
$L_2 = \int_0^1 x^{-\frac{1}{2}} \ln\ln x^{-1} dx = 0.231863031316825$
$L_3 = \int_0^1 (1+x)^{-2} \ln\ln x^{-1} dx = -0.06281647980603901$
$L_4 = \int_0^1 (1-\ln x)^{-1}(-\ln x)^{-\frac{1}{2}} dx = 1.343293421646734$
$L_5 = \int_0^1 \ln x \ln(1-x) dx = 0.3550659331517735$
$L_6 = \int_0^1 (1-x)^{-1} \ln x dx = -1.644934066848227$

Table 5

In Table 6 results are presented for tests carried out on the various integrals considered. The number of quadrature points N required to obtain a relative error of 10^{-10} in each of the integrals is depicted for values of n ranging from $n = 1$ to $n = 8$. Patterson quadrature is utilised with $N = 3, 7, 15, 31, \ldots, 511$ points, the N-point formula integrating exactly a polynomial of degree $(3N+1)/2$. (The corresponding Gaussian formula has polynomial precision $(2N-1)$). For comparison purposes, the last column shows the number of points used in the acceleration procedure of Evans, Hyslop and Morgan (1983) described earlier.

Number of Patterson Quadrature points (N) required to attain relative error $E = 1.0(-10)$ for various integrals using $x = t^n$transformation						
n	1	2	4	6	8	EHM
I_1	-	-	-	-	-	21*
I_2	63	15	15	15	31	27
I_3	-	127	31	31	31	111
I_4	-	15	15	15	15	127
J_1	127	31	31	31	31	57
J_2	31	3	7	15	15	29
J_3	511	63	15	15	15	59
J_4	127	31	15	15	31	51
J_5	-	3	3	3	3	45
J_6	-	511	31	15	15	133
J_7	-	127	31	15	15	89
J_8	-	-	127	31	31	75
J_9	-	31	31	31	31	79
J_{10}	-	31	31	15	31	79
J_{11}	-	255	63	31	31	141
K_1	255	3	3	7	7	45
K_2	-	511	31	3	15	45
K_3	-	-	511	3	31	45
K_4	15	7	15	15	31	17
K_5	-	127	31	15	15	97
K_6	-	511	63	31	15	141
K_7	15	15	15	31	31	29
L_1	-	15	15	31	31	127
L_2	-	-	63	31	31	135
L_3	-	127	31	31	31	110
L_4	-	31	15	31	31	127
L_5	63	31	15	15	31	54
L_6	-	63	31	15	31	88
* 6-figure accuracy only For doubly singular integrals the number of function evaluations is $2N$ see (5.5.5) EHM - Evans, Hyslop and Morgan(1983)						

Table 6

The $n = 1$ column corresponds to ignoring the singularity and it will be observed that even the highest order Patterson formula fails in many cases to produce convergence to the required accuracy. This is, indeed, a measure of the difficulty of the integrals considered. Nonetheless, the $x = t^n$ transformation rapidly smooths the integrands and convergence is attained in all cases (except I_1) with a maximum of $N = 31$ quadrature grid-points. It is necessary to use values of n as large as $n = 6$ to achieve convergence in general for the types of singularity tested here. Calculations were carried out for larger values of n ($n = 6$ and $n = 8$ results are shown here) and it was found that numerical instability due to round-off only began to appear for values approaching $n = 10$. There is slight evidence of this effect appearing at $n = 8$ for I_2 and for K_4, so that $n = 6$ may be a good practical limit.

To deal with singularities at both end-points, the power transformation is applied separately at each end of the interval. Thus in (27), to alleviate the singularity at $x = 0$ the transformation

$$x = u^n \tag*{31)}$$

is used and, if the integrand is still singular at $x = u = 1$, then the further substitution

$$1 - u = (1 - t)^m \tag*{32)}$$

is adopted. Hence the function $\psi(t)$ of equation(29) is defined by

$$x = \psi(t) = (1 - (1 - t)^m)^n \tag*{33)}$$

and the double singularity alleviating derivative $\dot{\psi}(t)$ by

$$\dot{\psi}(t) = nm(1 - t)^{m-1}(1 - (1 - t)^m)^{n-1} \quad (n, m \geq 1) \tag*{34)}$$

The transformation $x = (1-(1-t)^m)^n$ whilst proving more effective than $x = t^n$ for dealing with doubly singular integrals suffers from the defect of being markedly skew for larger values of n and m.

For this reason, the transformation based on

$$\dot{x}(t) = ct^{n-1}(1 - t)^{m-1} \tag*{35)}$$

is proposed, where c is a normalizing constant. It is clear that

$$c = 1/\beta(n, m) \tag*{36)}$$

where $\beta(n, m)$ is the beta function

$$\beta(n, m) = \int_0^1 t^{n-1}(1-t)^{m-1}dt \tag{37}$$

The transformation itself involves the incomplete beta function defined by

$$\beta_t(n, m) = \int_0^t t^{n-1}(1-t)^{m-1}dt \tag{38}$$

and is written in the form

$$x(t) = I_t(n, m) = \beta_t(n, m)/\beta(n, m) \tag{39}$$

which represents a distribution much used in statistical applications.

5. The tanh transformation

An alternative approach is to employ a transformation which moves the singular points of the integrand to infinity. A *low* order quadrature rule such as the trapezoidal rule then proves to be particularly effective, as the derivatives in the Euler-Maclaurin summation formula often fall off rapidly at infinity.

This approach was investigated by Schwartz (1969b) and subsequently studied by a number of authors including Takahasi and Mori(1973), Stenger(1973) and Haber (1977). One of the transformations suggested by Schwartz has become known as the 'tanh rule' and has proved to be very effective over a wide class of integrals, the method having been developed further by Squire (1976,1979). Despite the efficiency and simplicity of the Squire process, it suffers from the serious defect of being non-progressive.

Evans, Forbes and Hyslop (1984) present an alternative progressive version of the Squire procedure, together with various modifications to the tanh transformations. Critical comparisons are presented for a number of test integrals.

The most general singular integral considered is of the form

$$S = \int_a^b f(x)dx \tag{40}$$

where $f(x)$ may be singular at both end-points. Integrands possessing more singularities may be reduced to several integrands of this form by sub-division at the singularities. Generalizing the procedure of Squire (1979), the substitution

$$x = \frac{1}{2}(b+a) + \frac{1}{2}(b-a)\tanh g(t) \qquad 41)$$

is proposed in which $g(t) \to \pm\infty$ as $t \to \pm\infty$. For example, the choice $g(t) = t$ gives the classical tanh transformation employed by Squire. Consequently

$$S = \frac{1}{2}(b-a)\int_{-\infty}^{\infty} f[\frac{1}{2}(b+a) + \frac{1}{2}(b-a)\tanh g(t)]\mathrm{sech}^2 g(t)\dot{g}(t)dt \qquad 42)$$

in which $\dot{g}(t)$ represents the derived function dg/dt. With a suitable choice of $g(t)$, the $\mathrm{sech}^2 g(t)$ factor falls off extremely rapidly as $t \to \pm\infty$ and controls the behaviour of the transformed integrand even though the singular values of $f(a)$ and $f(b)$ are being approached. For this reason, when the midpoint rule is applied by setting

$$t = jh \qquad j = 0, \pm 1, \pm 2, \ldots \qquad 43)$$

the summation may be truncated at $j = \pm N$, say, yielding

$$S \simeq 2h(b-a)\sum_{j=-N}^{N} A_j f(x_j) \qquad 44)$$

where

$$x_j = \frac{1}{2}(b+a) + \frac{1}{2}(b-a)(1-q_j)(1+q_j)^{-1}, \qquad 45)$$

$$A_j = q_j(1+q_j)^{-2}\dot{g}(jh) \qquad 46)$$

and

$$q_j = \exp[-2g(jh)] \qquad 47)$$

As mentioned by Squire (1979), difficulty arises in evaluating $f(x_j)$ for the larger values of j. Hence again, the function $f(x)$ should be programmed as a two-parameter function $f(x, \delta)$ in which the 'residual' δ, representing the distance from the singularity, should also be specified. It is recommended therefore that x_j in equation (44) should be expressed as

$$x_j = a + (b-a)q_j(1+q_j)^{-1} \tag{48)}$$

in the vicinity of $x = a$ and as

$$x_j = b - (b-a)(1+q_j)^{-1} \tag{49)}$$

near $x = b$, so providing accurate numerical values for δ. Some simplification is possible when attention is confined to *odd* functions for $g(t)$, suitable examples being

$$g(t) = t^{\nu} \qquad (\nu = 1, 3, 5, \ldots) \tag{50)}$$

or

$$g(t) = c \sinh t. \tag{51)}$$

In such circumstances, it is easily seen that (44) reduces to

$$S \simeq 2h(b-a)\{A_0 f[\frac{1}{2}(b+a) + \sum_{j=1}^{N} A_j[f(a+\delta_j) + f(b-\delta_j)]\} \tag{52)}$$

in which

$$\delta_j = (b-a)q_j(1+q_j)^{-1} \tag{53)}$$

Equation (50) is the form adopted for computational purposes, remembering that the functions are programmed as two-parameter procedures. As an example, consider the integral

$$\int_{-1}^{1} (x-2)^{-1}(1-x)^{-\frac{1}{4}}(1+x)^{-\frac{3}{4}}\,dx \tag{54)}$$

which was one of the test cases treated by Takahasi and Mori. The integrand $f(x)$ is then programmed according to the algorithm

$$\begin{aligned} \text{IF } x < 0 \text{ THEN} f(x,\delta) &\leftarrow (x-2)^{-1}(1-x)^{-\frac{1}{4}}\delta^{-\frac{3}{4}} \\ \text{ELSE} f(x,\delta) &\leftarrow (x-2)^{-1}\delta^{-\frac{1}{4}}(1+x)^{-\frac{3}{4}} \end{aligned} \tag{55)}$$

thus avoiding cancellation effects as $x \to \pm 1$.

6. Generalized transformations

In the previous paragraph, the problem of applying a transformation

$$x = g(t) \tag{56}$$

to a quadrature

$$I = \int_{-1}^{1} f(x)dx \tag{57}$$

to yield

$$I = \int f(g(t))g'(t)dt \tag{58}$$

has been considered for various functions $g(t)$

Two purposes are served by the process. The first is to remove any singularities occurring in $f(x)$ by choosing $g'(t)$ to have strong zeros in the region of these singularities (usually at the end points). The resulting quadrature can then be very easily integrated by conventional means.

The second purpose is to generate a new quadrature with an integrand which vanishes with all its derivatives at the end points. It is well-known from the Euler-Maclaurin summation formula

$$\sum_{k=0}^{r} f_k = \frac{1}{h}\int_{x_0}^{x_r} f(x)dx + \frac{1}{2}(f_0 + f_r) + \sum_{i=1}^{m} \frac{B_{2i}}{(2i)!} h^{2i-1}[f_r^{(2i-1)} - f_0^{(2i-1)}] + E_m \tag{59}$$

where

$$E_m = r\frac{B_{2m+2}h^{2m+2}}{(2m+2)!} f^{(2m+2)}(\xi) \tag{60}$$

with $x_0 < \xi < x_r$. that the trapezoidal rule then behaves as a very high order technique Evans(1992). Hence it is possible to generate general purpose quadrature algorithms based on this principle. The development of these ideas has interested many authors. Schwartz (1969) proposed

$$s = \frac{1}{(1+e^{-t})} \quad \text{to map} \quad (0,1) \text{ to } (-\infty, \infty) \tag{61}$$

and later various other transformations were used.

Two classes emerge, the first mapping a finite interval into $(-\infty, \infty)$ and including variations on (61) such as

$$x = \tanh t \qquad 62)$$

which maps $(-1, 1)$ to $(-\infty, \infty)$ which was considered by Evans, Forbes and Hyslop (1984) and Squire (1976 and 1979).

The transformation based on

$$x = \tanh(c \sinh t) \qquad 63)$$

is called the double exponential formula after the behaviour of $x'(t)$ as $|t| \to \infty$. This was considered by Takahasi and Mori(1974) who also analysed the errors (1973). It is important to note that Evans, Forbes and Hyslop(1984) demonstrated that the same result can be achieved by replacing $c \sinh t$ by an odd function

$$g(t) = t^{\nu} \qquad (\nu = 1, 3, 5) \qquad 64)$$

Theoretically (60) gives a faster decrease as $|t| \to \infty$ but in practice (62) is perfectly adequate. The other common type of formula in this class is

$$x = \operatorname{erf}(t) \qquad 65)$$

and this was also considered by Takahasi and Mori(1973).

The second class involves mapping finite intervals into finite intervals and is based on

$$x = d \int_{-1}^{t} e^{-\left(\frac{c}{1-s^2}\right)} ds \qquad 66)$$

which maps $(-1, 1)$ into $(-1, 1)$ where d is a normalization constant and c a constant. This is the IMT rule originally proposed by Iri, Moriguti and Takasawa (1970) and later reviewed in a more general context by Murota and Iri (1982) and Mori (1985).

Aihie and Evans (1989) consider some generalizations of these formulae, including methods to implement the transformation formulae themselves, as these are no longer standard functions. The transformations are then tested on a range of singular and general quadrature examples and the results considered in the light of the error analysis of Takahasi and Mori (1973).

The formula (66) defines a transformation say $g_1(t)$ and yields a non-standard function for which efficient methods for computing this function are required so that unnecessary work is not caused in the ancillary parts of the algorithm. It would be feasible to use a suitable quadrature formula for (66) but this seems a circular process as it may well be equally feasible to apply the same rule to the original function. Hence there is the need for formulae to compute the transformation accurately. If these prove to be highly time consuming to implement, a conventional rational approximation can then be computed using, say, the exchange algorithm with accurate stable formulae as a basis.

Hence ignoring the normalization constant, $g_1(t)$ can be written

$$g_1(t) = \int_{-1}^{t} e^{-\left(\frac{c}{1-x^2}\right)} dx = \int_{\infty}^{\frac{1}{1-t^2}} \frac{e^{-cy}}{2y^2\sqrt{1-1/y}} dy \qquad 67)$$

by putting $y = 1/(1-x^2)$.

A useful series is obtained by expanding $(\sqrt{1-1/y})^{-1}$ to yield a series of exponential integrals. Hence

$$g_1(t) = \frac{1}{2}\frac{E_2(cz)}{z} + \frac{1}{2}\sum_{i=1}^{\infty} \frac{1.3.5\ldots(2i-1)}{2^i.i!.z^{i+1}} E_{i+2}(cz) \qquad 68)$$

where $x = 1/(1-t^2)$.

By using the recurrence relation $nE_{n+1}(z) = e^{-z} - zE_n(z)$, it becomes straight forward to implement (68) with only $E_2(z)$ requiring separate computation. This formula can be used effectively for $-1 \le t \le -0.5$ where z is large. For the range of t, $-0.5 \le t \le 0.0$ a series solution the the differential equation

$$1 = (1-t^2)^2 \hat{g}_1'(t) - \hat{g}_1(t)[2t(2+c-2t^2)] \qquad 69)$$

was used where

$$g_1(t) = (1-t^2)^2 e^{-\left(\frac{c}{1-t^2}\right)} \hat{g}_1(t) \tag{70}$$

and $\hat{g}_1(t)$ is given by $\sum_{i=0}^{\infty} a_i t^i$ where

$$\begin{aligned}
a_0 &= e^c g_1(0) \\
a_1 &= 1 \\
a_2 &= \frac{a_0}{2}(4+2c) \\
a_3 &= \frac{a_1}{3}(6+2c) \\
a_4 &= \frac{1}{4}[(8+2c)a_2 - 4a_0] \\
\text{and} \quad a_{n+1} &= [(2n+2+2c)a_{n-1} - (n+1)a_{n-3}]/(n+1)
\end{aligned}$$

Symmetry is used for $0 \le t \le 1.0$ to complete the required range.

A generalization of (65) yields the transformation

$$g_2(n,t) = c\int_0^t e^{-x^{2n}}\,dx \tag{71}$$

with

$$\frac{1}{c} = \int_0^\infty e^{-x^{2n}}\,dx$$

For $n = 1$ the standard error function is recovered. For larger n similar special formulae are developed to the previous case. For small t the series expansion

$$\int_0^t e^{-x^{2n}}\,dx = \sum_{m=0}^{\infty} \frac{(-1)^m t^{2nm+1}}{m!(2nm+1)} \tag{72}$$

can be used, but for increasing t this series suffers from stability problems because of large alternating terms causing cancellation before the terms of the series fall off eventually for large m. An alternative is to write

$$\int_0^t e^{-x^{2n}}\,dx = e^{-t^{2n}}\hat{g}_2(t) \tag{73}$$

and then $\hat{g}_2(t)$ satisfies

$$1 = \hat{g}_2(t) - 2nt^{2n-1}\hat{g}_2(t)$$

which has a series solution

$$g_2(t) = t + \sum_{m=1}^{\infty} \frac{(2n)^m t^{2nm+1}}{(2n+1)(4n+1)\dots(2mn+1)} \tag{74}$$

This series is effective for $0 \le t \le 5$ for $n = 1$, for $0 \le t \le 2.0$ for $n = 2$ and for $0 \le t \le 1.25$ for $n = 3$. For large t an asymptotic formula is used based on successive integration by parts. By putting $y = x^{2n}$ in

$$\int_t^{\infty} e^{-x^{2n}}\, dx$$

gives

$$\begin{aligned} I_{2n-1} &= \int_t^{\infty} \frac{e^{-y}}{2nx^{2n-1}} dy \\ &= \frac{e^{-t^{2n}}}{2nt^{2n-1}} - \int_t^{\infty} \frac{(2n-1)e^{-y}}{(2n)^2 x^{4n-1}} dy \\ &= \frac{e^{-t^{2n}}}{2nt^{2n-1}} - \frac{(2n-1)}{(2n)^2} I_{4n-1} \end{aligned} \tag{75}$$

to yield the asymptotic series

$$I_{2n-1} = e^{-t^{2n}} \sum_{m=1}^{\infty} \frac{(2n-1)\dots(2mn-1)(2n(m+1)-1)}{(2n)^m t^{2mn-1}} \tag{76}$$

The integration formulae require accurate values of the distance from the curve $x = g_2(t)$ to the asymptote $x = 1$ which (76) yields directly. On a low accuracy machine this can be a problem in computing $1 - g_2(t)$ from (74) as $g_2(t)$ tends quite rapidly to 1.0 to many figures. If this proves to be a problem in the upper range of t for which (74) needs to be used, a short range Gaussian quadrature can be used to give accurate values for the small area between $x = g_2(t)$ and its asymptote. Hence on a seven digit machine a 16-point Gauss-Legendre formula was used to evaluate

$$\int_t^{N} e^{-x^{2n}}\, dx$$

with

$$N = \exp(\ln(M \ln(10)/2n)$$

and $M = 38$, 10^{-38} being the smallest real representable on the machine.

7. Tests of generalized transformations

The test integrals listed in Tables 3, 4, and 5 have been considered. Table 8 shows a comparative performance of the new rules.

Relative performance of transformation methods to attain relative error $E = 1.0(-10)$ for various integrals.

	1	2	3	4	5	6	7
I_1	21	-	-	-	-	-	-
I_2	26	15	25	16	13	17	25
I_3	111	31	31	16	21	21	33
I_4	127	15	29	32	13	21	33
J_1	57	31	33	16	29	33	25
J_2	29	15	25	16	-	13	25
J_3	59	15	21	16	15	17	25
J_4	51	15	29	16	13	17	25
J_5	45	3	21	16	25	21	33
J_6	133	15	21	16	21	21	33
J_7	87	15	25	16	17	17	25
J_8	75	31	25	16	27	25	33
J_9	79	31	33	16	33	17	33
J_{10}	79	15	25	16	17	17	33
J_{11}	141	31	25	16	21	21	33
J_{12}	127	31	25	32	15	17	33
K_1	45	7	25	16	15	17	25
K_2	45	3	23	16	21	21	33
K_3	45	3	25	16	29	25	13
K_4	17	15	19	16	-	13	25
K_5	97	15	25	32	21	21	33
K_6	141	31	25	32	21	21	33
K_7	29	31	29	32	33	17	33
K_8	75	31	25	32	27	25	33
K_9	129	31	25	32	15	17	33
L_1	127	31	21	16	25	21	33
L_2	135	31	25	16	27	25	33
L_3	110	31	29	16	19	21	33
L_4	127	31	25	32	15	17	33
L_5	54	31	17	32	13	17	25
L_6	88	31	17	32	19	21	33

1. extrapolation: 2. beta 3. tanh: 4. IMT
5. erf,n=1: 6. erf,n=2 7. erf,n=3

Table 8

In general the rules compare very well with existing methods. The IMT type of rule is very difficult to apply because there is no way of knowing the appropriate choice of c before-hand and this experiment has demonstrated that there is no discernible trend in the way of choosing the value of c. Among the erf class, the erf function which is recovered for $n = 1$ failed in J_2, K_4 and L_8. The erf class with $n = 2$ appears to be the best in that the number of function evaluations is generally less than when $n = 3$. However the new rules compare favourably with other existing methods.

In addition the new rules are also robust for non-singular integrals, but they require too many function evaluations compared with conventional general purpose methods. So the recommendation is that where possible the new rules should be used for singular integrals only.

To conclude this discussion, some mention should be made of the routines available in the NAG library. There are two major routines recommended for singular integrals, D01ACF and D01APF. The first of these is essentially QUADPACK using 10-point Gauss-Legendre and 21-point Kronrod, and is similar to D01AJF, mentioned in Chapter 2. However this alternative routine allows for the user to supply break points in the range of integration which allows smoothing transformations to be applied. The routine D01APF is a Clenshaw-Curtis based routine with allowance for specific end-point singularities of either a logarithmic or algebraic variety. This special routine is employed on any subinterval which contains either of the end-points a and b, 12 and 24 point routines being used. For other interior subintervals the more usual prescription of 7-point Gauss-Legendre and 15-point Kronrod is used.

7. Applications

A very commonly employed application of quadrature rules occurs in the important methods for the numerical solution of partial differential equations, namely the finite element method (Zienkiewicz(1977)) and the boundary element method (Brebbia(1978), Brebbia and Dominguez(1989)). In each case, integrals are required over the elemental regions. In the boundary element method, the integrals for a two dimensional region are over the one dimensional boundary, and the Gauss-Legendre rule is commonly employed. A feature of these methods is that singularities arise in some of the required integrals, most commonly either logarithmic singularities or singularities such as $1/\sqrt{(a^2 + x^2)}$. A good practical application is the use of the boundary element method for modelling three dimensional cracks, Penzel(1993). The method depends on a fully discretised Galerkin scheme in which scalar products are found by a quadrature formula. Restrictions are found on which quadrature rules are sufficient to ensure the validity of the error analysis.

Within the realms of the finite element method itself many developments now require the evaluation of singular integrals. Hence the boundary point method, for example, involves integrating the singular fundamental solutions of Laplace's equation to obtain the algebraic elements, and is given in detail in Maz'ya(1993). In this method, a choice is made of a set of basis functions, each of which is concentrated near a particular boundary point, or falls off rapidly away from this point. These functions are used to approximate the solution. Both the boundary data and the potential densities are so approximated, and the resulting algebraic system depends on the co-ordinates of a finite number of boundary points. For non-stationary problems, stable, explicit methods arise.

A further example is that of the solution of parabolic equations with weakly singular positive type memory terms as described by Slodicka(1993). Singular integrals arise in most of the practically useful cases. Many authors still use the crude process of 'ignoring the singularity' in their finite and boundary element codes. It is demonstrated here using typical examples that considerable savings and increased accuracy can be achieved by using more modern methods. The use of the tanh transformation and the midpoint rule not only gives such savings, the algorithm itself is very simple. There is no management needed of files of Gaussian coefficients, and a progressive approach will yield the required accuracy efficiently.

References

Aihie V.U. and Evans G.A. "The tanh transformation for double integrals with singularities- a progressive rule" *Int. J. Comp. and Appl. Math.* **24**,349-354,(1988)

Aihie V.U. and Evans G.A. "Variable transformation methods and their use in general and singular quadratures" *Int. J. Comp. Maths.* **27**, No.2, 91-101, (1989)

Aihie V.U. and Evans G.A. "A comparison of the error function and the tanh transformation as Progressive rules for double and triple singular integrals" *Int. J. Comp. and Appl. Math.* **30**, No. 2,145-154, (1990)

Brebbia C. A. "The boundary element method for engineers" *Pentech Press, London*, (1978)

Brebbia C. A. and Dominguez J. "Boundary elements, an introductory course" *Computation mechanics publications, McGraw-Hill*, (1989)

Chisholm J. S. R., Genz A. and Rowlands G. E. "Accelerated convergence of sequences" *J.Comp.Phys***10**,284-307,(1972)

Clenshaw C. W. and Curtis A. R. "A method for numerical integration on an automatic computer" *Num.Math.***12**,197-205,(1960)

Cohen A. M. "Cautious Romberg extrapolation" *Int. J. Comp. Maths.* **8**, 137-147, (1980)

Davis P. J. and Rabinowitch P. "Methods of numerical integration" *Academic Press, New York*, (1984)

Evans G. A. "The high order use of the trapezoidal rule in numerical quadrature" *Int. J. Math. Ed.* **23** No.4, 525-536, (1992)

Evans G. A. "Practical numerical integration" *John Wiley and Sons, Chichester*, (1993)

Evans G. A., Forbes R. C. and Hyslop J. "Polynomial transformations for singular integrals" *Int. J. Comp. Math.* **14**, No.4, 157-170, (1983)

Evans G. A., Forbes R. C. and Hyslop J. "The implementation of the tanh transformation in the evaluation of singular integrals" *Int. J. Comp. Math.* **15**,339-358,(1984)

Evans G. A., Forbes R. C. and Hyslop J. "Polynomial transformations for singular integrals" *Int. J. Comp. Math.* **14**, No.4, 157-170, (1983)

Evans G. A., Hyslop J. and Morgan A. P. G. "An extrapolation procedure for the evaluation of singular integrals" *Int. J. Comp. Math.* **12**, 251-265,(1983)

Fox L. and Hayes L. "On the definite integration of singular integrals" *SIAM rev.* **12**, 449-457, (1970)

Gauss C.F. "Methodus nova integralium valores per approximationem inveniendi" *C. F. Gauss Werke, Göttingen: Königliche Gesellschaft der Wissenschaften* **3**, 163-196, (1866)

Gradshteyn I. S. and Ryzhik I. M. "Table of integrals, series and products" *Academic Press, New York*,(1965)

Haber S. "The tanh rule for numerical integration" *SIAM J. Numer. Anal.* **14**,668-685,(1977)

Harris C. G. and Evans W. A. B. "Extension of numerical quadrature formulae to cater for end-point singular behaviour over finite intervals" *Int. J. Comp. Maths.* **6**,219-227,(1977)

Iri M., Moriguti S. and Takasawa Y. "On certain quadrature formulae" *Kokyuroko Res.Inst.Math. Sci.,Kyoto Univ.* **91**,82-118,(1970)

Levin D. "Development of non-linear transformations for improving convergence of sequences" *Int. J. Comp. Maths.*

Ma J., Rokhlin V. and Wandzura S. "Generalized Gaussian quadrature rules for systems of arbitrary functions" *SIAM J. Num. Anal.* **33**, No.3, 971-996, (1996)

Maz'ya V. "Approximate approximation" *The mathematics of finite elements and its applications* ed. Whiteman J. R., Wiley and Sons, (1993)**3**,371-388, (1973)

Mori M. "Quadrature formulae obtained by variable transformations and the D. E. rule" *J. Comp. Appl. Math.* **12** and **13**,119-130,(1985)

Murota K. and Iri M. "Parameter tuning and repeated application of the IMT-type transformation in numerical quadrature" *Num. Math.* **38**,347-363,(1982)

Patterson T. N. L. "The optimum addition of points to quadrature formulae" *Math.Comp.* **22**, 847-856, (1968)

Penzel F. "Boundary elements for 3-D cracks" *The mathematics of finite elements and its applications* ed. Whiteman J. R., Wiley and Sons, (1993)

Schwartz C. "Numerical integration of analytic functions" *J. Comp.Phys.* **4**,19-29,(1969)

Simpson T. "Mathematical dissertations on Physical and Analytic subjects", London (1743)

Shanks D. "Non-linear transformations of divergent and slowly convergent sequences" *J. Math. Phys.* **34**,1-42,(1955)

Slodicka M. "Parabolic equations with weakly singular positive type memory terms" *The mathematics of finite elements and its applications* ed. Whiteman J. R., Wiley and Sons, (1993)

Squire W. "A quadrature method for finite intervals" *Int. J. Num. Meth. Engng.* **10**,708-712,(1976)

Squire W. "A comment on quadrature in the presence of end-point singularities" *Int. J. Comp. Maths.* **7**,239-241,(1979)

Stenger F. "Integration formulae based on the trapezoidal formula" *J. Inst. Maths. Applic.* **12**,102-114,(1973)

Takahasi H. and Mori M. "Quadrature formulas obtained by variable transformation" *Numer. Math.* **21**,206-219,(1973)

Takahasi H. and Mori M. "Double exponential formulae for numerical integration" *Publ. Res. Inst. Math. Sci., Kyoto Univ* **9**,721-741,(1974)

Wynn P. "On a device for computing the $e_m(s_n)$ transformation" *Maths. Comp.* **10**,91-96,(1956)

Zienkiewicz O. C. "The finite element method" *McGraw-Hill*, (1977)

A H SIDDIQI[1], M K AHMAD[2] AND A MUKHEIMER[3]

Current developments in fractal image compression

1. Introduction

Image Compression Techniques and methods are of vital importance in high technologies especially in Information Technology. These techniques and methods are related to communicating and storing images and data (information) in the shortest possible time (in minimum space on hard disk or CD-ROM) and retrieval with a permissible distortion. Tools and Techniques of Fourier Analysis like Fourier Transform, Convolution, Shannon Sampling theorems and Walsh - Fourier methods have been successfully used for a long time. Matrix transform, Optimization and Variational techniques were also employed to study image and data compression. Wavelet theory entered in this area in the late eighties and started replacing the role of DCT(JPEG) which was dominating the scene till that time in view of its high compression ratio and the quality of information after retrieval. An interesting account of this development with comprehensive references can be found in chapter 5 [37]. For more technical details we refer to Wickerhauser [58].

The concept of a fractal and the discipline of Fractal Geometry was introduced by Benôit Mandelbrot in the early eighties to study irregular shapes like coastal lines, mountains, clouds or rain fall. Fractals are complicated looking sets like Cantor set, Sierpinski gasket, Sierpinski carpet, von-Koch curve, Julia set, but they arise out of simple algorithms. By now it is a well established discipline and a comprehensive and updated bibliography can be found in, for example, Barnsley [4], Barnsley and Hurd [6], Fisher [12], and web sources mentioned in this article.

A conference in the third week of June 1997 in Arcachon in the south west of France with Mandelbrot as the honorary chairman and all other leading workers in the field in the organising committee has treated the following theoretical and application aspects. Theoretical aspect: Dimension theory, Multi-fractal Analysis, IFS theory, Harmonic Analysis of Fractals, Dynamical systems, Wavelet and Fractals and Stochastic Analysis on Fractals. Applications: Signal Processing, Image Processing, Finance, Fractal Wave interactions, Computer Networks, Traffic Analysis, Processes on Fractal surfaces and Mechanical Aspects of Fractals.

Barnsley [4,5] established a close connection between Functional Analysis, Fractals and Multimedia by demonstrating that Fractals can be defined in terms of fixed points of mappings defined on an appropriate metric space into itself and image compression can be studied through this methodology achieving marvelous results. He has already commercialized his achievements in the form of the top selling multimedia encyclopedia *Encarta*, published by Microsoft corporation, includes on one CD-ROM seven thousand colour photographs which may be viewed interactively on a computer screen. There are diverse images like that of buildings, musical instruments, people's faces, baseball bats, ferns. The most surprising thing is that even many functional analysts do not know that the *Banach Contraction Fixed Point Theorem* (contraction mapping theorem) is the key to this exciting development. This development is known as the IFS theory (Iterated Function System theory) for image compression. Stewart [53] has explained in a simple language the far-reaching consequences of this theory which will drastically reduce the expenses on communication through fax machines manufactured based on this technique and methodology. It has been established that for a fairly large class of images IFS theory provides better compression ratio and quality of images after retrieval (see for example Fisher [12,13] and recent papers on web resources) compared to most popular methods till now like DCT(JPEG) and Wavelet. It is also expected that the combination of fractal and wavelet techniques may still yield better results. This is a fast growing field in which besides mathematicians, computer and informatics scientists, physicists, chemists and engineers are actively involved and so it seems difficult to keep track of every new results. However we have tried our level best to include updated literature. The main objective of this paper is to discuss salient features of the IFS theory incorporating our work in this area. The proof of those results will appear separately. Although compression ratio for many types of data and images have been examined by IFS theory, we have mentioned our investigations for some wood blocks and a singer and a bride of our sub-continent (see Figures 1-6). In our case, wavelet method gives slightly better result than IFS method which gives far superior result than JPEG (see Table 2.).

2. IFS Theory

Let (X, d) be a complete metric space and let $T : X \longrightarrow X$ be a mapping of X into itself. The iterates of T are mappings $T^{0n} : X \longrightarrow X$, defined by $T^{00}(x) = x, T^{01}(x) = T(T^{00}(x)) = T(x), T^{02}(x) = T(T^{01}(x))$ or $T^{02} =$

$T^{00}oT^{01}, \cdots, T^{0n}(x) = T(T^{0(n-1)}(x))$ or $T^{0n} = ToT^{0(n-1)}$. A mapping F on the set of real numbers R into itself is called an affine transformation if it is of the form $F(x) = ax + b$ for all $x \in R$, where a and b are constants. A mapping G on R^2 into itself is called an *affine transformation* if it can be written as

$$G(x, y) = (ax + by + e, cx + dy + f)$$

where a, b, c, d, e and f are real numbers. G can also be written as:

$$G\begin{pmatrix} x \\ y \end{pmatrix} = \begin{pmatrix} a & b \\ c & d \end{pmatrix}\begin{pmatrix} x \\ y \end{pmatrix} + \begin{pmatrix} e \\ f \end{pmatrix} = Ax + B$$

where $A = \begin{pmatrix} a & b \\ c & d \end{pmatrix}$ is 2×2 matrix and B is the column vector $\begin{pmatrix} e \\ f \end{pmatrix}$. The mapping T is called a Lipschitz mapping if there exists a constant $\alpha \geq 0$ such that $d(T(x), T(y)) \leq \alpha d(x, y)$ for all $x, y \in X$. T is called *contraction mapping* if $0 \leq \alpha < 1$. α is called the contractivity factor of T. A Lipschitz continuous function T is called *eventually contractive* if there is a number n such that T^{0n} is a contraction map. Let Y be a non-empty subset of a compact metric space (X, d) then a mapping S on Y into X is called a *local (partitioned) contraction mapping* on (X, d) if there is a number $s, 0 \leq s < 1$, such that $d(S(x), S(y)) \leq sd(x, y)$ for all $x, y \in Y$. A complete metric space (X, d) equipped with n contraction mappings $w_i : X \longrightarrow X, i = 1, 2, \cdots, n$, denoted by $\{X, d, w_i, i = 1, 2, \cdots\}$ is called an *iterated function system (IFS)*. A complete metric space (X, d) equipped with n eventually contractive mappings $w_i : X \longrightarrow X, i = 1, 2, \cdots, n$, denoted by $\{X, d, w_i, i = 1, 2, \cdots, n\}$ is called an *eventually iterated function system* . A local (partitioned) iterated function system (LIFS) is a compact metric space (X, d) equipped with n local contractive mappings $w_i : Y \longrightarrow X$, $Y \subseteq X$. A *recurrent iterated function system* (RIFS) is a collection $w_1, w_2, \cdots w_n$ of n Lipschitz maps in a complete metric space X and $n \times n$ matrix (a_{ij}) satisfying $\sum_j a_{ij} = 1$ for all i. The Iterated Function System (IFS) theorem [4] states that:

For an IFS $\{X, d, w_i\}$ with contractivity factors $\alpha_i, i = 1, 2, \cdots n$, the mapping W defined on $H(X)$ = space of all compact subsets of X, into itself by $W(B) = \cup_{i=1}^n w_i(B)$ is a contraction mapping on the complete metric space $(H(X), h(.,.))$, where

$$h(A, B) := \max\{d(A, B), d(B, A)\},$$

$$d(A, B) := \max\{d(x, B) \mid x \in A\} ,$$

$$d(x, B) := \min\{d(x, y) \mid y \in B\}$$

with the contractivity factor $\alpha = \max_{1 \leq i \leq n} \alpha_i$, that is,

$$h(W(B), (W(C)) \leq \alpha h(B, C)$$

for all $B, C \in H(X)$. Consequently, W has a unique fixed point say $A \in H(X)$ which satisfies the relation $A = W(A) = \cup_{i=1}^{n} w_i(A)$, and is given by

$$A = \lim_{n \to \infty} W^{0n}(B) \quad \text{for any} \quad B \in H(X) .$$

This fixed point A is called the *attractor or deterministic fractal or fractal.* $h(.,.)$ is known as the Hausdorff metric.

Let $\{R^2, d, w_i, i = 1, 2, \cdots n\}$ be an IFS where w_i's are given by

$$w_i \begin{pmatrix} x \\ y \end{pmatrix} = \begin{pmatrix} a_i & b_i \\ c_i & d_i \end{pmatrix} \begin{pmatrix} x \\ y \end{pmatrix} + \begin{pmatrix} e_i \\ f_i \end{pmatrix}, i = 1, 2, \cdots n .$$

Then the following table is known as the IFS code

w	a	b	c	d	e	f	p
w_1	a_1	b_1	c_1	d_1	e_1	f_1	p_1
w_2	a_2	b_2	c_2	d_2	e_2	f_2	p_2
$\vdots$	$\vdots$	$\vdots$	$\vdots$	$\vdots$	$\vdots$	$\vdots$	$\vdots$
w_n	a_n	b_n	c_n	d_n	e_n	f_n	p_n

Table 1. IFS code

where

$$p_i \simeq \frac{|detA|}{\sum_{i=1}^{N} |detA_i|} = \frac{|a_i d_i - b_i c_i|}{\sum_{i=1}^{N} |a_i d_i - b_i c_i|}$$

for $i = 1, 2, \cdots n$, the symbol $\simeq$ means "approximately equal to". The numbers p_i's can be interpreted as probabilities for finding the attractor of an IFS using the chaos game algorithm. An image can be treated as a closed bounded (compact) subset of R^2. The following result, known as the *Collage Theorem,* is very important for designing IFS whose attractors or fractals are close to a given image. Let (X, d) be a complete metric space and S be a given image, that is, $S \in H(X)$ and let $\epsilon \geq 0$ be also given. Choose an IFS $\{X, d, w_i\}, i = 1, 2, \cdots n$, with contractivity factors $\alpha_i, i = 1, 2, \cdots n$, such that

$$h(S, \cup_{i=1}^{n} w_i(S)) \leq \epsilon .$$

Then

$$h(S, A) \leq \frac{1}{1-s} h(S, W(S)) \leq \frac{\epsilon}{1-s} ,$$

where

$$W(s) = \cup_{i=1}^{n} w_i(S)) \quad \text{and} \quad s = \max_{1 \leq i \leq n} \alpha_i$$

and A is the attractor of the IFS. This theorem precisely tells us that if we can find an IFS code so that the Hausdorff distance between S and $W(S)$ is very small, the attractor of W will be very close to the target image S. There are algorithms for finding attractors of given IFS like the Chaos Game Algorithm, the Photocopy Machine Algorithm (see Barnsley and Hurd [6] and Lu [30]).

Basic technique of image compression through IFS is to find out appropriate affine contraction mappings $w_1, w_2, \cdots w_n$ such that the condition of *Collage Theorem* is satisfied namely S is very close to $W(S)$ and so instead of communicating / storing the image we can communicate / store the fractal or attractor of IFS, that is, coefficients in the IFS code. For details of this method we refer to Barnsley [4,5], Barnsley and Hurd [6], Lu [30], Jacquin [24,25], Fisher [12], Mazel and Hayes [33]. Most of the results concerning IFS have been extended for eventually, recurrent, and local (partitioned) IFSs enhancing the compression ratio and the quality of retrieved pictures (see Fisher [12], Saupe, Hamzaoui, Hartenstein [46], Hart [22], Barnsley, Elton and Hardin [7]).

3. Inverse Problem For Image (Function)

The problem of representing a given image (or a function) by the IFSs or their variations is a typical inverse problem. This involves finding the IFS parameters of an image that is exactly generated via an IFS. In the recent years the *Iterated Function System with probabilities* (IFSP), *Iterated Fuzzy Set Systems* (IFZS) and IFS *with grey level maps* (IFSM) have been introduced and the corresponding inverse problems have been investigated. Such an inverse problem is related to the problem of finding the image / function as the fixed point element of a given iteration algorithm of the type IFS, IFSP and IFSM on function spaces like L_p, $H^{m,p}$ (Sobolev space of order m), weighted Sobolev space and Sobolev space on an arbitrary metric space. In purely abstract mathematical terms it comprises the following steps:

(i) Finding a suitable metric space X in which to represent the image (function)

(ii) Finding an appropriate metric $d(.,.)$ on X

(iii) Finding an appropriate contraction map T on X into itself.

The fact that such problems have more than one solution motivated the search of different kinds of optimality. This problem has been studied in the recent years by Forte and Vrscay in series of papers (see [14-17]) in the setting for functions belonging to L_p and the space of distributions. Siddiqi, Ahmad and Mukheimer (see [49-51]) have studied the problem in the setting of Sobolev and Besov spaces.

We present here a resume of these results.
Let $F^*(X)$ denote the set of all functions $u : X \longrightarrow [0,1]$ which are (i) upper semi continuous on (X, d) (ii) normalized, that is, for each $u \in F^*(X)$ there exists an $x_0 \in X$ such that $u(x_0) = 1$. For $u \in F^*(X)$ and for each $\alpha \in [0,1]$ the α-level sets of u are defined by

$$[u]^\alpha = \{x \in X \mid u(x) \geq \alpha\}, \alpha \in (0,1]$$

$$[u]^0 = \overline{\{x \in X \mid u(x) > 0\}}\,.$$

It is clear that $[u]^\alpha \in H(X)$. For each $u, v \in F^*(X)$ we can define a metric

$$d_\infty(u,v) = \sup_{0 \leq \alpha \leq 1} h([u]^\alpha, [v]^\alpha)\,.$$

$(F^*(X), w_i, \Phi)$, where $w_i : X \longrightarrow X$ are contraction maps, $\Phi = \{\phi_1, \phi_2, \cdots \phi_n\}$ and ϕ_i's $i = 1, 2, \cdots n$ are mappings of [0,1] into itself ($\phi_i : [0,1] \longrightarrow [0,1]$), each of them being (a) non-decreasing (b) right continuous (c) $\phi_i(0) = 0$ for all i (d) for at least one i, $\phi_i(1) = 1$, is called the *Iterated Fuzzy set system (IFZS)*.

Let contraction mappings $w_1, w_2, \cdots, w_n$ be associated with probabilities $p_1, p_2, \cdots, p_n$ with $\sum_{i=1}^n p_i = 1$. Further more, let $\mathcal{B}(X)$ denote the $\sigma-$algebra of Borel subsets of X and $\mathcal{M}(X)$ denote the set of all probability measure on $\mathcal{B}(X)$. $(\mathcal{M}(X), w_i, T)$, where T is defined by the relation

$$(T\nu)(S) = (M\nu)(S) = \sum_{i=1}^{n} p_i(\nu(w_i^{-1}(S)))$$

for a $\nu \in \mathcal{M}(X)$ and each $S \in H(X)$ is called the *Iterated Function system with probabilities (IFSP)*. $\mathcal{M}(X)$ is a metric space with respect to the metric

$$d_H(\mu, \nu) = \sup_{f \in Lip_1(X)} \left| \int_X f d\mu - \int_X f d\nu \right|$$

where

$$Lip_1(X) = \{f : X \longrightarrow R \mid |f(x_1) - f(x_2)| \leq d(x_1, x_2) \ \forall \ x_1, x_2 \in X\} .$$

T defined above is called the Markov operator on $(\mathcal{M}(X), d_H(.,.))$. For a measure μ on $\mathcal{B}(X)$ and for any integer $p \geq 1$, let $L_p(X, \mu)$ denote the vector space of all real valued functions u such that u^p is integrable on $(\mathcal{B}(X), \mu)$. $L_p(X, \mu)$ is a complete metric space with respect to the metric induced by the L_p norm, that is,

$$d(u, v) = \|u - v\|_{L_p} = \left(\int_X |u(x) - v(x)|^p d\mu(x) \right)^{1/p} .$$

$(L_p(X, \mu), w_i, \Phi)$, where $\Phi = \{\phi_1, \phi_2, \cdots \phi_n\}$ with $\phi_i : R \longrightarrow R$, known as the grey level maps, is called the *Iterated Function system with grey level maps (IFSM)*.

An *operator* T can be defined on IFSM as

$$(Tu)(x) = \sum_{i=1}^{n}{}' \phi_i(u(w_i^{-1}(x))) .$$

The prime ($'$) signifies that the sum operates on all those terms for which $w_i^{-1}(x)$ is defined. If $w_i^{-1}(x) = 0$ for all $i = 1, 2, \cdots, n$ then $(Tu)(x) = 0$.

For $X \subset R^n$, let $m^{(n)} \in \mathcal{M}(X)$ denote the Lebesgue measure on $\mathcal{B}(X)$. The Indicator function of a subset A of X denoted by $I_A(x)$ is defined by

$$I_A(x) = \begin{cases} 1, & x \in A \\ 0, & otherwise \end{cases}$$

$$Lip(R) = \{\phi : R \longrightarrow R \mid |\phi(t_1) - \phi(t_2)| \leq \beta |t_1 - t_2|\}$$

$\forall\ t_1, t_2 \in R$ and for some $\beta \in [0, \infty)$. It can be verified that for any $u \in L_p(X, \mu)$, $1 \leq p < \infty$, and $\phi \in Lip(R)$, $1 \leq i \leq n$, T is a mapping on $L_p(X, \mu)$ into itself. In fact, T becomes a contraction mapping under certain assumptions and hence has a unique fixed point as $L_p(X, \mu)$ is a complete metric space.

Affine IFSM on $L_p(X, \mu)$ is that IFSM on $L_p(X, \mu)$ where ϕ_i are given by $\phi_i(t) = \alpha_i t + \beta_i, \quad t \in R, \quad i = 1, 2, \cdots n$. Let $X = [0, 1]$ and $\mu = m^{(1)}$ with $w_i(x) = s_i x + a_i$ and $\alpha_i = |s_i| < 1, \quad 1 \leq i \leq n$. If T is contractive with fixed point $\bar{u}$, then

$$\bar{u}(x) = \sum_{i=1}^{n}{}' \alpha_i \bar{u}\left(\frac{x - a_i}{s_i}\right) + \beta_i I_{w_i(X)}(x) = \sum_{i=1}^{n}{}' \left[\alpha_i \psi_i(x) + \beta_i \chi_i(x)\right].$$

This means that $\bar{u}$ may be expressed as a linear combination of both piece wise constant functions $\chi_i(x)$ as well as functions $\psi_k(x)$ which are obtained by dilations and translations of $\bar{u}(x)$ and $I_X(x) = 1$ respectively. This reminds us of the role of scaling functions in wavelet theory.

The Collage Theorem mentioned earlier can be rephrased as follows:

Let (X, d) be a complete metric space, and for a given $x \in X$ there exists a contraction map $W : X \longrightarrow X$ with contractivity factor α such that $d(x, W(x)) < \epsilon$. Then

$$d(x, \bar{x}) < \frac{\epsilon}{1 - \alpha},$$

where $\bar{x}$ is the fixed point of W $(W(\bar{x}) = \bar{x})$.

In view of this result the inverse problem for approximation of functions in $L_p(X, \mu)$ by IFSM may be stated as follows:

Given a target function $v \in L_p(X, \mu)$ and a $\delta > 0$, find an IFSM $(L_p(X, \mu), w_i, \phi_i)$ with associated operator T such that $\|v - Tv\|_p \leq \delta$.

For $\mu \in M(X)$, a family $\mathcal{A}$ of subsets $A = \{A_i\}$ of X is called μ-dense in a family $\mathcal{N}$ of subsets B of X if for every $\epsilon > 0$ and any $B \in \mathcal{N}$ there exists a collection $A \in \mathcal{A}$ such that $A \subseteq B$ and $\mu(B \setminus A) < \epsilon$. Let $\{w_i\}$ be an infinite sequence of contraction maps on X into itself. We say that $\{w_i\}$ generates a "μ-dense and non-overlapping"- to be abbreviated as "$\mu - d - n$" family $\mathcal{A}$ of subsets of X if for every $\epsilon > 0$ and every $B \subseteq X$ there exists a finite set of integers $i_k \geq 1, \quad 1 \leq k \leq n$, such that

(i) $A = \cup_{k=1}^{n} w_{i_k}(X) \subseteq B$

(ii) $\mu(B \setminus A) < \epsilon$ and

(iii) $\mu(w_{i_k}(X) \cap w_{i_l}(X)) = 0$ if $k \neq l$

If $\{w_i\}$ satisfies the above conditions on (X, d), then $\inf_{1 \leq i < \infty}\{\alpha_i\} = 0$, where α_i's are the contractivity factors of w_i's, independent of μ. If $X = [0,1]$ and μ is the Lebesgue measure, then the wavelet type functions

$$w_{ij}(x) = 2^{-i}(x + j - 1), \quad i = 1, 2, \cdots, j = 1, 2, \cdots, 2^i$$

can form a "$\mu - d - n$" family.

For each $i^* \geq 1$, the set of maps $\{w_{i^*j}, \quad j = 1, 2, \cdots, 2^{i^*}\}$ provides a set of 2^{-i^*} contractions of $[0,1]$ which tile $[0,1]$. In 1995 Forte and Vrscay obtained the following result which provided the solution of the inverse problem.

Theorem A[16] Let $v \in L_p(X, \mu), \quad 1 \leq p < \infty$, then

$$\lim_{n \to \infty} \inf \|v - T^n v\|_p = 0,$$

provided the sequence of contraction maps w_i generates a "$\mu - d - n$" family $\mathcal{A}$ of subsets of X and w_i's are also one-to-one, where

$$(T^n v)(x) = \sum_{i=1}^{n}{}' \phi_i(u(w_i^{-1}(x))) .$$

This theorem has been studied for local IFSM and special cases like $p = 2$ and ϕ_i's affine maps. Forte and Vrscay have also carried out approximation of the target image "Lena", a 512×512 pixel grey scale image, with each fixed pixel having 256 possible values (8 bits, with values from 0 to 255, which are rescaled to values in [0,1]). This type of approximation is being studied by us for our sub-continent's singer and bride.

As mentioned above the inverse problem in the setting of *Sobolev, weighted Sobolev, Orlicz Sobolev spaces, and Sobolev space on an arbitrary metric space* have been studied recently. For presenting a typical result the following definitions and notations are required. We will confine ourselves to $X = [a, b]$ being a closed interval in one dimension, however most of the results can be extended to n-dimensional case ($X \subseteq R^n$) or even in the case where X is a complete metric space. For detailed information about these spaces we refer to Adam [1], Kufner [27], Kufner and Sändig [28], Griffel [19], and Hajlasz *et al.* [20,21]. The *support* of a function $f : R \longrightarrow R$ is the closure of $\{x \mid f(x) \neq 0\}$ denoted by supp(f). f is said to have *bounded support* if supp$(f) \subset [a, b]$. It is said to have *compact support* if the closure of the set of points at which f does not vanish is compact. f is called *infinitely differentiable or smooth* if its derivatives of all orders exist and are continuous. A *test function* is a smooth function with compact (in one dimension only, bounded) support. The set of all test functions over R / $[a, b]$ is denoted by $\mathcal{D}(R)$ / $\mathcal{D}(a, b)$. A linear function F on $\mathcal{D}(R)$ is a map $F : \mathcal{D}(R) \longrightarrow R$ such that

$$F(\phi + \psi) = F(\phi) + F(\psi) \quad \text{and} \quad F(\alpha\phi) = \alpha F(\phi)$$

for all $\phi, \psi \in \mathcal{D}(R)$ and $\alpha \in R$.

$$F(\phi) = \phi(0) \quad \text{and} \quad G(\phi) = \int_a^b f(x)\phi(x)dx\,,$$

where f is a real valued integrable function, are examples of linear functionals while H defined as

$$H(\phi) = \int_a^b |\phi(t)|^2 dt$$

is not linear. A sequence of test functions $\{\phi_n\}$ is said to converge to ϕ ($\phi_n \longrightarrow \phi$) if

(i) there is an interval $[a, b]$ containing supp(ϕ) and supp(ϕ_n) for all n, and

(ii) for any k, $\phi_n^{(k)}(x) \longrightarrow \phi^{(k)}(x)$ as $n \to \infty$, uniformly for $x \in [a, b]$, where $\phi^{(k)}$ denotes the k-th derivative of ϕ.

A functional F on $\mathcal{D}(R)$ is called *continuous* if it maps a convergent sequence of $\mathcal{D}(R)$ into a convergent sequence in R, that is, if $F(\phi_n) \longrightarrow F(\phi)$ whenever $\phi_n \longrightarrow \phi$ in $\mathcal{D}(R)$. A continuous linear functional on $\mathcal{D}(R)$ is called a *distribution or generalized function* and the space of all distributions is denoted by $\mathcal{D}'(R)$. $F(\phi)$ or (F, ϕ) will denote the image of ϕ under F or action of F

at ϕ. The *delta distribution* is defined by $(\delta, \phi) = \phi(0)$ for all ϕ in $\mathcal{D}(R)$. A function $f : R \longrightarrow R$ is called *locally integrable* if

$$\int_a^b |f(x)| dx$$

exists for all real numbers a and b. $\log|x|$ and $|x|^a$ for $-1 < a \leq 0$ are locally integrable, but functions with stronger singularities, such as $|x|^a$, for $a \leq -1$, are not. To every locally integrable function f there corresponds a distribution f defined as

$$F(\phi) = (F, \phi) = \int_a^b f(x)\phi(x) dx \; .$$

F is said to be generated by the function f, some times denoted by F_f but often this distribution is denoted by simply f. Such distributions are called *regular* distributions. All other distributions are called *singular*. The space of all distributions is a *vector space*. It is also a *metric space*. The product of two distributions is not necessarily a distribution unless one of them is a smooth function. The *Heaviside function* H is defined by

$$H(x) = \begin{cases} 0, & x < 0 \\ \frac{1}{2}, & x = 0 \\ 1, & x > 0 \end{cases}$$

and an almost identical function H_1 is defined by

$$H_1(x) = \begin{cases} 0, & x \leq 0 \\ 1, & x > 0 \end{cases}$$

H and H_1 are locally integrable and they generate the same regular distribution

$$H_f(\phi) = \int_0^\infty \phi(x) dx$$

and

$$H_{1f}(\phi) = \int_0^\infty \phi(x) dx \; .$$

There is no distinction between H and H_1 in distribution theory and this reflects the fact that from the point of view of the physicists the distinction between them is artificial. In fact one could never distinguish between H and

H_1 experimentally. Two distributions F and G are called equal on (a, b) if $(F, \phi) = (G, \phi)$ for all $\phi \in \mathcal{D}(a, b)$ such that $\text{supp}(\phi) \subset (a, b)$. The *derivative* of a distribution f denoted by $\mathcal{D}_F$ or F' is defined by

$$(\mathcal{D}_F, \phi) = -(F, \phi')$$

or $(D_F(\phi) = F(\phi'))$ for all $\phi \in \mathcal{D}$. $|x|$ is not differentiable in the classical sense but if we treat it as a distribution being locally integrable, its distributional derivative is the *sign function* defined as

$$sgn(x) = \begin{cases} -1, & x < 0 \\ 1, & x > 0 \end{cases}$$

The distributional derivative of the Heaviside and the Dirac delta functions are respectively the Dirac delta function and $-\phi'(0)$. If a locally integrable function is also differentiable in the classical sense then the two derivatives coincide. It is clear that every distribution is differentiable in the sense of the definition mentioned above. A sequence of distributions $\{F_n\}$ is said to be *convergent* if the sequence of numbers $\{F_n(\phi)\}$ is convergent for every test function ϕ. A distribution F is an integral of a distribution G if $\mathcal{D}_F = G$. For any distribution F there are infinitely many distributions G such that $\mathcal{D}_G = F$, and difference between any two of them is a constant.

The space of all distributions generated by Cauchy sequences in $\mathcal{D}(a, b)$ is denoted by $\mathcal{L}_2(a, b)$. $\mathcal{L}_2(a, b)$ is a *Hilbert space* with respect to the inner product defined by

$$\langle F, G \rangle_{\mathcal{L}_2} = \lim_{n \to \infty} \int_a^b \phi_n(x) \psi_n(x) dx = \lim_{n \to \infty} \langle \phi_n, \psi_n \rangle_{\mathcal{L}_2} ,$$

where $\{\phi_n\}$ and $\{\psi_n\}$ are Cauchy sequences generating the distributions F and G. For any Cauchy sequence $\{\phi_n\}$ of test functions with compact support, there is a distribution F defined by $(F, \psi) = \lim_{n \to \infty}(F, \phi_n)$ for all $\psi \in \mathcal{D}(a, b)$. Such a distribution F is called the distribution generated by $\{\phi_n\}$. It can be shown that the space of all square integrable functions in the sense of Lebesgue over (a, b), i.e., $L_2(a, b)$, is a subspace of the Hilbert space $\mathcal{L}_2(a, b)$ with respect to the inner product defined above.

Sobolev space $\mathcal{H}^1(a, b)$ of distributions of order 1 is a subspace of $\mathcal{L}_2(a, b)$ of those elements whose distributional derivatives also belong to $\mathcal{L}_2(a, b)$, that is,

$$\mathcal{H}^1(a, b) = \{F \in \mathcal{L}_2(a, b) \mid D_F \in \mathcal{L}_2(a, b)\}$$

$\mathcal{H}^1$ is a Hilbert space with respect to the inner product,

$$\langle F, G\rangle_{\mathcal{H}^1} = \langle F, G\rangle_{\mathcal{L}_2(a,b)} + \langle DF, DG\rangle_{\mathcal{L}_2(a,b)}$$

$$= \lim_{n\to\infty} \int_a^b \phi_n(x)\psi_n(x)dx + \lim_{n\to\infty} \int_a^b \phi_n'(x)\psi_n'(x)dx$$

or

$$\|F\| = \|F\|_{\mathcal{L}_2(a,b)} + \|D_F\|_{\mathcal{L}_2(a,b)} \,.$$

An analogue of theorem A for elements of $\mathcal{H}^1(a,b)$ is the following.

Theorem B [49] Let $F \in \mathcal{H}^1(a,b)$ then

$$\lim_{n\to\infty} \inf \|F - T^n F\|_{\mathcal{H}^1} = 0,$$

where $\{\mathcal{H}^1(a,b), w_i, \phi_i\}$ is an IFSM,

$$X = \mathcal{H}^1(a,b), \quad w_i : \mathcal{H}^1(a,b) \longrightarrow \mathcal{H}^1(a,b), \quad \phi_i : R \longrightarrow R$$

satisfies a weak Lipschitz condition on the space of all distributions that is there exists a $k \geq 0$ such that for all $\psi \in \mathcal{D}(a,b)$,

$$\left|\int_a^b \Big[(\phi_i F_1)(x) - (\phi_i F_2)(x)\Big]\psi(x)dx\right| \leq k \left|\int_a^b \Big[F_1(x) - F_2(x)\Big]\psi(x)dx\right|$$

for all $F_1, F_2 \in \mathcal{D}'(a,b)$ and w_i's are one-to-one contraction mappings generating a $\mu - d - n$ family $\mathcal{A}$ of subsets of X. In the case of one-to-one affine maps:

$$w_i(x) = A_i(x) + b_i \;,$$

where A_i's are $n \times n$ matrices with non-zero determinant, b_i's are n-vectors; the theorem is valid if the Jacobians of the w_i's are sufficiently small.

4. Comparison of JPEG, Wavelet (EPIC) and IFS techniques for image compression

A comparison of performance in terms of compression ratio (CR), and visual appearance (PSNR) of IFS (Fractal), Wavelet (EPIC) and JPEG techniques for some wood blocks and a singer and bride of Indian sub-continent are given in the following table. For images see Figures 1-6.

Image	*Compression techniques*					
	Fractal		*EPIC*		*JPEG*	
	CR	*PSNR*	*CR*	*PSNR*	*CR*	*PSNR*
Wood 1	4.36:1	31.59	4.83:1	34.33	5:1	28.31
Wood 1	18.78:1	27.52	18.41:1	28.73	20:1	25.83
Wood 2	4.69:1	29.15	4.78:1	33.98	5:1	26.95
Wood 2	18.74:1	25.03	17.97:1	27.12	20:1	23.77
Singer	4.36:1	33.27	5.10:1	36.99	5:1	30.49
Singer	18.81:1	27.30	18.22:1	29.56	20:1	25.62
Bride	4.36:1	32.33	5.16 :1	40.35	5:1	30.64
Bride	18.79:1	27.02	18.68:1	30.65	20:1	26.06

Table 2. CR and PSNR

Wavelet technique is slightly better in terms of the visual quality of images after compression followed by Fractal and JPEG.

5. Some mathematical inquisition

(i) As we have seen that the IFS theory is based on the properties of the metric space X, and contraction maps of X into itself. The Banach contraction principle for existence of the unique fixed point plays important role. Some generalizations of the Banach contraction principle have been used to develop new kinds of IFS. The Banach contraction principle has been extended for set valued maps (see Nadler [35] and Istrâtescu [23]). Natural question arise, whether one can introduce the concept of an iterated multivalued system replacing single valued contraction maps by multi-valued (set valued) maps. The concept of multivalued contraction map is already introduced in the paper of Nadler [35] and he has extended the theorem, key to fractal image compression, in the following form:

Let (X, d) be a complete metric space and $F : X \longrightarrow H(X)$ be a multivalued contraction map on X into $H(X)$, the space of all compact subsets of X, then F has a fixed point, that is, there exists an element $u \in X$ such that $u \in F(u)$.

In view of the remark on p.477 [35] if $F : X \longrightarrow H(X)$ then $\hat{F}$ defined as $\hat{F}(B) = \cup\{F(x) \mid x \in B\}$ for all $B \in H(X)$ has a unique fixed point, say A (the attractor in our case). Further more the set of all fixed points of F, say S is contained in A. Is it possible to use this set for image compression? A result analogous to the Collage theorem may be stated as,

If $B \in H(X)$ and $h(B, F(B)) \leq \epsilon$ then

$$h(B, S) \leq \frac{\epsilon}{1-\alpha}$$

where S is the set of all fixed points of F and α contractivity factor of F .

Is this result true? Can we find inf and sup of $h(S, A)$ where A is the fractal?

(ii) Fisher [12, p.48] and Barnsley and Hurd [6] have discussed the concept of *Partitioned (Local) Iterated Function System (PIFS or LIFS)* and it has been pointed out by Fisher that a complete theory remains to be developed. There is a closely related concept in mathematics literature (see Edelstein [11] or Istrâtescu [23]) namely the concept of (ϵ, λ)- uniformly locally contractive maps. A mapping T of a metric space (X, d) into itself is called (ϵ, λ)- uniformly locally contractive if for every $x \in X$ there exists ϵ and λ, $\epsilon > 0, 0 \leq \lambda < 1$, independent of x such that

$$d(T(x), T(y)) < \lambda \, d(x, y) \quad \text{whenever} \quad d(x, y) < \epsilon \, .$$

In other words, T is *uniformly locally contractive* if T is contractive in a small neighbourhood, that is, T satisfies contractive conditions for sufficiently close points. The Banach contraction fixed point theorem has been extended in the following form [11] :

If (X, d) is ϵ- chainable complete metric space, then every $(\epsilon - \lambda)$- uniformly locally contractive map on X into itself has a unique fixed point.

Here ϵ- chainable means that for every $a, b \in X$ there exists an ϵ- chain that is finite set of points $a = x_0, x_1, \cdots, x_n = b$ (n may depend on both a and b) such that $d(x_{i-1}, x_i) < \epsilon$, $1 = 1, 2, \cdots, n$.
One may like to know

(a) What are the extensions of the IFS and Collage theorem for (ϵ, λ)-uniformly locally contractive maps?

(b) Is it possible to apply these results for image and data compression?

(iii) Banach contraction fixed point theorem has an extension for the probability metric space (PMS) see Istrâtescu [23] for details. Is it possible to develop an IFS theory in the setting of PMS?

(iv) Non-Archimedean Functional Analysis has been a popular topic of research in eighties (see Narici and Beckenstein [36], Van Rooij [57]). A natural question arise what could be the IFS theory in this strange terrain? What could be the inverse function theorem for

$$L_p(X, \mu) = \left\{ f : X \longrightarrow K \mid \|f\|_p = \left[\int_X |f(x)|^p d\mu \right]^{1/p} \right\}$$

where K is a field with non-archimedean valuation?

(v) The concept of 2-metric spaces introduced by Gähler [18] has attracted the attention of many mathematicians and an analogue of the Banach contraction mapping has been proved by Rhodes [41]. As we know axioms of metric are axiomization of the notion of distance on the real line while conditions for 2-metric are axiomization of the properties of the area of a triangle in the plane. More precisely 2-metric $d(.,.,.)$ on X is a function defined on $X \times X \times X = X^3$ such that

(ia) For every $x \neq y$ there exists $z \in X$ such that $d(x, y, z) \neq 0$

(ib) $d(x, y, z) = 0$ if any two of the three elements are equal

(ii) $d(x, y, z) = d(x, z, y) = d(z, x, y)$ for all $x, y, z \in X$

(iii) $d(x, y, z) \leq d(x, y, u) + d(x, u, z) + d(u, y, z)$ for all $u \in X$

Is it possible to define a Hausdorff metric with the help of 2-metric? Is it possible to define a 2-metric on a class of subsets of a metric space with some properties?

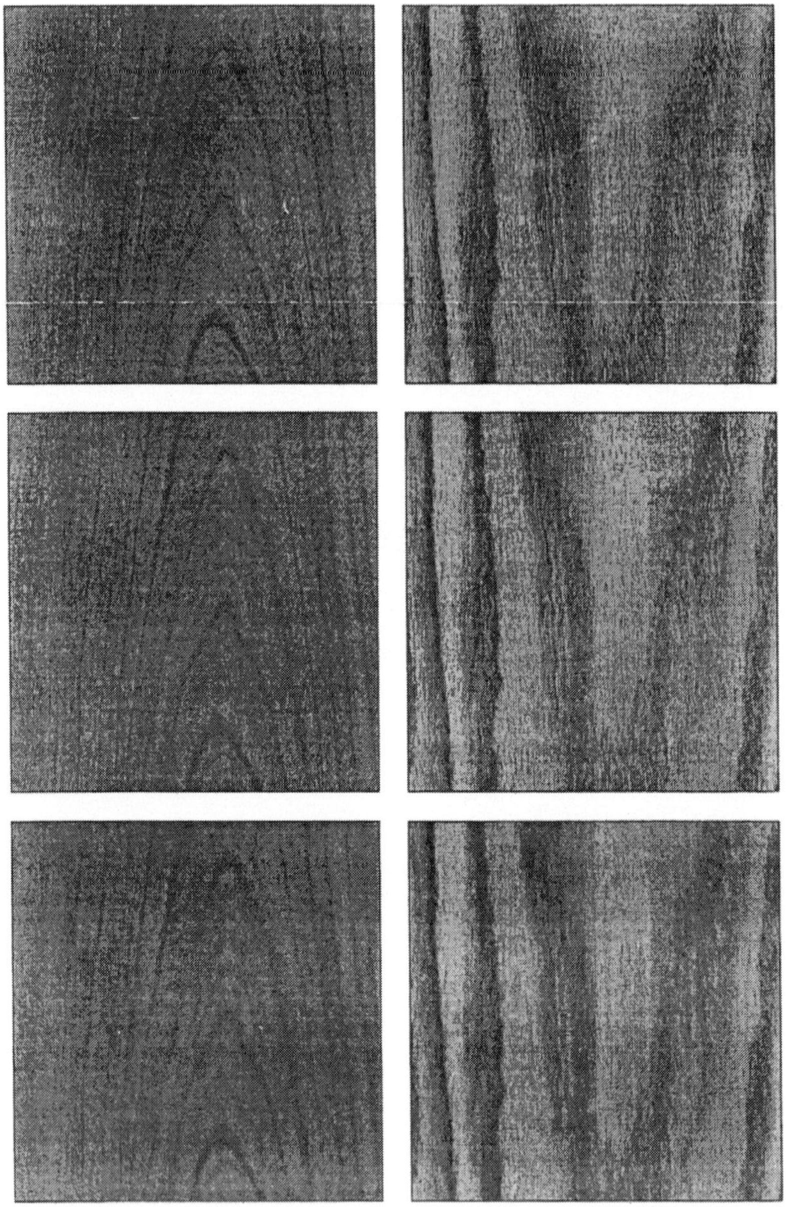

Figure 1: COMPRESSION OF WOOD BLOCKS BY **FRACTAL**

Top: Original wood blocks, **Middle:** compressed blocks, compression: 4.36:1 & 4.69:1 resp., **PSNR:** 31.59 & 29.15 dB resp., **Bottom:** compressed blocks, compression: 18.78:1 & 18.74:1 resp., **PSNR:** 27.52 & 25.03 dB resp.

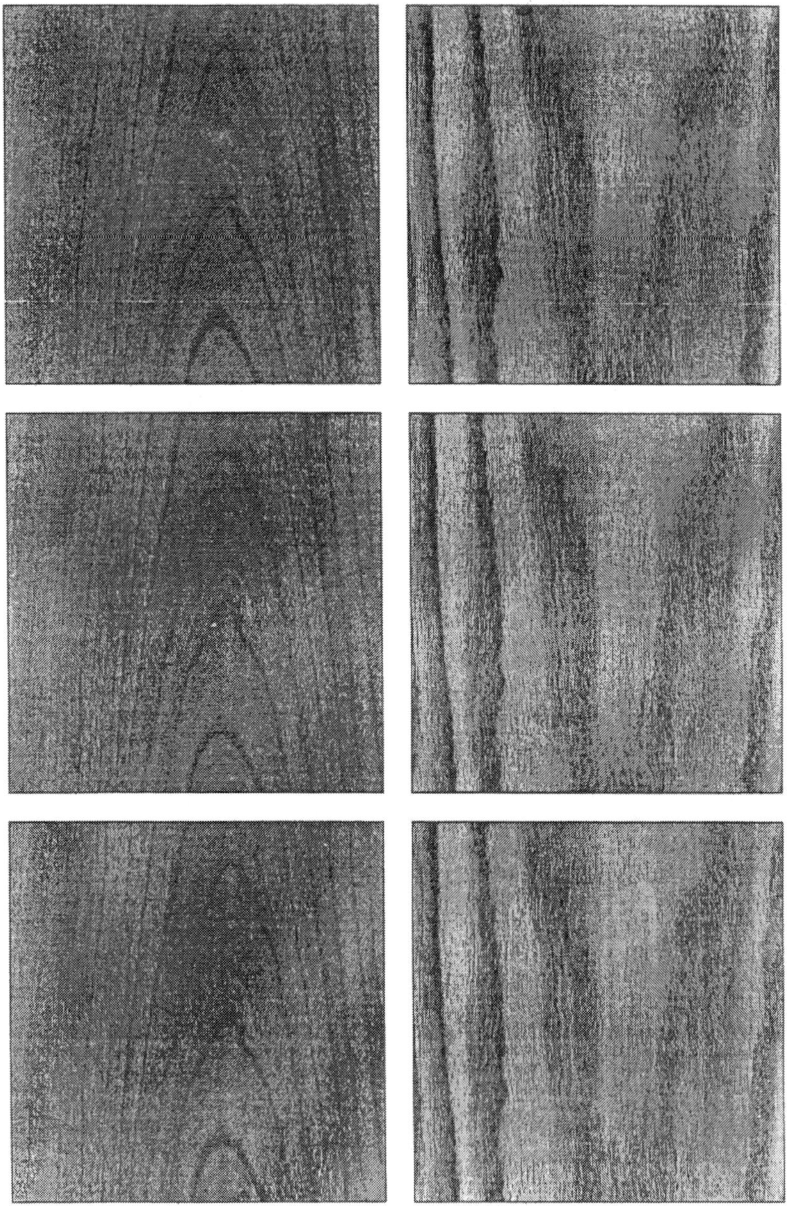

Figure 2: COMPRESSION OF WOOD BLOCKS BY **EPIC**
Top: Original wood blocks, **Middle:** compressed blocks, compression: 4.83:1 & 4.78:1 resp., **PSNR:** 34.33 & 33.98 dB resp., **Bottom:** compressed blocks, compression: 18.41:1 & 17.97:1 resp., **PSNR:** 28.73 & 27.12 dB resp.

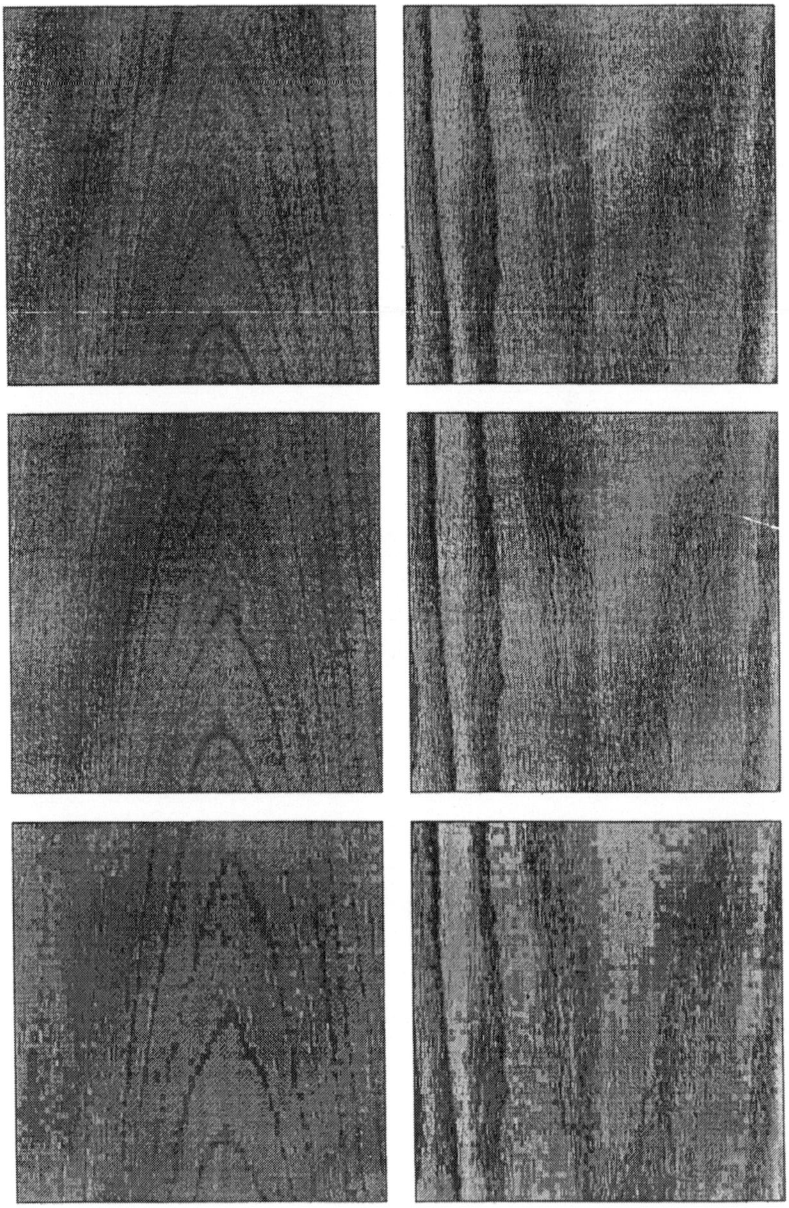

Figure 3: COMPRESSION OF WOOD BLOCKS BY **JPEG**
Top: Original wood blocks, **Middle:** compressed blocks, compression: 5:1 in both, **PSNR:** 28.31 & 26.95 dB resp., **Bottom:** compressed blocks, compression: 20:1 in both, **PSNR:** 25.83 & 23.77 dB resp.

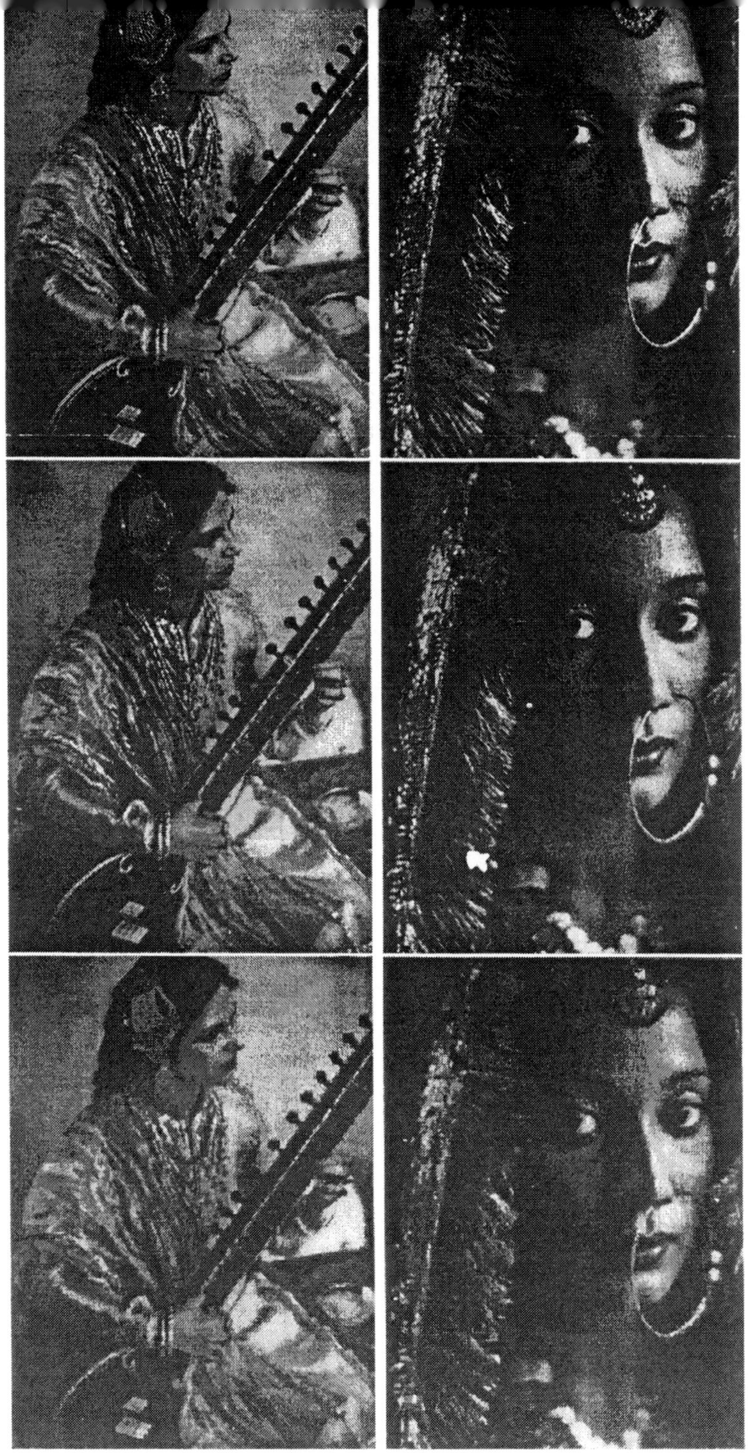

Figure 4: COMPRESSION OF SINGER AND BRIDE BY **FRACTAL**
Top: Original images, **Middle:** compressed blocks, compression: 4.36:1 in both, **PSNR:** 33.27 & 32.33 dB resp., **Bottom:** compressed blocks, compression: 18.81:1 & 18.79:1 resp., **PSNR:** 27.30 & 27.02 dB resp.

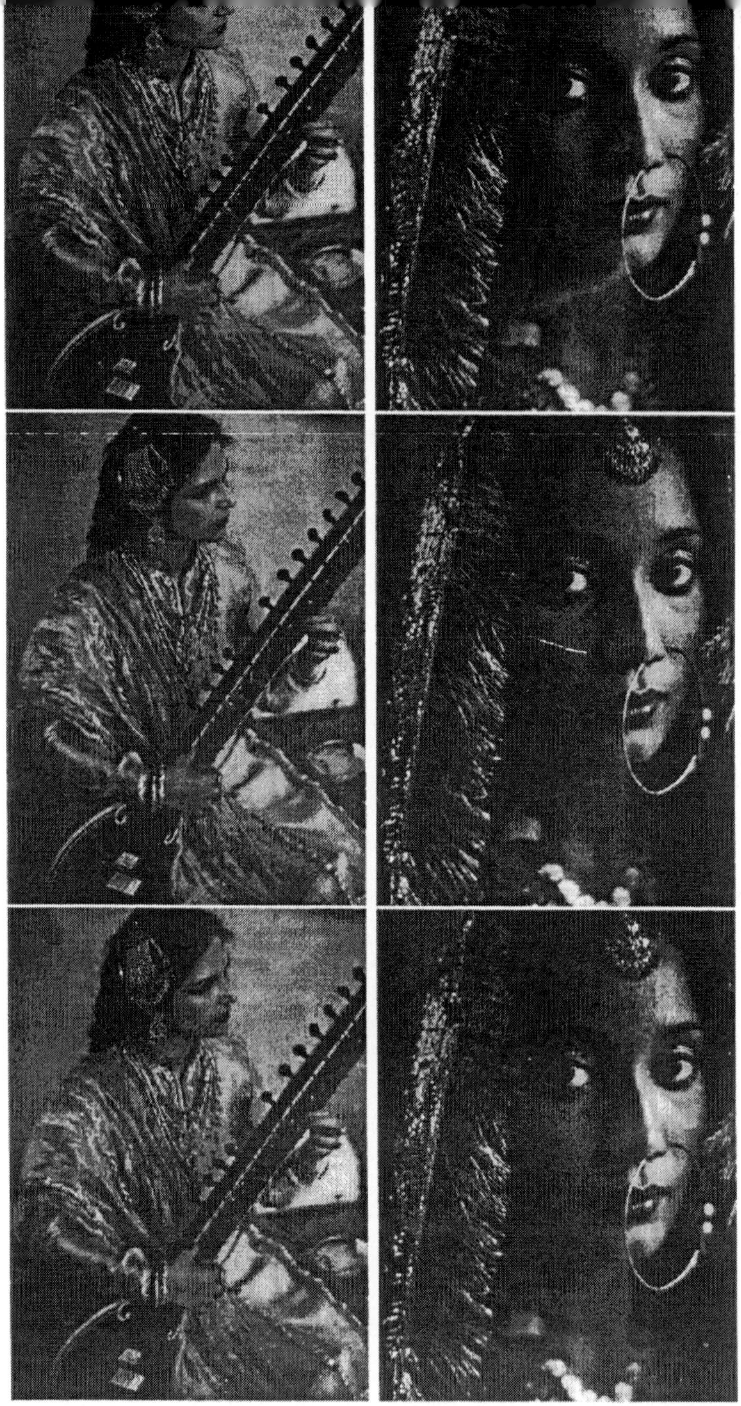

Figure 5: COMPRESSION OF SINGER AND BRIDE BY **EPIC**
Top: Original images, **Middle:** compressed blocks, compression: 5.1:1 & 5.16:1 resp., **PSNR:** 36.99 & 40.35 dB resp., **Bottom:** compressed blocks, compression: 18.22:1 & 18.68:1 resp., **PSNR:** 29.56 & 30.65 dB resp.

Figure 6: COMPRESSION OF SINGER AND BRIDE BY JPEG
Top: Original images, **Middle:** compressed blocks, compression: 5:1 in both, **PSNR:** 30.49 & 30.64 dB resp., **Bottom:** compressed blocks, compression: 20:1 in both, **PSNR:** 25.62 & 26.06 dB resp.

References

[1] Adams, R. A., *Sobolev Spaces,* Academic Press, New York, 1975.

[2] Ali, M. and Clarkson, T., *Using linear fractal interpolation functions to compress video images,* Fractals 2,3(1994), 417-421.

[3] Arnedo, A., Bacry, E. and Muzy, J.F., *Solving the inverse fractal problem from wavelet analysis,* Europhysics Letters, 25, 7(1994), 479-484.

[4] Barnsley, M.F., *Fractals Everywhere,* Academic Press, San Diego, CA, 1988.

[5] Barnsley, M.F., *Fractal Image Compression,* Notices of the AMS 43(1996), 657-662.

[6] Barnsley, M.F. and Hurd, Lyman P., *Fractal Image Compression,* AK Peters, Ltd., Wellesley, Massachusetts, 1993.

[7] Barnsley, M.F., Elton, J.H. , and Hardin,D.P., *Iterated Function systems,*, Constr. Approx., 5(1989), 3-31.

[8] Cabrelli, C.A., Forte, B., Motler, U.M., and Vrscay, E.R., *Iterated fuzzy set systems, a new approach to the inverse problem for fractal and other sets,* J. Math. Anal. and Appl. 17(1993), 465-486.

[9] Cabrelli, C.A. and Molter, U.M., *Generalized Self-Similarity,* Wavelets and Image Analysis, Preprint, 1996.

[10] Cheung, K.M. and Shahshahani, M., *A Comparison of the Fractal and JPEG Algorithms*, TDA Progress Report 42, 107(1991), 21-26.

[11] Edelstein, M., *An extension of the Banach's contraction principle*, Proc. Amer. Math. Soc. 12 (1961), 7-10.

[12] Fisher, Yuval (editor), *Fractal Image Compression: Theory and Application,* Springer Verlag,New York, 1995.

[13] Fisher, Y., Rogovind, D., and Shen, T.P., *A comparison of fractal methods with DCT(JPEG) and wavelet,* Proceeding from SPIE Neural and Stochastic Methods in Image and Signal Processing, 2304, 16 (1994).

[14] Forte,B. and Vrscay, E. R. , *Solving the inverse problem for function/image approximation using iterated function systems I*, Theoretical basis, Fractals 2,3 (1994), 325–334.

[15] Forte,B. and Vrscay, E. R. , *Solving the inverse problem for function/image approximation using iterated function systems II*, Algorithm and computations, Fractals 2,3 (1994), 335–346. ·

[16] Forte,B. and Vrscay, E. R. , *Solving the Inverse Problem for Function and Image Approximation using iterated function systems,* Dynamics of Continuous, Discrete and Impulsive Systems 1(1995), 177-231.

[17] Forte,B. and Vrscay, E. R. , Theory of generalized fractal transforms, in Conf. Proceedings of the NATO ASI on Fractal Image Encoding and Analysis, July 1995, Trondheim, Fisher(ed.), Springer Verlag, N.Y.

[18] Gähler, S., *2- metrische Räume und ihre topologische struktur,* Math. Nachr. 26(1963/64), 115-148.

[19] Griffel, D.H., *Applied Functional Analysis,* Ellis Horwood Ltd., New York, 1981.

[20] Hajlasz, P., *Sobolev Spaces on an Arbitrary Metric Space,* Potential Analysis, 5(1996), 403-415.

[21] Hajlasz, P. and Martio, O., *Traces of Sobolev Functions on Fractal Type Sets and Characterization of Extension Domains,* Journal of Functional Analysis 143(1997), 221-246.

[22] Hart, J., *Fractal image compression and the inverse problem of recurrent iterated function systems,* IEEE, Computer Graphics and Applications, 16, 4(1996),25-32.

[23] Istrâtescu, V.I., *Fixed Point Theory,* D. Reidel Publishing Company, Dordrecht, Boston, London, 1981

[24] Jacquin, Arnard E., *Image Coding Based on a Fractal Theory of Iterated Contractive Image Transformations,* IEEE Transactions on Image Processing, 1(1992), 18-30.

[25] Jacquin , Arnard E., *Fractal image coding: A review*, Proceedings of the IEEE, 81, 10(1993), 1451-1465.

[26] Köhn, C., *Bildanalyse und Bilddatencompression,* fraktale verfahren und klassische Methoden im praktischen Einsatz, Hanser, München, Wien, 1996.

[27] Kufner, A., *Weighted Sobolev Spaces,* Teubner-Texte zur Mathematik, Vol 31, Leipzig, 1980.

[28] Kufner, A. and Sändig, A.M., *Some Applications of Weighted Sobolev Spaces,* vol. 100, Teubner-Texte zur Mathematik, Leipzig, 1987.

[29] Liu, Y., *Solution of fractal equation,* Proceeding from IS & T / SPIE 1995 Symposium on Electronic Imaging, Science & Technology, vol. 2418: Still image compression, 1995.

[30] Lu, G., *Fractal image compression,* Signal Processing, Image Communication 5(1993), 327-343.

[31] Lu, G., Yew, T.L., *Image compression using partitioned iterated function systems,* Proceedings from SPIE Image and Video Compression, vol 2186(1994).

[32] Lu, G., Yew, T.L., *Image compression using quadtree partitioned iterated function systems,* Electronic Letters, 30, 1(1994), 23-24.

[33] Mazel, D.S. and Hayes, M.H., *Using iterated function systems to model discrete sequences,* IEEE Transactions on Signal Processing 40, 7(1992), 1724-1734.

[34] Mass, P. and Stark, H.G., *Surveys of Mathematics in Industry,* Wavelets and digital image processing, 4(1994/95), 195-235.

[35] Nadler, S.B. Jr., *Multivalued contraction mappings,* Pacific J. Math. 30(1969), 475-488.

[36] Narici, L., Beckenstein, E., *Strange Terrain - Nonarchimedian Spaces* Amer. Math. Monthly, 88(1981), 667-676.

[37] Neunzert, H., and Siddiqi, A.H., *Topics in Industrial Mathematics* Narosa-springer, India, to appear in 1997.

[38] Obrador, P., Caso, G., and Kuo, C.C., *A fractal based methods for textured image compression,* Proceeding from IS & T / SPIE 1995 Symposium on Electronic Imaging, Science & Technology, vol. 2418: Still image compression 1995.

[39] Oien, G.E., Hamzaoui, R., and Saupe, D., *On the limitations of fractal image texture coding,* Proceedings of NORSIG'96 1996, IEEE Nording Signal Processing Symposium, Espoo, Sept. 1996.

[40] Popescu, D. and Yan H., *Fractal based method for colour image compression,* Journal of Electronic Imaging 4,1(1995), 23-30.

[41] Rhodes, B.E., *Contraction Type Mapping on a 2-metric space,* Math. Nachr., 91(1979), 151-155.

[42] Rinaldo, R. and Zakhor, A., *Inverse and approximation problem for two dimensional fractal sets,* IEEE Tans. Image Processing, 3,6(1994), 802-822.

[43] Saupe, D., *A new view of fractal image compression as convolution transform coding,* IEEE Signal Processing Letters, 3,7(1996).

[44] Saupe, D. and Hamzaoui, R., *A review of the fractal image compression literature,* Computer Graphics, 28,4(1994), 268-279.

[45] Saupe, D. and Hamzaoui, R., *Complexity reduction methods for fractal image compression,* I.M.A. Conf. Proc. on Image Processing, Mathematical Method and Applications, Sept. 1994, J.M. Blackledge(Ed.), Oxford Press, 1996.

[46] Saupe, D., Hamzaoui, R. and Hartenstein, H., *Fractal image compression - An introductory overview,* Fractal Models for Image Synthesis, Compression and Analysis, D. Saupe, J. Hart (edits.), ACMSIGGRAPH'96 Course Notes 27, New Orleans, Louisiana, Aug. 1996.

[47] Saupe, D. and Hartenstein, H., *Lossless acceleration of fractal image compression by fast convolution,* Proc. ICIP-96, IEEE International conference on Image Processing, Lausane, Sept. 1996.

[48] Siddiqi, A.H., *Functional Analysis with Applications,* 4th print, Tata McGraw Hill, 1994.

[49] Siddiqi, A.H. and Mukheimer, A., *Inverse problem for function and Image approximation in Sobolev space and its generalization,* to be published.

[50] Siddiqi, A.H. and Ahmad, M.K., *Certain remarks on local IFS theory and multivalued maps,* to be published.

[51] Siddiqi, A.H. and Ahmad, M.K., *IFS in the strange - terrain Non-Archimedean spaces,* to be published.

[52] Simon, B., *Explicit link between local fractal transform and multiresolution transform,* Proc. ICIP-95 IEEE International conference on image processing, Washington, D.C., 1995.

[53] Stewart, I., *Four encounters with Sierpinski Gasket,* Mathematical Intelligencer, 17(1995), 52-64.

[54] Triebal, H., *Theory of Function Spaces,* Monograph in Mathematics, vol. 78, Birkhauser, Basel, 1983.

[55] Triebal, H., *Spaces of Besov-Hardy-Sobolev Type,* vol. 15, Teubner Texte Math., Teubner, Leipzig, 1978.

[56] Van de Walle, A., *Merging Fractal image compression and Wavelet transform Methods,* NATO ASI Conf. Fractal Image Encoding and Analysis, Trondheim, July 1995, to appear in a special issue of Fractals.

[57] Van Rooij, A., *Non Archimedean Functional Analysis,* Marcel Dekker, New York, 1978.

[58] Wickerhauser, M.V., *Adapted Wavelet Analysis from Theory to Software,* A.K. Peters, Wellesley, Massachusetts, 1994

Web resources:

http://www.informatik.uni-freiburg.de/~hamzaoui/ftp.html
http://inls.ucsd.edu/y/Fractals/
http://www.iterated.com/
http://dmsun4.bath.ac.uk/
http://www.eurecom.fr/~image/
http://links.uwaterloo.ca/
http://www-syntim.inria.fr/fractales/
http://spanky.triumf.ca/
http://www.mitsubishi.co.jp/jp/fractal/fractal.html

The first two authors would like to express their profound gratitude to Prof. Dr. Helmut Neunzert for encouragement in carrying out this work. They would also like to thank the German Academic Exchange Services (DAAD) for providing financial support to complete the paper.

1 Department of Mathematics
Aligarh Muslim University
Aligarh - 202002
India

Email: mmt01/ahs@amu.nic.in

and

Industrial Mathematics Group
Department of Mathematics
University of Kaiserslautern
D-67663 Kaiserslautern
Germany
siddiqi@mathematik.uni-kl.de

2 Industrial Mathematics Group
Department of Mathematics
University of Kaiserslautern
D-67663 Kaiserslautern
Germany
Email: ahmad@itwm.uni-kl.de

3 Department of Mathematics
Aligarh Muslim University
Aligarh - 202002, India

and

King Abdul Aziz University
Riyadh,
Kingdom of Saudi Arabia

M A MOHAMED ALI[1] AND A K PANI

Mixed finite element methods for compressible miscible displacement problems in reservoir studies

1 Introduction

Numerical simulations of enhanced oil recovery processes such as steam flooding, in-situ combustion, chemical flooding and miscible displacement by carbon dioxide are essentially aiming to predict whether a given field is economically feasible to produce or not. Here, we shall discuss finite element methods for simulating the flow of two-component single-phase, compressible miscible displacement of one fluid (say oil, called resident fluid) by another (say CO_2, called invading fluid) in a porous medium. Assuming that the reservoir $\Omega \subset R^2$ is of unit thickness and homogeneous in the vertical direction, the flow can be described by the following system of strongly coupled partial differential equations , see Douglas and Roberts [7]. For a given $T > 0$, find pressure 'p' of the total fluid and concentration 'c' of the invading fluid such that

$$\begin{aligned}
&(a) \quad d(c)\frac{\partial p}{\partial t} + \nabla \cdot u \equiv d(c)\frac{\partial p}{\partial t} - \nabla \cdot (a(c)\nabla p) = q, \quad (x,t) \in \Omega \times J,\\
&(b) \quad \phi\frac{\partial c}{\partial t} + u \cdot \nabla c + b_1(c)\frac{\partial p}{\partial t} - \nabla \cdot (D(u)\nabla c) = (\hat{c} - c)q, \ (x,t) \in \Omega \times J\\
&\qquad \text{with boundary conditions}\\
&(c) \quad u \cdot \nu = 0, \ \ and \ (D(u)\nabla c - uc) \cdot \nu = 0 \quad , x \in \partial\Omega, \ \ t \in J, \qquad (1.1)
\end{aligned}$$

[1]On leave from Faculty of Science, Suez Canal University, Ismailia, Egypt.
AMS Classification: 65M60.

and initial conditions

$$(d) \quad c(x,0) = c_0(x) \; and \; p(x,0) = p_0(x), \qquad x \in \Omega.$$

Here, ν is the outward normal to the boundary $\partial\Omega$, J=(0,T], $a(c) = a(x,c) = k(x)/\mu(c)$, with μ concentration-dependent viscosity, k is the permeability and $\phi(x)$ denotes the porosity of the medium. Moreover, the diffusion-dispersion coefficient $D(u) = D(x,u)$ arises from two sources: molecular-diffusion (due to random motion of the molecules) and anisotropic velocity dependent mechanical dispersion (due to mechanical mixing), and is given by

$$D(u) = \phi(x)\{d_m I + |u|(d_l E(u) + d_t E^T(u))\},$$

where d_m, d_l and d_t are, respectively, the constant coefficients of molecular diffusion, longitudinal and transverse dispersion, see Douglas [5] and Peaceman [17]. Here $|u|$ denotes the standard Euclidean norm of the vector u. Further, $E(u)$ is a 2×2 matrix $[E_{ij}]_{1\leq i,j\leq 2}$ of projection in the direction of the flow with $E_{ij} = \frac{1}{|u|^2}u_i u_j$, $i,j = 1,2$, and $E^T(u)$ is the matrix of projection in the direction orthogonal to the flow, i.e; $E^T = I - E$. The function q(x,t) is the volumetric flow rate and $\hat{c}$ is a function specified at the injection points and is equal to the resident fluid concentration at a production point, and $c = c_1 = 1 - c_2$, where c_1 and c_2 are the concentrations of the two flow components, $0 \leq c \leq 1$ and $u = -a(c)\nabla p$ is the Darcy velocity. Further, $d(c) = d(x,c) = \phi(x)(z_1\, c_1 + z_2\, c_2)$ and $b_1(c) = b_1(x,c) = \phi(x)c_1\{z_1 - \sum_{j=1}^{2} z_j\, c_j\}$, where z_i is the "constant compressibility" factor for the ith component.

Now, using (1.1a), the concentration equation (1.1b) can be rewritten as

$$\phi\frac{\partial c}{\partial t} - \nabla.(D(u)\nabla c) + u\cdot\nabla c - b(c)\nabla\cdot u = g(c), \quad (x,t) \in \Omega \times J, \qquad (1.1b')$$

where $b(c) = b_1(c)/d(c)$ and $g(c) = [(\tilde{c} - c) + b(c)]q$.

Equation (1.1a) is called the "pressure equation" and $(1.1b')$ called the " concentration equation" and they are coupled by the dependence of the viscosity $\mu = \mu(c)$ on the concentration c and by the dependence of $D(u)$ as well as the transport term $u \cdot \nabla c$, explicitly on the Darcy velocity u. Because of nonlinearity and strong coupling, these equations are hard to solve analytically and hence, one resorts to numerical methods.

Earlier, Douglas and Roberts [7] have derived the above mentioned model and have also discussed convergence analysis, when a finite element method is applied to the problem $(1.1(a) - 1.1(d))$. Subsequently, Chou and Li [4] have considered the modified equations $(1.1(a), (b'), (c)$, and $1.1(d))$ and obtained optimal estimates in L^2 and almost optimal error estimates in L^∞-norm under much weaker condition than one given by Douglas and Roberts. In both these papers the diffusion -dispersion coefficients D is independent of the Darcy velocity u. However, it is well noted in the literature of reservoir studies that the dispersion due to mechanical mixing plays a vital role than the molecular diffusion and hence, can not be neglected. The aim of this article is to derive optimal error estimates for both semidiscrete and fully discrete schemes, when D depends on u.

The principal variable of physical interest in $(1.1a)$ and $(1.1b')$ is the concentration, as it shows how much of the reservoir is swept by the solvent or more precisely how much oil is recovered. Standard finite difference or finite element procedures for solving (1.1a) determines p as a set of cell averages, nodal points, or piecewise smooth functions. The resulting p is differenced or differentiated and then multiplied by a possibly rough function k/μ to determine the velocity u, which in turn reduces the accuracy of the approximation of the concentration c, i.e; the simulation may give less accurate prediction. In order to approximate the velocity u more accurately, mixed finite element methods are proposed in the literature, see [6], [8] and [9] for incompressible miscible displacement problems, and for the compressible miscible displacements [7] and [4]. Since $(1.1b')$ is strongly convection dominated, standard computational methods suffer from grid orientation effect. However the finite element methods with either modified method of characteristics or C^0-interior penalty reduces these difficulties substantially. In this article, we shall not pursue these techniques and for more details see, Mohamed Ali [12]

In Section 2 of the present paper, we shall extend the analysis of [7] and [4] to the dispersion-dependent compressible miscible displacement in porous media. The parabolic mixed method which we shall employ here for the pressure equation (1.1a) is a generalization of the method introduced and analyzed by Johnson and Thomée [10]. A standard Galerkin method will be used to approximate the concentration equation $(1.1b')$. Optimal error estimates in L^2- and H^1- norms have been derived for the semidiscrete case in the first part of the second section under certain strong restriction on the finite element mesh parameters. However, it is possible to weaken the restrictions on the parameters and this will form a part of a separate paper. Since the error analysis for the discrete-in-time case is not available in the literature, we shall in the second part of Section 2 prove some preliminary results on a priori error estimates of backward Euler method.

It is well known that the mixed Galerkin formulation should satisfy the *LBB* (Ladyzhenskaya - Babuška - Brezzi, see [11],[1],[2]) consistency condition on the approximating subspaces as a stability requirement. This is restrictive for the choice of the approximation spaces severely. Consequently, some of the best known and widely used finite element spaces say C^0- piecewise linear elements can not be used.

Attempts have been made in the past to avoid LBB conditions for mixed methods. For an example, Carey *et al.* [3] and Pehlivanov *et al.* [19] have proposed and analyzed least-squares mixed method for two point boundary value problems. In [3], extensive numerical computations have been conducted and comparisons have been made with classical mixed methods. It is observed that this method has superior rate of convergence to the classical one. Subsequently, least squares method is applied to elliptic problems, see Pehlivanov *et al.* [20], [18] and [21]. Recently, Pani [16] has proposed an alternate mixed finite element method for parabolic equation which can be thought of as a non symmetric version of least-squares procedure. This method called as H^1-Galerkin mixed method is not restrictive in the sense that the approximating finite element spaces need not satisfy the LBB condition. Further, C^0-piecewise linear polynomials can be used for approximating both p and u. Although slightly higher regularity is assumed, but the order of convergence in velocity is better than the classical method.

In section 3 of this paper, we shall extend the analysis of Pani [16] to the compressible miscible displacement problems (1.1) and shall indicate the results without proofs.

2 Mixed Finite Element Procedure

In this section, we shall discuss mixed finite element procedure for the pressure equation (1.1a) combined with the standard Galerkin finite element method for the concentration equation (1.1b').

2.1 Weak Formulation and Approximation Procedure

Using the form of the Darcy velocity, we split (1.1a) into the following system of two equations

$$\begin{aligned} &(a) \quad d(c)\frac{\partial p}{\partial t} + \nabla \cdot u = q, \quad (x,t) \in \Omega \times J, \\ &(b) \quad \alpha(c)u - \nabla p = 0, \quad (x,t) \in \Omega \times J, \\ &(c) \quad p(x,0) = p_0(x), \quad x \in \Omega, \end{aligned} \tag{2.1}$$

where $\alpha(c) = a^{-1}(c)$.

Let $H(div,\Omega) = \{v : v \in L^2(\Omega)^2, \nabla.v \in L^2(\Omega)\}$, $V = \{v : v \in H(div,\Omega) : v.\nu = 0$ on $\partial\Omega\}$, and $W = L^2(\Omega)$. Then the weak form of the pressure equation (1.1a) is equivalent to the following saddle point formulation [7] and [4]: Find a solution $\{u,p\}$: $J \to V \times W$ such that

$$(d(c)\tfrac{\partial p}{\partial t}, w) + (\nabla \cdot u, w) = (q, w), \quad w \in W, \tag{2.2a}$$

$$(\alpha(c)u, v) - (\nabla \cdot v, p) = 0, \quad v \in V, \tag{2.2b}$$

where (.,.) denotes the inner product in $L^2(\Omega)$ or in $L^2(\Omega)^2$.

Now, using Green's formula and the boundary conditions (1.1c), we have

$$-(b(c)\nabla \cdot u, z) = (b(c)u, \nabla z) + (\frac{\partial b}{\partial c}u \cdot \nabla c, z) + (\frac{\partial b}{\partial x}.u, z), \quad z \in H^1(\Omega).$$

Thus, a weak form of $(1.1(b'))$ is to seek a solution $c : J \to H^1(\Omega)$ which satisfies

$$\begin{aligned} (\phi c_t, z) + (D(u)\nabla c, \nabla z) + (e(c)u \cdot \nabla c, z) + (b(c)u, \nabla z) \\ +(B(c) \cdot u, z) = (g(c), z), \qquad z \in H^1(\Omega). \end{aligned} \tag{2.2c}$$

where

$$e(c) = \frac{\partial b}{\partial c}(c) + 1 \text{ and } B(c) = \frac{\partial b}{\partial x}(c).$$

For Galerkin approximation, let $h = max(h_c, h_p)$, where h_c and h_p are positive. Let $V_h = V_h^k \subset V$ and $W_h = W_h^k \subset W$, be Raviart-Thomas-Nedlec spaces, see Johnson and Thomée [10] for constructing of these spaces. Roughly, these spaces have to satisfy $div \cdot V_h = W_h$, i.e; there exists $v_h = (v_1, v_2)$ in V_h for every $w_h \in W_h$ such that $\frac{\partial v_1}{\partial x_1} + \frac{\partial v_2}{\partial x_2} = w_h$. These spaces are first constructed by Raviart and Thomas [22] for rectangular elements, then modified by Nedlec [15] on triangular elements. Since the outer boundaries of petroleum reservoir are not known exactly and the regularity of the solutions of the differential equations is limited by the presence of sources and sinks, we shall, in particular, assume in the analysis that the boundary condition $v \cdot \nu = 0$ is represented exactly in V_h.

The approximation properties for $V_h \times W_h$ are given by

$$
\begin{aligned}
&(a) \quad \inf_{v_h \in V_h} \|v - v_h\| \equiv \inf_{v_h \in V_h} \|v - v_h\|_{L^2(\Omega)^2} \leq K\|v\|_{k+1} h_p^{k+1}, \\
&(b) \quad \inf_{v_h \in V_h} \|\nabla.(v - v_h)\| \quad \leq \quad K\{\|v\|_{k+1} + \|\nabla \cdot v\|_{k+1}\} h_p^{k+1}, \\
&\qquad\qquad v \in V \subset H^{k+1}(\Omega)^2, \\
&(c) \quad \inf_{w_h \in W_h} \|w - w_h\| \leq K\|w\|_{k+1} h_p^{k+1}, \quad w \in H^{k+1}(\Omega).
\end{aligned} \tag{2.3}
$$

Let $M_h \subset W^{1,\infty}(\Omega)$ denote a standard finite element space, which shall be used to approximate concentration and be such that

$$
\inf_{z_h \in M_h} \{\|z - z_h\| + h_c\|z - z_h\|_1\} \leq K\|z\|_{l+1} h_c^{l+1}, \quad z \in H^{l+1}(\Omega). \tag{2.4}
$$

Further, set all standard inverse relations hold for elements in M_h.

The semidiscrete finite element approximation for problem (2.2) is to seek $\{u_h, p_h, c_h\} : J \to V_h \times W_h \times M_h$ such that

$$
\begin{aligned}
&(a) \quad (d(c_h)\frac{\partial p_h}{\partial t}, w) + (\nabla \cdot u_h, w) = (q, w), \qquad w \in W_h \\
&(b) \quad (\alpha(c_h)u_h, v) - (\nabla \cdot v, p_h) = 0, \qquad v \in V_h, \\
&(c) \quad (\phi\frac{\partial c_h}{\partial t}, z) + (D(u_h)\nabla c_h, \nabla z) + (e(c_h)u_h \cdot \nabla c_h, z) \\
&\qquad +(B(c_h) \cdot u_h, z) + (b(c_h)u_h, \nabla z) = (g(c_h), z), \quad z \in M_h,
\end{aligned} \tag{2.5}
$$

and the initial approximation for $p_h(0), u_h(0)$, and $c_h(0)$, are specific functions to be defined later according to the requirements of the analysis to come.

The above mentioned semidiscrete scheme yields a system of nonlinear ordinary differential equations. However, replacing time derivatives by difference quotions, we obtain fully discrete schemes and the whole procedure yields a system of (nonlinear) algebraic equations. Then one may apply Newton or Newton like methods to solve the resulting system. As in incompressible miscible displacement problems, see Russell and Wheeler [23], it is possible to obtain a system of linear algebraic equations using sequential methods (linearized methods). We shall not pursue this further in the present paper.

We shall define a fully discrete Galerkin procedure which is based on backward differencing in time (see Thomée [24]). Let $\Delta t > 0, N = T/\Delta t \in Z$, and $t^n = n\Delta t, n = 0, 1, 2, \ldots, N$. Also, let $\psi^n = \psi(x, t^n)$ and $\partial_t \psi^n = (\psi^{n+1} - \psi^n)/\Delta t$. Denote an approximation of p by $P : \{0 = t^0, t^1, \cdots, t^N = T\} \to W_h$, an approximation of u by $U : \{t^0, t^1, \cdots, t^N\} \to V_h$, and an approximation of c by $C : \{t^0, t^1, \cdots, t^N\} \to M_h$. Assume that P^n, U^n, and C^n are known, we determine P^{n+1}, U^{n+1}, and C^{n+1} as solutions of

$$\begin{aligned}
&(a) \quad C^0 = \tilde{C}(0), P^0 = \tilde{P}(0), \text{ and } U^0 = \tilde{U}(0),\\
&(b) \quad (d(C^{n+1})\partial_t P^n, w) + (\nabla \cdot U^{n+1}, w) = (q^{n+1}, w), \quad w \in W_h,\\
&(c) \quad (\alpha(C^{n+1})U^{n+1}, v) - (\nabla \cdot v, P^{n+1}) = 0, \quad v \in V_h, \qquad (2.6)\\
&(d) \quad (\phi \partial_t C^n, z) + (D(U^{n+1})\nabla C^{n+1}, \nabla z) + (e(C^{n+1})U^{n+1} \cdot \nabla C^{n+1}, z)\\
&\qquad + (b(C^{n+1})U^{n+1}, \nabla z) + (B(C^{n+1}) \cdot U^{n+1}, z) = (g(C^{n+1}), z), \quad z \in M_h,
\end{aligned}$$

where $\tilde{C}, \tilde{P}$, and $\tilde{U}$ are appropriate projections to be defined later.

2.2 Error Estimates for the Semidiscrete Case

It is convenient for our error analysis to project the solution of the problem (1.1) onto the finite element spaces. First, as in [7] and [4], we shall introduce auxiliary functions $\{\tilde{p}, \tilde{u}, \tilde{c}\}$ as projections of $\{p, u, c\}$ onto $W_h \times V_h \times M_h$ defined through

$$\begin{aligned}
&(a) \quad (\nabla \cdot (u - \tilde{u}), w) = 0, \qquad w \in W_h,\\
&(b) \quad (\alpha(c)(u - \tilde{u}), v) - (\nabla \cdot v, (p - \tilde{p})) = 0, \quad v \in V_h, \qquad (2.7)\\
&(c) \quad (D(u)\nabla(c - \tilde{c}), \nabla z) + (e(c)u \cdot \nabla(c - \tilde{c}), z) + \lambda(c - \tilde{c}, z) = 0, \quad z \in M_h.
\end{aligned}$$

Here λ is chosen in such a way that the bilinear form in (2.7) is H^1-coercive. Below, we shall use the following notations for our future use. Let $\eta = p - \tilde{p}, \xi = \tilde{p} - p_h, \rho = u - \tilde{u}, \sigma = \tilde{u} - u_h, \theta = c - \tilde{c}$, and $\pi = \tilde{c} - c_h$. Then $p - p_h = \eta + \xi$, $u - u_h = \rho + \sigma$ and $c - c_h = \theta + \pi$.

Following Douglas and Roberts [7], we have the following estimates for ρ and η

$$\begin{aligned}
&(a) \quad \|\rho\|_{H(div,\Omega)} + \|\eta\| \le K\|p\|_{k+3} h_p^{k+1},\\
&(b) \quad \|\frac{\partial \rho}{\partial t}\|_{H(div,\Omega)} + \|\frac{\partial \eta}{\partial t}\| \le K\{\|p\|_{k+3} + \|\frac{\partial p}{\partial t}\|_{k+3}\} h_p^{k+1}. \qquad (2.8)
\end{aligned}$$

By standard results on the finite element method for elliptic problems see Chou and Li [4] and Wheeler [25], it is possible to obtain the following estimates involving θ:

$$(a) \quad \sum_{j=0}^{2} \|\frac{\partial^j \theta}{\partial t^j}\| + h_c\|\nabla \theta\| \le K\{\sum_{j=0}^{2} \|\frac{\partial^j c}{\partial t^j}\|_{l+1}\} h_c^{l+1}, \qquad (2.9)$$

$$(b) \quad \|\theta\|_{0,\infty} \leq K h_c^{l+1} (log h_c^{-1})^r \|c\|_{l+1,\infty},$$

where

$$r = \begin{cases} 0 & \text{if } l > 1 \\ 1 & \text{when } l = 1. \end{cases}$$

Note that the presence of u in D does not pose any special problem for these estimates. Now, from (2.2a,b),(2.5a,b), and (2.7a,b) we obtain the error equations in ξ and σ as

$$\begin{aligned} (a) \quad & (d(c)\frac{\partial \xi}{\partial t}, w) + (\nabla \cdot \sigma, w) = -(d(c)\frac{\partial \eta}{\partial t}, w) \\ & \qquad + ([d(c_h) - d(c)]\frac{\partial p_h}{\partial t}, w), \quad w \in W_h \\ (b) \quad & (\alpha(c_h)\sigma, v) - (\nabla \cdot v, \xi) = -([\alpha(c) - \alpha(c_h)]\tilde{u}, v), \quad v \in V_h. \end{aligned} \tag{2.10}$$

Further, from (2.2c), (2.5c), and (2.7c) we have the following equation involving π

$$\begin{aligned} (\phi\frac{\partial \pi}{\partial t}, z) \; &+ \; (D(u_h)\nabla\pi, \nabla z) = (g(c) - g(c_h), z) - (\phi\frac{\partial \theta}{\partial t}, z) + \lambda(\theta, z) \\ &- \; (b(c)u - b(c_h)u_h, \nabla z) - (e(c)u \cdot \nabla\tilde{c} - e(c_h)u_h \cdot \nabla c_h, z) \\ &- \; (B(c) \cdot u - B(c_h) \cdot u_h, z) - ([D(u) - D(u_h)]\nabla\tilde{c}, \nabla z). \end{aligned} \tag{2.11}$$

For our error analysis, we shall now assume temporarily that there is a constant $K^* \geq 2K_1$ with

$$\|\tilde{c}\|_{L^\infty(L^\infty)} + \|\frac{\partial \tilde{p}}{\partial t}\|_{L^2(L^\infty)} \leq K_1$$

such that

$$\|c_h\|_{L^\infty(L^\infty)} + \|\frac{\partial p_h}{\partial t}\|_{L^2(L^\infty)} \leq K^*. \tag{2.12}$$

Theorem 2.1 *Chose $p_h(0) = \tilde{p}(0)$ and $c_h(0) = \tilde{c}(0)$, so that $\xi(0) = 0$ and $\pi(0) = 0$. Then there exists a positive constant K such that*

$$\|p - p_h\|_{L^\infty(L^2)} + \|c - c_h\|_{L^\infty(L^2)} + \|u - u_h\|_{L^2(L^2)} \leq K(K^*)(h_c^{l+1} + h_p^{k+1}).$$

Besides, if

$$\min\{h_c^{-1}h_p^{k+1}, h_p^{-2}h_c^{l+1}\} \to 0 \quad as \quad h \to 0, \tag{2.13}$$

then for sufficiently small h_p and h_c, and $k \geq 2$ the constant K can be chosen independent of K^*.

Proof. Choose $w = \xi$ in (2.10a) and $v = \sigma$ in (2.10b) and add the resulting equations to have

$$\begin{aligned}(d(c)\frac{\partial \xi}{\partial t}, \xi) + (\alpha(c_h)\sigma, \sigma) &= -(d(c)\frac{\partial \eta}{\partial t}, \xi) + ([d(c_h) - d(c)]\frac{\partial p_h}{\partial t}, \xi) \\ &\quad -([\alpha(c) - \alpha(c_h)]\tilde{u}, \sigma).\end{aligned}$$

Note that

$$(d(c)\frac{\partial \xi}{\partial t}, \xi) = \frac{1}{2}\frac{d}{dt}(d(c)\xi, \xi) - \frac{1}{2}(\frac{\partial d}{\partial c}(c)\frac{\partial c}{\partial t}\xi, \xi).$$

Then integrating with respect to time from zero to t, we have

$$\begin{aligned}\|\xi\|^2 &+ 2\alpha_* \int_0^t \|\sigma\|^2 \le 2\int_0^t (d(c)\frac{\partial \eta}{\partial t}, \xi)ds \\ &+2\int_0^t ([d(c_h) - d(c)]\frac{\partial p_h}{\partial t}, \xi)ds + \int_0^t (\frac{\partial d}{\partial c}(c)\frac{\partial c}{\partial t}\xi, \xi)ds \\ &+2\int_0^t ([\alpha(c) - \alpha(c_h)]\tilde{u}, \sigma)ds \\ &= I_1 + I_2 + I_3 + I_4. \end{aligned} \tag{2.14}$$

The terms on the right side of (2.14) will be estimated as follows: For I_1 and I_3, an application of Cauchy-Schwarz inequality yields

$$|I_1| + |I_3| \le K\int_0^t [\|\eta_t(s)\|^2 + \|\xi(s)\|^2]ds.$$

For I_2, use (2.12) with Cauchy-Schwarz inequality to obtain

$$\begin{aligned}|I_2| &\le K\int_0^t [\|\frac{\partial p_h}{\partial t}\|_{L^\infty}\|c - c_h\|\|\xi\|]ds \\ &\le \frac{1}{2}\max_{0\le s\le t}\|\xi(s)\|^2 + K(K^*)\int_0^t \|c - c_h\|^2 ds.\end{aligned}$$

Set

$$\|\xi(t^*)\| = \max_{0\le s\le t}\|\xi(s)\|, \quad for\ t^* \in [0, t].$$

For I_4, we have

$$|I_4| \le K\int_0^t \|c - c_h\|^2 ds + \epsilon\int_0^t \|\sigma\|^2 ds$$

Use the above estimates in (2.14) with (2.8)-(2.9) and choose ϵ such that $(\alpha_* - \epsilon) > 0$ to have

$$\|\xi\|^2 + \int_0^t \|\sigma\|^2 ds \le K(K^*)(h_c^{2l+2} + h_p^{2k+2}) + K(K^*)\int_0^t [\|\pi\|^2 + \|\xi\|^2]ds + \frac{1}{2}\|\xi(t^*)\|^2.$$

The above result is also true when t is replaced by t^*, i.e;

$$\begin{aligned}\|\xi(t)\|^2 &\le \|\xi(t^*)\|^2 \\ &\le K(K^*)(h_c^{2l+2}+h_p^{2k+2})+K(K^*)\int_0^t[\|\pi\|^2+\|\xi\|^2]ds.\end{aligned}$$

In fact, we obtain

$$\|\xi\|^2+\int_0^t\|\sigma\|^2ds\le K(K^*)(h_c^{2l+2}+h_p^{2k+2})+K(K^*)\int_0^t[\|\pi\|^2+\|\xi\|^2]ds.$$

Apply Gronwall's Lemma to have

$$\|\xi\|^2+\int_0^t\|\sigma\|^2ds\le K(K^*)(h_c^{2l+2}+h_p^{2k+2})+K(K^*)\int_0^t\|\pi\|^2ds. \tag{2.15}$$

Now, choose $z=\pi$ in (2.11) and obtain

$$\begin{aligned}\frac{1}{2}\frac{d}{dt}\|\pi\|^2+D_*\|\nabla\pi\|^2 &\le -[(\phi\frac{\partial\theta}{\partial t},\pi)+\lambda(\theta,\pi)]+(g(c)-g(c_h),\pi)\\ &- (e(c)u.\nabla\tilde{c}-e(c_h)u_h.\nabla c_h,\pi)-(b(c)u-b(c_h)u_h,\nabla\pi)\\ &- (B(c).u-B(c_h).u_h,\pi)-([D(u)-D(u_h)]\nabla\tilde{c},\nabla\pi)\\ &= L_1+L_2+\cdots+L_6.\end{aligned} \tag{2.16}$$

For L_1 and L_2 , use Cauchy-Schwarz inequality to have

$$|L_1|+|L_2|\le K(\|\theta\|^2+\|\theta_t\|^2)+K\|\pi\|^2.$$

For L_3, we rewrite it as

$$\begin{aligned}L_3 &= ([e(c)-e(c_h)]u.\nabla\tilde{c},\pi)+(e(c_h)[u-u_h].\nabla\tilde{c},\pi)\\ &+(e(c_h)[u_h-\tilde{u}].\nabla\pi,\pi)+(e(c_h)\tilde{u}.\nabla\pi,\pi),\end{aligned}$$

and then apply (2.12) to obtain

$$\begin{aligned}|L_3| &\le K(\|\nabla\tilde{c}\|_{0,\infty})(\|c-c_h\|+\|u-u_h\|)\|\pi\|\\ &+K(\|\pi\|_{0,\infty})(\|\rho\|+\|\pi\|)\|\nabla\pi\|\\ &\le K(K^*,\epsilon)(\|\theta\|^2+\|\sigma\|^2+\|\rho\|^2+\|\pi\|^2)+\epsilon\|\nabla\pi\|^2.\end{aligned}$$

For L_4, rewrite it as

$$L_4=-([b(c)-b(c_h)]u+b(c_h)[u-u_h],\nabla\pi),$$

and hence,

$$|L_4|\le K(\epsilon)(\|\theta\|^2+\|\sigma\|^2+\|\rho\|^2+\|\pi\|^2)+\epsilon\|\nabla\pi\|^2.$$

Similarly, L_5 and L_6 can be bounded by

$$|L_5| + |L_6| \leq K(\epsilon)(\|\theta\|^2 + \|\sigma\|^2 + \|\rho\|^2 + \|\pi\|^2) + \epsilon\|\nabla\pi\|^2.$$

Integrate (2.16) with respect to time from 0 to t and use the estimates of $L_1, L_2, \ldots, L_6$ to obtain

$$\begin{aligned}\|\pi\|^2 + 2(D_* - 3\epsilon)\int_0^t \|\nabla\pi\|^2 ds &\leq K(K^*)(h_c^{2l+1} + h_p^{2k+2}) \\ &+ K(K^*)\int_0^t \|\pi\|^2 ds + K(K^*)\int_0^t \|\sigma\|^2 ds.\end{aligned}$$

Here, we have used the estimates (2.8)-(2.9). Choose ϵ so that $(D_* - 3\epsilon) > 0$ and apply (2.15) for the last term to have

$$\|\pi\|^2 + \int_0^t \|\nabla\pi\|^2 ds \leq K(K^*)\{h_c^{2l+1} + h_p^{2k+2} + \int_0^t \|\pi\|^2 ds\}.$$

Again an application of Gronwall's lemma yields

$$\|\pi\|^2 + \int_0^t \|\nabla\pi\|^2 ds \leq K(K^*, T)(h_c^{2l+1} + h_p^{2k+2}). \tag{2.17}$$

Using the above estimate for (2.15), it is easy to check that

$$\|\xi\|^2 + \int_0^t \|\sigma\|^2 \leq K(K^*, T)(h_c^{2l+2} + h_p^{2k+2}). \tag{2.18}$$

To verify (2.12), we need an estimate $\|\xi_t\|_{L^2(L^\infty)}$. From the error equation (2.10a) with $w = \xi_t$, we have

$$(d(c)\frac{\partial\xi}{\partial t}, \frac{\partial\xi}{\partial t}) = -(\nabla.\sigma, \frac{\partial\xi}{\partial t}) - (d(c)\frac{\partial\eta}{\partial t}, \frac{\partial\xi}{\partial t}) + ([d(c_h) - d(c)]\frac{\partial p_h}{\partial t}, \frac{\partial\xi}{\partial t}).$$

Integrate the above equation with respect to time from 0 to t, and use (2.12) to obtain

$$d_*\int_0^t \|\frac{\partial\xi}{\partial t}\|^2 ds \leq K\int_0^t [\|\nabla.\sigma\|^2 + \|\eta_t\|^2] ds + \frac{3}{2}\epsilon\int_0^t \|\frac{\partial\xi}{\partial t}\|^2 ds + K(K^*)\|c - c_h\|^2_{L^\infty(L^2)}$$

Using inverse hypothesis for W_h, it follows that

$$\|\nabla\cdot\sigma\| \leq K\|\sigma\|_{H(div,\Omega)} \leq K h_p^{-1}\|\sigma\|,$$

and hence,

$$(d_* - \frac{3}{2}\epsilon)\int_0^t \|\frac{\partial\xi}{\partial t}\|^2 ds \leq K\int_0^t [h_p^{-2}\|\sigma\|^2 + \|\eta_t\|^2] ds + K(K^*)\|c - c_h\|^2_{L^\infty(L^2)}.$$

Choose ϵ so that $(d_* - \frac{3}{2}\epsilon) > 0$, then use (2.8)-(2.9), and (2.18) to obtain

$$\int_0^t \|\frac{\partial\xi}{\partial t}\|^2 ds \leq K(K^*)(h_p^{2k} + h_p^{-2}h_c^{2l+2}).$$

Again an use of inverse hypothesis for V_h yields

$$\|\frac{\partial \xi}{\partial t}\|_{L^2(L^\infty)} \le K(K^*)(h_p^{k-1} + h_p^{-2}h_c^{l+1}).$$

Therefore,

$$\begin{aligned}
\|\frac{\partial p_h}{\partial t}\|_{L^2(L^\infty)} &\le \|\frac{\partial \tilde{p}}{\partial t}\|_{L^2(L^\infty)} + \|\frac{\partial \xi}{\partial t}\|_{L^2(L^\infty)} \\
&\le K_1 + K(K^*)\{h_p^{k-1} + h_p^{-2}h_c^{l+1}\} \\
&\le 2K_1 \le K^*.
\end{aligned}$$

To verify $\|c_h\|_{L^\infty(L^\infty)}$ in (2.12), we note that from (2.13)

$$\begin{aligned}
\|c_h\|_{L^\infty(L^\infty)} &\le \|\tilde{c}\|_{L^\infty(L^\infty)} + \|\pi\|_{L^\infty(L^\infty)} \\
&\le K_1 + K(K^*)(h_c^l + h_c^{-1}h_p^{k+1}) \\
&\le 2K_1 \le K^*,
\end{aligned}$$

for sufficiently small h, and consequently K can be chosen independent of K^*. Using the estimates (2.8)-(2.9) and (2.18), the triangle inequality completes the proof of the theorem. □

Remark 2.1 *Now the condition (2.13) is satisfied if we choose $h_p^{k+1} = h_c^{l+1}$, i.e; for*

$$h_p^{k+1}h_c^{-1} = h_p^{(k+1)l/(l+1)}, \qquad h_c^{l+1}h_p^{-2} = h_c^{(l+1)(k-1)/(k+1)}.$$

In fact the first term always tends to zero, while the second tends to zero if $k \ge 2$. If $h_p = O(h_c)$ then k as well as $l \ge 2$.

However, it is possible to prove the optimal error estimates when k and $l \ge 1$ and a weaker condition on the mesh parameters to satisfy $h_c^{-1}h_p^{k+1} \to 0$ as $h \to 0$. A more detailed analysis, can be found in Mohamed Ali and Pani [14].

2.3 A Priori Error Estimates for the Fully Discrete Scheme

In this subsection, we shall discuss a priori bounds for the errors $c^n - C^n, p^n - P^n$, and $u^n - U^n$. The results to come will be similar to those in the continuous case. Here, we shall use the same notations as in section 2 and assume that (2.7), (2.8), and (2.9) hold at each time level $t = t^n$.

Now, from (2.2a), (2.6b), and (2.7a), we obtain the error equation in pressure as

$$\begin{aligned}
(d(c^{n+1})\partial_t \xi^n, w) &+ (\nabla \cdot \sigma^{n+1}, w) = -(d(c^{n+1})[\frac{\partial p^{n+1}}{\partial t} - \partial_t p^n], w) \\
&- (d(c^{n+1})\partial_t \eta^n, w) - ([d(c^{n+1}) - d(C^{n+1})]\partial_t P^n, w). \quad (2.19)
\end{aligned}$$

Similarly, from (2.2b),(2.6c), and (2.7b), the error equation in the velocity can be written as

$$(\alpha(C^{n+1})\sigma^{n+1}, v) - (\nabla.v, \xi^{n+1}) = -([\alpha(c^{n+1}) - \alpha(C^{n+1})]\tilde{u}^{n+1}, v). \qquad (2.20)$$

Also, from (2.2c), (2.6d), and (2.7c), we have the following error equation in the concentration

$$\begin{aligned}(\phi\partial_t\pi^n, z) &+ (D(U^{n+1})\nabla\pi^{n+1}, \nabla z) = (g(c^{n+1}) - g(C^{n+1}), z) + \lambda(\theta^{n+1}, z)\\ &- (\phi[\frac{\partial c^{n+1}}{\partial t} - \partial_t c^n], z) - (\phi\partial_t\theta^n, z)\\ &- (e(c^{n+1})u^{n+1}.\nabla\tilde{c}^{n+1} - e(C^{n+1})U^{n+1}.\nabla C^{n+1}, z)\\ &- (B(c^{n+1}).u^{n+1} - B(C^{n+1}).U^{n+1}, z) \qquad (2.21)\\ &- (b(c^{n+1})u^{n+1} - b(C^{n+1})U^{n+1}, \nabla z)\\ &- ([D(u^{n+1}) - D(U^{n+1})]\nabla\tilde{c}^{n+1}, \nabla z).\end{aligned}$$

Theorem 2.2 *Suppose $k \geq 2, l \geq 1$, and that the space and time discretization parameters h_p, h_c and Δt satisfy the relations*

$$max(h_p^{-2}h_c^{l+1}\ ,\ h_c^{-1}h_p^{k+1}\ ,\ h_p^{-2}\Delta t\ ,\ h_c^{-1}\Delta t) \to 0 \ \ as\ h \to 0.$$

Then, there exists a positive constant K Such that for sufficiently small h and Δt

$$\begin{aligned}\|c^J - C^J\| + \|p^J - P^J\| &+ \Delta t\sum_{n=0}^{J}\|u^n - U^n\|^2\\ &\leq K(h_c^{2l+2} + h_p^{2k+2} + (\Delta t)^2)\ \ ,\ J = 0, 1, \cdots, N.\end{aligned}$$

Proof. Choose $w = \xi^{n+1}$ in (2.19) and $v = \sigma^{n+1}$ in (2.20), then add to have

$$\begin{aligned}&(d(c^{n+1})\partial_t\xi^n, \xi^{n+1}) + (\alpha(C^{n+1})\sigma^{n+1}, \sigma^{n+1}) =\\ &\quad -[(d(c^{n+1})[\frac{\partial p^{n+1}}{\partial t} - \partial_t p^n], \xi^{n+1}) + (d(c^{n+1})\partial_t\eta^n, \xi^{n+1})]\\ &\quad -([d(c^{n+1}) - d(C^{n+1})]\partial_t P^n, \xi^{n+1}) - ([\alpha(c^{n+1}) - \alpha(C^{n+1})]\tilde{u}^{n+1}, \sigma^{n+1}).\\ &\quad = I_1 + I_2 + I_3. \qquad (2.22)\end{aligned}$$

Note that

$$\begin{aligned}(d(c^{n+1})\partial_t\xi^n, \xi^{n+1}) &= \frac{1}{2}\partial_t(d(c^n)\xi^n, \xi^n) + \frac{\Delta t}{2}(d(c^{n+1})\partial_t\xi^n, \partial_t\xi^n)\\ &\quad - \frac{1}{2\Delta t}([d(c^{n+1}) - d(c^n)]\xi^n, \xi^n),\end{aligned}$$

and hence, summing from n=0 to J-1, we have

$$\sum_{n=0}^{J-1}(d(c^{n+1})\partial_t\xi^n, \xi^{n+1}) \geq \frac{1}{2\Delta t}\{(d(c^J)\xi^J, \xi^J) - (d(c^0)\xi^0, \xi^0)\} - \frac{1}{2\Delta t}\sum_{n=0}^{J-1}(\partial_t[d(c^n)]\xi^n, \xi^n)\Delta t.$$

Multiply the resulting equation by $2\Delta t$ to have

$$d_*\|\xi^J\|^2 + 2\alpha_*\sum_{n=0}^{J}\|\sigma^n\|^2\Delta t \leq 2\sum_{n=0}^{J-1}[I_1^n + I_2^n + I_3^n]\Delta t + \sum_{n=0}^{J-1}(\partial_t[d(c^n)]\xi^n, \xi^n)\Delta t. \tag{2.23}$$

Now, we shall bound the terms on the right-side of (2.23) as follows : For I_1^n, using Taylor series expansion we have

$$\frac{\partial p^{n+1}}{\partial t} - \partial_t p^n = -\frac{1}{\Delta t}\int_{t^n}^{t^{n+1}}(t^{n+1} - s)\frac{\partial^2 p}{\partial t^2}(s)ds,$$

and hence,

$$\|\frac{\partial p^{n+1}}{\partial t} - \partial_t p^n\|^2 \leq K(\int_{t^n}^{t^{n+1}}\|\frac{\partial^2 p}{\partial t^2}(s)\|^2 ds)\Delta t.$$

Therefore,

$$\sum_{n=0}^{J-1}|I_1^n| \leq K(\Delta t)^2 + K\sum_{n=0}^{J-1}\|\xi^n\|^2\Delta t + K\Delta t\|\xi^J\|^2 + K\sum_{n=0}^{J-1}\|\partial_t\eta^n\|^2\Delta t.$$

we shall now, assume temporarily that there exists a positive constant $K^* \geq 2K_1$ with

$$\|\tilde{c}^J\|_{L^\infty(L^\infty)} + \sum_{n=0}^{J-1}\|\partial_t\tilde{p}^n\|_{L^\infty}^2\Delta t \leq K_1$$

such that

$$\|C^J\|_{L^\infty(L^\infty)} + \sum_{n=0}^{J-1}\|\partial_t P^n\|_{L^\infty}^2\Delta t \leq K^*, \quad J = 1, 2, \cdots, N. \tag{2.24}$$

Now, I_2^n is bounded by

$$\sum_{n=0}^{J-1}|I_2^n|\Delta t \leq K\sum_{n=0}^{J-1}(\|\partial_t P^n\|_{L^\infty}\|c^{n+1} - C^{n+1}\|\|\xi^{n+1}\|)\Delta t$$
$$\leq \frac{d_*}{4}\max_{0\leq n\leq J}\|\xi^n\|^2 + K(K^*)\sum_{n=0}^{J-1}\|c^{n+1} - C^{n+1}\|^2\Delta t.$$

For I_3^n, use Cauchy-Schwarz inequality to have

$$\sum_{n=0}^{J-1} |I_3^n|\Delta t \leq \epsilon \sum_{n=0}^{J} \|\sigma^n\|^2 \Delta t + K(\epsilon) \sum_{n=0}^{J-1} \|c^{n+1} - C^{n+1}\|^2 \Delta t.$$

Further,

$$\sum_{n=0}^{J-1} |(\partial_t[d(c^n)]\xi^n, \xi^n)|\Delta t \leq K \sum_{n=0}^{J-1} \|\xi^n\|^2 \Delta t.$$

Assume that for $0 \leq m \leq J$

$$\|\xi^m\| = \max_{0 \leq n \leq J} \|\xi^n\|.$$

Using the above estimates with (2.8)-(2.9) in (2.23), we obtain

$$\begin{aligned}(d_* - 2K\Delta t)\|\xi^J\|^2 \;&+\; 2(\alpha_* - \epsilon) \sum_{n=0}^{J} \|\sigma^n\|^2 \Delta t \leq K(K^*)(h_c^{2l+2} + h_p^{2k+2} + (\Delta t)^2) \\ &+\; K(K^*) \sum_{n=0}^{J-1} [\|\pi^{n+1}\|^2 + \|\xi^n\|^2]\Delta t + \frac{d_*}{2}\|\xi^m\|^2.\end{aligned}$$

Choose Δt such that $(\frac{d_*}{2} - K\Delta t) > 0$ and ϵ small so that $(\alpha_* - \epsilon) > 0$. Further note that the above inequality remains valid if we replace J by m. As $\|\xi^J\| \leq \|\xi^m\|$, with $0 \leq m \leq J$, we have using discrete Gronwall's Lemma

$$\begin{aligned}\|\xi^J\|^2 + \sum_{n=0}^{J} \|\sigma^n\|^2 \Delta t \;&\leq\; K(K^*)(h_c^{2l+2} + h_p^{2k+2} + (\Delta t)^2) \\ &+ K(K^*) \sum_{n=0}^{J-1} \|\pi^{n+1}\|^2 \Delta t. \qquad (2.25)\end{aligned}$$

To estimate π^n, choose $z = \pi^{n+1}$ in (2.21), and obtain

$$\begin{aligned}(\phi\partial_t\pi^n, \pi^{n+1}) + (D(U^{n+1})\nabla\pi^{n+1}, \nabla\pi^{n+1}) &= [(g(c^{n+1}) - g(C^{n+1}), \pi^{n+1}) \\ &+\lambda(\theta^{n+1}, \pi^{n+1}) - (\phi[\frac{\partial c^{n+1}}{\partial t} - \partial_t c^n], \pi^{n+1}) - (\phi\partial_t\theta^n, \pi^{n+1})] \\ &-(e(c^{n+1})u^{n+1}.\nabla\tilde{c}^{n+1} - e(C^{n+1})U^{n+1}.\nabla C^{n+1}, \pi^{n+1}) \\ &-(B(c^{n+1}).u^{n+1} - B(C^{n+1}).U^{n+1}, \pi^{n+1}) \\ &-[(b(c^{n+1})u^{n+1} - b(C^{n+1})U^{n+1}, \nabla\pi^{n+1}) \\ &+([D(u^{n+1}) - D(U^{n+1})]\nabla\tilde{c}^{n+1}, \nabla\pi^{n+1})]. \\ &= Z_1^n + Z_2^n + \cdots + Z_4^n. \qquad (2.26)\end{aligned}$$

We now estimate the terms on the right-hand side of (2.26). For Z_1^n, use Taylor's expansion and Cauchy-Schwarz inequality to have

$$\begin{aligned}|Z_1^n| \;&\leq\; K(\int_{t^n}^{t^{n+1}} \|\frac{\partial^2 c}{\partial t^2}\|^2 ds)\Delta t + \frac{K}{\Delta t}\int_{t^n}^{t^{n+1}} \|\theta_t\|^2 ds \\ &+K[\|c^{n+1} - C^{n+1}\|^2 + \|\theta^{n+1}\|^2].\end{aligned}$$

For Z_2^n, we rewrite it as

$$\begin{aligned} Z_2^n &= -([e(c^{n+1}) - e(C^{n+1})]u^{n+1} \cdot \nabla \bar{c}^{n+1}, \pi^{n+1}) \\ &\quad -(e(C^{n+1})[u^{n+1} - U^{n+1}] \cdot \nabla \bar{c}^{n+1}, \pi^{n+1}) \\ &\quad -(e(C^{n+1})\sigma^{n+1} \cdot \nabla \pi^{n+1}, \pi^{n+1}) \\ &\quad -(e(C^{n+1})\tilde{u}^{n+1} \cdot \nabla \pi^{n}, \pi^{n+1}), \end{aligned}$$

and hence, using (2.24), we obtain

$$\begin{aligned} |Z_2^n| &\leq \epsilon \|\nabla \pi^{n+1}\|^2 + K(K^*)\|\sigma^{n+1}\|^2 \\ &\quad + K[\|c^{n+1} - C^{n+1}\|^2 + \|u^{n+1} - U^{n+1}\|^2]. \end{aligned}$$

For the estimation of Z_3^n, we rearrange it as

$$\begin{aligned} Z_3^n &= ([B(c^{n+1}) - B(C^{n+1})] \cdot u^{n+1}, \pi^{n+1}) \\ &\quad +(B(C^{n+1}) \cdot [u^{n+1} - U^{n+1}], \pi^{n+1}), \end{aligned}$$

and hence,

$$|Z_3^n| \leq K[\|c^{n+1} - C^{n+1}\|^2 + \|u^{n+1} - U^{n+1}\|^2].$$

Similarly, for Z_4^n, we obtain

$$|Z_4^n| \leq 3\epsilon \|\nabla \pi^{n+1}\|^2 + K[\|c^{n+1} - C^{n+1}\|^2 + \|u^{n+1} - U^{n+1}\|^2].$$

Multiply (2.26) by $2\Delta t$ and sum up from n=0 to J-1, then use the above estimates with the estimates (2.8)-(2.9) to have

$$\begin{aligned} (1 - K\Delta t)\|\pi^J\|^2 &+ 2(D_* - 4\epsilon) \sum_{n=0}^{J} \|\nabla \pi^n\|^2 \Delta t \leq K(h_c^{2l+2} + h_p^{2k+2} + (\Delta t)^2) \\ &+ K(K^*) \sum_{n=0}^{J-1} \|\pi^n\|^2 \Delta t + K(K^*) \sum_{n=0}^{J} \|\sigma^n\|^2 \Delta t. \end{aligned}$$

Choose ϵ so that $(D_* - 4\epsilon) > 0$, and Δt so that $(1 - K\Delta t) > 0$. Then an application of discrete Gronwall's Lemma with (4.7) yields

$$\|\pi^J\|^2 + \sum_{n=0}^{J} \|\nabla \pi^n\|^2 \Delta t \leq K(K^*)(h_c^{2l+2} + h_p^{2k+2} + (\Delta t)^2). \tag{2.27}$$

Using the above inequality in (2.25), it is easy to see that

$$\|\xi^J\|^2 + \sum_{n=0}^{J} \|\sigma^n\|^2 \Delta t \leq K(K^*)(h_c^{2l+2} + h_p^{2k+2} + (\Delta t)^2). \tag{2.28}$$

To verify (2.24), we need an estimate $\sum_{n=0}^{J-1}\|\partial_t\xi^n\|_{L^\infty}^2\Delta t$. Choose $w=\partial_t\xi^n$ in (2.19), to have

$$\begin{aligned}(d(c^{n+1})\partial_t\xi^n,\partial_t\xi^n) &+ (\nabla\cdot\sigma^{n+1},\partial_t\xi^n) = -(d(c^{n+1})[\frac{\partial p^{n+1}}{\partial t}-\partial_t p^n],\partial_t\xi^n)\\ &- (d(c^{n+1})\partial_t\eta^n,\partial_t\xi^n)-([d(c^{n+1})-d(C^{n+1})]\partial_t P^n,\partial_t\xi^n).\end{aligned}$$

Sum over time from n=0 to J-1 and then multiply by Δt to obtain

$$\begin{aligned}(d_*-3\epsilon)\sum_{n=0}^{J-1}\|\partial_t\xi^n\|^2\Delta t &\le K\sum_{n=0}^{J-1}[\|\nabla\cdot\sigma^{n+1}\|^2+\|\partial_t\eta^n\|^2+\Delta t]\Delta t\\ &+K\max_{0\le n\le J}\|c^n-C^n\|^2(\sum_{n=0}^{J-1}\|\partial_t P^n\|_{L^\infty}^2\Delta t).\end{aligned}$$

Choose ϵ so that $(d_*-3\epsilon)>0$, then use (2.24) and the estimates (2.8)-(2.9) and (2.27)-(2.28) to have

$$\begin{aligned}\sum_{n=0}^{J-1}\|\partial_t\xi^n\|^2\Delta t &\le K(K^*)K\sum_{n=0}^{J-1}h_p^{-2}\|\sigma^{n+1}\|^2\Delta t+K(K^*)\{h_c^{2l+2}+h_p^{2k+2}+(\Delta t)^2\}\\ &\le K(K^*)\{h_p^{2k}+h_p^{-2}h_c^{2l+2}+h_p^{-2}(\Delta t)^2\}.\end{aligned}$$

Note that inverse hypothesis for W_h has been used. Again an use of inverse hypothesis for V_h yields

$$\sum_{n=0}^{J-1}\|\partial_t\xi^n\|_{L^\infty}^2\Delta t \le h_p^{-2}\sum_{n=0}^{J-1}\|\partial_t\xi^n\|^2\Delta t,$$

and hence,

$$\sum_{n=0}^{J-1}\|\partial_t\xi^n\|_{L^\infty}^2\Delta t\le K(K^*)\{h_p^{2k-2}+h_p^{-4}h_c^{2l+2}+h_p^{-4}(\Delta t)^2\}.$$

Therefore,

$$\begin{aligned}\sum_{n=0}^{J-1}\|\partial_t P^n\|_{L^\infty}^2\Delta t &\le \sum_{n=0}^{J-1}\|\partial_t\tilde{p}^n\|_{L^\infty}^2\Delta t+\sum_{n=0}^{J-1}\|\partial_t\xi^n\|_{L^\infty}^2\Delta t\\ &\le K_1+K(K^*)\{h_p^{2k-2}+h_p^{-4}h_c^{2l+2}+h_p^{-4}(\Delta t)^2\}.\\ &\le 2K_1\le K^*,\end{aligned}$$

for $k\ge 2, h_c=o(h_p^{\frac{2}{l+1}})$ and $\Delta t=o(h_p^2)$. To verify $\|C^J\|_{L^\infty(L^\infty)}$ in (2.24), we observe that

$$\begin{aligned}\|C^J\|_{L^\infty(L^\infty)} &\le \|\tilde{c}^J\|_{L^\infty(L^\infty)}+\|\pi^J\|_{L^\infty(L^\infty)}\\ &\le K_1+Kh_c^{-1}\|\pi^J\|_{L^\infty(L^2)}\\ &\le K_1+Kh_c^{-1}(h_c^{l+1}+h_p^{k+1}+\Delta t)\\ &\le 2K_1\le K^*,\end{aligned}$$

for $h_p = o(h_c^{\frac{1}{k+1}}), \Delta t = o(h_c)$ and $l \geq 1$. Consequently, K can be chosen independent of K^*. Using the estimates (2.8)-(2.9) with (2.27)-(2.28), the triangle inequality now completes the proof of the theorem. □

3 H^1-Galerkin Mixed Finite Element

In this section, we shall briefly discuss an H^1-Galerkin mixed finite element procedure for the compressible miscible displacement problems (1.1) in porous media.

3.1 Weak Formulation and Galerkin Procedure

Now split the pressure equation (1.1a) as

$$\begin{aligned} &(a) \quad u = -a(c)\nabla p, \\ &(b) \quad d(c)\frac{\partial p}{\partial t} + \nabla \cdot u = q. \end{aligned} \tag{3.1}$$

Let $W = \{w \in (L^2(\Omega))^2 : \nabla \cdot w \in L^2(\Omega)$ and $w \cdot \nu = 0$ 0n $\partial\Omega\}$. Multiply (3.1a) and (3.1b) by ∇v and $\nabla \cdot w$, respectively, then integrate over Ω to obtain

$$\begin{aligned} &(a) \quad (a(c)\nabla p, \nabla v) = -(u, \nabla v), \qquad v \in H^1, \\ &(b) \quad (\frac{\partial p}{\partial t}, \nabla \cdot w) + (\beta(c)\nabla \cdot u, \nabla \cdot w) = (\beta(c)q, \nabla \cdot w), \quad w \in W, \end{aligned} \tag{3.2}$$

where $\beta(c) = \frac{1}{d(c)}$. Using (3.1a), we shall write (3.2b) as

$$(\alpha(c)\frac{\partial u}{\partial t}, w) + (\beta(c)\nabla \cdot u, \nabla \cdot w) = (\beta(c)q, \nabla \cdot w) - (\frac{\partial \alpha}{\partial c}(c)\frac{\partial c}{\partial t}u, w), \quad w \in W.$$

where $\alpha(c) = \frac{1}{a(c)}$. Now, the weak formulation of the problem (1.1) may be stated as : Find $\{p, u, c\} : J \to H^1 \times W \times H^1$ as the solution of

$$\begin{aligned} &(a) \quad (a(c)\nabla p, \nabla v) = -(u, \nabla v), \qquad v \in H^1, \\ &(b) \quad (\alpha(c)\frac{\partial u}{\partial t}, w) + (\beta(c)\nabla \cdot u, \nabla \cdot w) = (\beta(c)q, \nabla \cdot w) - (\frac{\partial \alpha}{\partial c}(c)\frac{\partial c}{\partial t}u, w), \quad w \in W, \\ &(c) \quad (\phi\frac{\partial c}{\partial t}, z) + (D(u)\nabla c, \nabla z) + (e(c)u \cdot \nabla c, z) \\ &\qquad +(B(c) \cdot u, z) + (b(c)u_h, \nabla z) = (g(c), z), \quad z \in H^1(\Omega), \end{aligned} \tag{3.3}$$

where

$$e(c) = \frac{\partial b}{\partial c}(c) + 1, \quad B(c) = \frac{\partial b}{\partial x}(c), \quad and \quad g(c) = [(\tilde{c} - c) + b(c)]q.$$

For Galerkin approximation, let $h = (h_p, h_c)$ where h_p and h_c are positive. Let V_h^k, W_h^r and M_h^l be the standard finite element subspaces of H^1, W and H^1, respectively which satisfy the following approximation properties: For $1 \le p \le \infty$ and for positive integers k, r and l

$$\inf_{v_h \in V_h^k}\{\| v - v_h \|_{L^p} + h_p \| v - v_h \|_{W^{1,p}}\} \le K h_p^{k+1} \quad \| v \|_{W^{k+1,p}}, \ v \in H_0^1 \cap W^{k+1,p},$$

$$\inf_{w_h \in W_h^r}\{\| w - w_h \|_{L^p} + h_p \| w - w_h \|_{W^{1,p}}\} \le K h_p^{r+1} \quad \| w \|_{W^{r+1,p}}, \ w \in W \cap W^{r+1,p},$$

$$\inf_{z_h \in M_h^l}\{\| z - z_h \|_{L^p} + h_c \| z - z_h \|_{W^{1,p}}\} \le K h_c^{l+1} \quad \| z \|_{W^{l+1,p}}, \ z \in H^1 \cap W^{l+1,p}.$$

Note that we can use C^0- piecewise linear polynomial spaces for V_h^k and W_h^r, i.e; when $k = 1 = r$.

Then, an H^1- Galerkin mixed finite element approximation is to determine $\{p_h, u_h, c_h\} : J \to V_h^k \times W_h^r \times M_h^l$ such that

$$\begin{aligned}
&(a) \quad (a(c_h)\nabla p_h, \nabla v_h) = -(u_h, \nabla v_h), \qquad v_h \in V_h^k, \\
&(b) \quad (\alpha(c_h)\frac{\partial u_h}{\partial t}, w_h) + (\beta(c_h)\nabla \cdot u_h, \nabla \cdot w_h) = (\beta(c_h)q, \nabla \cdot w_h) \\
&\qquad\qquad -(\frac{\partial \alpha}{\partial c}(c_h)\frac{\partial c_h}{\partial t}u_h, w_h), \quad w_h \in W_h^r, \qquad\qquad (3.4) \\
&(c) \quad (\phi\frac{\partial c_h}{\partial t}, z) + (D(u_h)\nabla c_h, \nabla z) + (e(c_h)u_h \cdot \nabla c_h, z) \\
&\qquad\qquad +(B(c_h) \cdot u_h, z) + (b(c_h)u_h, \nabla z) = (g(c_h), z), \quad z \in M_h,
\end{aligned}$$

where $u_h(0), p_h(0)$ and $c_h(0)$ are specific functions to be defined later.

We now define the elliptic projections $\tilde{p}_h$ and $\tilde{c}_h$ of p and c onto V_h^k and M_h^l respectively, by

$$\begin{aligned}
&(a) \quad (a(c)\nabla(p - \tilde{p}_h), \nabla v_h) = 0, \qquad v_h \in V_h^k, \\
&(b) \quad (D(u)\nabla(c - \tilde{c}), \nabla z) + (e(c)u \cdot \nabla(c - \tilde{c}), z) + \lambda(c - \tilde{c}, z) = 0, \quad z \in M_h.
\end{aligned}$$

Here λ is chosen in such a way that the bilinear form is H^1-coercive. Further, let $\tilde{u}_I$ be the interpolant of u onto W_h^r.

Let $\eta = p - \tilde{p}_h$, $\rho = u - \tilde{u}_I$ and $\theta = c - \tilde{c}_h$. Following Wheeler [25], it is easy to obtain the following estimates for η and θ

$$\|\eta\|_j + \|\eta_t\|_j \le K h_p^{k+1-j}\{\|p\|_{k+1} + \|p_t\|_{k+1}\}, \ j = 0, 1,$$

and

$$\|\theta\|_j + \|\theta_t\|_j \le K h_c^{l+1-j}\{\|c\|_{l+1} + \|c_t\|_{l+1}\}, \ j = 0, 1.$$

Since $\tilde{u}_I$ is an interpolant of u, we have from standard estimates of interpolation errors

$$\|\rho\| + h_p\|\rho\|_{H(div} \leq K h_p^{r+1}\|u\|_{r+1},$$
$$\|\rho_t\| + h_p\|\rho_t\|_{H(div} \leq K h_p^{r+1}\{\|u\|_{r+1} + \|u_t\|_{r+1}\}.$$

Now we shall state without proof the main result of this section in a form of a Theorem. For a detailed proof, see Mohamed Ali and Pani [13].

Theorem 3.1 *Choose* $u_h(0) = \tilde{u}_I(0)$ *and* $c_h(0) = \tilde{c}_h(0)$*, so that* $\sigma(0) = 0$ *and* $\pi(0) = 0$*, and the mesh parameters satisfy the relations*

$$max(h_c^{-1}h_p^r,\ h_p^{-1}h_c^{l+1}) \rightarrow 0 \ \ as\ \ h \rightarrow 0.$$

Then there exists a constant K *independent of* h_c *and* h_p *such that*

$$\begin{aligned}\|c - c_h\|_{L^2(L^\infty)} + \|u - u_h\|_{L^2(L^\infty)} \ &+ \ \|p - p_h\|_{L^2(L^\infty)} \\ &\leq \ K(c,u,p,T)\{h_c^{l+1} + h_p^r + h_p^{k+1}\}.\end{aligned}$$

Note that when $r = k + 1$, we obtain exactly the same result as in theorem 2.1 without using the LBB condition. However, sharper estimates for the velocity vector can be obtained. The detailed analysis will be discussed in a separate paper, see Mohamed Ali and Pani [13].

Acknowledgement. The first author acknowledges the support of the Government of India through ICCR, New Delhi and would like to thank the Government of Egypt through Mission Department, Cairo for granting him leave.

References

1. Babuška, I., *The Finite Element Method with Lagrange Multipliers*, Numer. Math., **20** (1973), pp. 179-192.

2. Brezzi, F., *On the Existence, Uniqueness and Approximation of Saddle-Point Problems Arising from Lagrange-Multipliers*, RAIRO Anal. Numer., **8** (1974), pp. 129-151.

3. Carey, G. F. and Shen, Y., *Convergence Studies of Least-Squares Finite Elements for First-Order Systems*, Comm. Appl. Numer. Methods., **5** (1989), pp. 427-434.

4. Chou, S.-H. and Li, Q., *Mixed Finite Element Methods for Compressible Miscible Displacement in Porous Media*, Math. Comp., **57** (1991), pp. 507-527.

5. Douglas, J. Jr., *Finite Difference for Two-phase Incompressible Flow in Porous Media*, SIAM J. Numer. Anal., **20** (1983), pp. 681-696.

6. Douglas, J. Jr., Ewing, R. E. and Wheeler, M. F., *The Approximation of the Pressure by a Mixed Method in the Simulation of Miscible Displacement*, RAIRO Anal. Numer., **17** (1983), pp. 17-33.

7. Douglas, J. Jr. and Roberts, J. E., *Numerical Methods for a Model for Compressible Miscible Displacement in Porous Media*, Math. Comp., **41** (1983), pp. 441-459.

8. Douglas, J. Jr., Wheeler, M. F., Darlow, B. L., and Kendall, R. P., *Self-Adaptive Finite Element Simulation of Miscible Displacement in Porous Media*, Comp. Methods Appl. Mech. Eng., **47** (1984), pp. 131-159.

9. Ewing, R. E. and Wheeler, M. F., *Galerkin Methods for Miscible Displacement Problems in Porous Media*, SIAM J. Numer Anal., **17** (1980), pp. 351-365.

10. Johnson, C. and Thomée, V., *Error Estimates for Some Mixed Finite Element Methods for Parabolic Type Problems*, RAIRO Anal. Numer., **15** (1981), PP. 41-78.

11. Ladyzhenskaya, O. A., *The Mathematical Theory of Viscous Incompressible Flows*, Gordon and Breach , London (1969).

12. Mohamed Ali, M. A., *Numerical Methods for Enhanced Oil Recovery in Reservoir Simulation*, Ph.D. Thesis, Department of Mathematics, IIT, Bombay, (1997).

13. Mohamed Ali, M. A. and Pani, A. K., *An H^1- Galerkin Mixed Finite Element Method for Compressible Miscible Displacement Problems in Porous Media*, (under preparation).

14. Mohamed Ali, M. A. and Pani, A. K., *Mixed Finite Element Galerkin Method for Compressible Miscible Displacement Problems in Porous Media*, Research Report, MRR-02-97, Dept. Maths, IIT, Bombay.

15. Nedlec, J. C., *Mixed Finite Elements in R^3*, Numer. Math., **35** (1980), pp. 315-341.

16. Pani, A. K., *An H^1-Galerkin Mixed Finite Element Method for Parabolic Partial Differential Equations*, Research Report MRR03-95, Department of Mathematics, IIT, Bombay. To appear in SIAM J. Numer. Anal.

17. Peaceman, D. W., *Improved Treatment of Dispersion in Numerical Calculations of Multidimensional Miscible Displacement*, Soc Petroleum Eng. J., **6** (1966), pp. 213-216.

18. Pehlivanov, A. I. and Carey, G. F., *Error Estimates for Least-Squares Mixed Finite Elements*, Math. Model. Numer. Anal., **28** (1994), pp. 499-516.

19. Pehlivanov, A. I., Carey, G. F., Lazarov, R. D. and Shen, Y., *Convergence Analysis of Least-Squares Mixed Finite Elements*, Computing, **51** (1993), pp. 111-123.

20. Pehlivanov, A. I., Carey, G. F. and Lazarov, R. D., *Least-Squares Mixed Finite Elements for Second-Order Elliptic Problems*, SIAM J. Numer. Anal., **31** (1994), pp. 1368-1377.

21. Pehlivanov, A. I., Carey, G. F., Lazarov, R. D. and Vassilevski, P. S., *Least-Squares Mixed Finite Element Methods for Non-Selfadjoint Elliptic Problems : I. Error Estimates*, Numer. Math., **72** (1996), pp. 501-522.

22. Raviart, P.A. and Thomas, J.M., *A Mixed Finite Element Method for 2nd Order Elliptic Problems*, Mathematical Aspects of Finite Element Method, Lecture Notes in Mathematics, **606**, Springer-Verlag, Berlin (1977), pp. 292-315.

23. Russell, T. F. and Wheeler, M. F., *Finite Element and Finite Difference Methods for Continuous Flows in Porous Media*, The Mathematics of Reservoir Simulation, R. E. Ewing (ed.), SIAM, Philadelphia (1983), pp. 35-107.

24. Thomée, V., *Galerkin Finite Element Methods for Parabolic Problems*, Lecture Notes in Mathematics, **1054**, Springer Verlag, New York (1984).

25. Wheeler, M. F., *A priori L^2-Error Estimates for Galerkin Approximations to Parabolic Differential Equations*, SIAM J. Numer. Anal., **10** (1973), pp. 723-749.

Department of Mathematics
Indian Institute of Technology, Bombay,
Powai, Mumbai - 400 076 (India).

Z ASLAN[(1)] AND H YAVUZ[(2)]

Mathematical approach based on Q.spline to define the taper curve of Scotch pine (*Pinus sylvestris*) and Crimean pine (*Pinus nigra*)

The taper curve of a tree is a central concept in forest mensuration. It gives mathematical model of a tree which can be used as a basis by all the tasks of forest inventory and mensuration. The form of a tree may have a simple appearence but the construction of its mathematical model is a demanding task. Inspite of a long line interest and a large number of publications no solution has appeared satisfying all the relevant conditions. Scotch pine (Pinus sylvestris) and Crimean pine (Pinus nigra) stems were used in this study. Diameters were measeured at fourteen different proportional heights from the ground, namely 1, 2.5, 5, 7.5, 10, 15, 20, 30, 40, 50, 60, 70, 80 and 90 percent. Moreover the diameter at 1.3m was registered. The diameters were measured to the nearest mm by crosswise calipering. The quality of the taper curve depends on both the number and placement of the interpolating points. There is not much choice in this study as far as placement is concerned, because we used certain measured at fixed places. Results of quadratic spline model have been evaluated for two different data sets.

1. INTRODUCTION

A taper curve can be constructed in several ways by using spline functions as explained in the work of Sloboda [1975 and 1977], Lahtinen and Laasasenaho [1979], Lappi [1986], Lahtinen [1988, 1990 and 1993] and Lahtinen and Laurila [1990]. These taper curves are called taper splines.

Spline functions form a class of approximating functions which can be adopted for several different purposes. Polynomial splines are easiest to handle. A polynomial spline of degree n consists of polynomial pieces of degree at most n which are joined together at so called breakpoints so that the resulting function has continuous derivatives up to the order $n-1$. A clear exposition on polynomial splines is to be found in De Boor [1978] and Schumaker [1981].

The most common polynomial spline is the quadratic splines. It is continuosly differentiable function consisting of polynomial pieces of degree at most two. An interpolating quadratic spline can be uniquely determined by using one initial condition if the interpolating points coincide with breakpoints. It can be evaluated by very simple algorithms [Lahtinen and Laasasenaho 1979].

2. MATERIAL AND METHODS

We used the same tree material as Yavuz [1995]. A more detail description of the materal can be found in Yavuz [1995]. For each tree the tree height was recorded and the height of the stump determined. Diameters were measured at 14 proportional heights from the ground, namely 1, 2.5, 5, 7.5, 10, 15, 20, 30, 40, 50, 60, 70, 80 and 90 percent. The diameter at 1.3m was registered. The diameter at the top was always supposed to be 0.1 cm.

In the test material we divided each tree into seven parts at the relative heights 5, 10, 20, 40, 60 and 80. The each part of tree stem can be formulated by a quadratic polynomial (QO):

$$Y_i = a_i + b_i x + c_i x^2 \tag{1}$$

the ith section or part volume of a tree can be calculated by equation (2).

$$V_i = \int_{x_{i-1}}^{x_i} Y_i^2(x)\, dx$$

$$V_i = \pi \int_{x_{i-1}}^{x_i} (a_i + b_i x + c_i x^2)^2\, dx$$

$$V_i = \pi |a_i^2 x + a_i b_i x^2 + \frac{1}{3}(b_i^2 + 2a_i c_i)x^3 + \frac{1}{2} b_i c_i x^4 + \frac{1}{5} c_i^2 x^5 |_{x_{i-1}}^{x_i} \tag{2}$$

As a result, the total stem volume of a tree can be obtained by addition of all the section volumes (Equation 3). In this study, the number n, of section, of each tree is 7. These section limits are stump to 5%, 5% to 10%, 10% to 20%, 20% to 40%, 40% to 60%, 60% to 80% and 80% to 100% respectively.

$$V = V_1 + V_2 + ... + V_n \tag{3}$$

Table 1. Percentage heights used in this study

Point set	The percentage heights													
15	1	2.5	5	7.5	10	15	20	30	40	50	60	70	80	90	100
8A	1	-	5	-	10	-	20	-	40	-	60	-	80	-	100
8B	-	2.5	-	7.5	-	15	-	30	-	50	-	70	-	90	100
8C	1	2.5	-	7.5	-	15	-	30	-	50	-	70	-	-	100
5A	1	-	-	7.5	-	-	20	-	-	-	60	-	-	-	100
5B	1	-	5	-	10	-	-	-	40	-	-	-	-	-	100
5C	-	2.5	-	7.5	-	-	20	-	-	-	60	-	-	-	100
4A	1	-	-	7.5	-	-	-	-	40	-	-	-	-	-	100
4B	-	2.5	-	-	10	-	-	-	-	50	-	-	-	-	100

3. RESULTS

We started the investigation with eight interpolating points. Three different combinations, called 8A, 8B and 8C were chosen as table 1 shows. Each of them contained 7 measured diameters and fixed top diameter.

The quadratic taper spline QO8A does not express the butt (from stump to 5% height) quite naturally but otherwise volume errors are small. In point set 8B the lowest height is 2.5%. This effects that the corresponding spline QO8B has very small mean error on the butt. Unfortunately the mean volume error on the next interval [5, 10] is respectively higher. Point set 8C attempts to eliminate problem with the butt by having an additional breakpoints there. This results is clear failure. The corresponding quadratic taper spline QO8C has greater volume errors on the butt than the other two alternatives (Table 2 and Table 3).

In the above section we saw that the taper spline with seven measured diameters had essentially the same qualities as that with 14 measured diameters. This encouraged us to reduce number of measured diameters further. The taper splines were again constructed to approximate the quadratic taper spline QO15.

Table 2

Mean total and partial volume differences expressed as percentages

Mean Volume Differences								
Scotch Pine								
Compared taper splines	St-100	St-5	5-10	10-20	20-40	40-60	60-80	80-100
QO8A-QO15	0.03	0.27	0.09	0.05	0.06	-0.12	-0.09	-0.70
QO8B-QO15	0.03	0.14	0.18	0.02	0.08	-0.08	-0.09	0.98
QO8C-QO15	0.04	0.25	0.18	-0.01	0.00	0.08	-0.20	1.11
QO5A-QO15	0.03	0.88	0.36	-0.18	0.23	0.36	0.08	1.11
QO5B-QO15	0.27	0.64	0.12	0.97	0.28	-0.75	0.38	2.29
QO5C-QO15	-0.90	0.11	0.03	-0.66	-0.41	0.018	0.11	1.18
QO4A-QO15	0.68	-0.46	0.64	-0.68	0.76	0.014	0.98	4.13
QO4B-QO15	0.22	0.14	0.27	1.17	2.01	2.13	3.87	6.17
Crimean Pine								
Compared taper splines	St-100	St-5	5-10	10-20	20-40	40-60	60-80	80-100
QO8A-QO15	-0.01	0.92	0.01	-0.88	1.29	0.83	-2.89	-3.15
QO8B-QO15	-0.02	0.77	0.01	-0.91	1.30	0.87	-2.90	-2.41
QO8C-QO15	0.09	0.85	0.01	-0.91	1.30	0.87	-2.90	-2.41
QO5A-QO15	0.04	0.98	0.22	-0.98	1.33	0.93	-6.66	-8.41
QO5B-QO15	0.79	0.27	0.18	-1.12	1.38	-1.11	-6.73	-8.81
QO5C-QO15	0.78	0.18	0.01	-0.98	1.32	-1.03	-6.66	-8.41
QO4A-QO15	0.85	1.26	-0.12	-1.25	1.13	-0.91	-6.43	-8.12
QO4B-QO15	0.51	-0.18	0.14	-0.77	1.81	-0.203	-5.95	-7.69

We chose three different five point combinations, namely the point sets 5A, 5B and 5C defined in Table 1. Each of them contained 4 measured diameters and the fixed top diameter. We again start with volume errors which are presented in Table 2. For the quadratic taper spline QO5A the mean percentile total volume is practically the same as for the quadratic taper spline QO15. The mean percentile partial volume errors remained, however, quite large at the butt. The small number of points apparently gives quite good volume estimates on the interval from 20% height to 60% height.

Table 3. Standard deviation of relative (%) to total partial differences

Mean Volume Differences								
Scotch Pine								
Compared taper splines	St-100	St-5	5-10	10-20	20-40	40-60	60-80	80-100
QO8A-QO15	0.22	0.67	1.34	1.66	2.88	2.96	5.17	4.14
QO8B-QO15	0.28	0.67	1.34	1.04	3.64	2.96	5.17	5.93
QO8C-QO15	0.34	0.66	1.35	0.82	1.24	1.11	9.92	22.73
QO5A-QO15	1.48	0.76	0.30	1.16	1.18	1.66	10.88	23.82
QO5B-QO15	1.76	1.76	1.10	2.36	2.13	1.11	9.92	22.76
QO5C-QO15	2.39	0.67	1.68	3.61	3.72	1.43	10.68	23.40
QO4A-QO15	2.83	1.18	1.62	2.36	4.23	3.74	9.88	23.88
QO4B-QO15	2.90	2.05	1.97	2.28	6.33	3.18	7.81	21.33
Crimean Pine								
Compared taper splines	St-100	St-5	5-10	10-20	20-40	40-60	60-80	80-100
QO8A-QO15	0.02	0.45	0.46	2.35	3.43	2.18	7.65	8.35
QO8B-QO15	0.02	0.44	0.48	2.31	3.28	2.21	7.98	8.66
QO8C-QO15	0.97	0.51	0.66	3.41	2.31	7.88	8.64	8.64
QO5A-QO15	2.00	0.76	0.81	2.81	3.74	2.86	17.60	22.20
QO5B-QO15	2.40	0.75	0.97	2.91	3.77	2.89	18.10	22.30
QO5C-QO15	2.10	1.10	0.99	2.33	3.48	2.74	17.60	22.30
QO4A-QO15	2.20	1.30	1.50	3.31	2.98	2.40	17.10	21.50
QO4B-QO15	2.40	1.00	1.90	2.05	4.67	3.53	15.70	20.40

The point set 5B is more concentrated to the butt than the previous set. The mean volume error of the quadratic taper spline QO5B is also much smaller at the butt than one of the quadratic taper spline QO5A. On the whole the quadratic taper spline QO5B is weaker in volume predictions than the quadratic taper spline QO5A. This can also be seen in standard deviations (Table 3). Thus we can not recommend the use of the alternative QO5B.

Lifting the lowest breakingpoints in point set 5A to the height of 2.5% produces taper spline QO5C. It gives clearly better volume estimates for the butt than the quadratic taper spline QO5A without any evident negative effect to the other parts. The mean total volume error is, however, slighty greater for QO5C than QO5A even if the standard deviation are smaller (Table 2 and Table 3).

Finally we made some experiments with four breakpoints. There were the sets 4A and 4B in Table 1. Both contained three measured diameters and the fixed top diameter. Table 2 shows that the mean total volume error is for both point taper splines clearly larger than the errors for the five point quadratic taper splines QO5A and QO5B. These errors are not very large at the butt, especially in the case of quadratic taper spline QO4B. In the middle parts of the stem the results are undoubtly inferior to the taper splines with more points. The standard deviations are also slightly larger (Table 3).

4. CONCLUSIONS

When seven measured diameters were used then the quadratic taper splines QO8A was quite good approximation to the quadratic taper spline QO15. They have in our sample tree material the same mean total volumes and there are in partial volumes differences only at the butt. On the whole the eight point taper spline is still a good taper curve.

The quadratic taper curve splines QO5A and QO5C with four measered diameters still offer satisfactory approximations to the quadratic taper spline QO15. The mean total volume is in our sample material practically the same as for the spline QO15 but there are differences in the mean partial volumes. It must be kept in mind that a taper spline with four measured diameters can not give a true shape to a singular tree.

A quadratic spline with three measured diameters can be recommended to a taper curve only for trees of regular shape. Also in this case there exist systematic errors. The four point taper spline still gives reasonable total volumes. Therefore it can be used in volume estimation also for arbitrary trees.

5. ACKNOWLEDGEMENTS

The authors wish to acknowledge the International Centre for Theoretical Physics,

Trieste. The authors would like to thank Professor Abul Hasan Siddiqi for providing valuable advise in completion of this paper.

References

[1] De Boor, C., A practical guide to Splines. Springer-Verlag, New York, 1978.

[2] Lahtinen, A., and J. Laasasenaho, On the construction of taper curves by using spline functions. Commun. Institute Forest Fennica 95 (1979), 1 - 63.

[3] Lahtinen, A., On the construction of monotony preserving taper curves. Acta Forestalia Fennica, 1988, 34.

[4] Lahtinen, A., Shape preserving interpolation by quadratic splines. Journal of Computational and Applied Mathematics, 29 (1990), 15 - 24.

[5] Lahtinen, A. and K. Laurila, The monotony preserving QO-spline as a taper curve. Scand. Journal of Forest Research, 5 (1990), 277 - 283.

[6] Lahtinen, A., On the construction of shape preserving taper curves. Silva Fennica, 27 (1993), 29 - 45.

[7] Lappi, J., Mixed linear models for analysing and predicting stem form variation of Scotch pine. Commun. Institute Forest Fennica, 134 (1986), 1 - 69.

[8] Schumaker, L., Spline functions, Basic theory. John Wiley and Sons., New York 1981.

[9] Sloboda, B., Anwendung der Spline-Interpolation bei der Wiedergabe der Baumschaftform, Allg. FZJ., 1975.

[10] Sloboda, B., Spline und Bitterlich'sche interpolation bei der Wiedergabe der Baumschaftform, Mitt. Forstl. Bundes-Versuchsanstalt, Wien 1977, 61 - 70.

[11] Yavuz, H., Development of compatible stem taper, total tree volume and volume ratio equation systems for Scotch and Crimean pine in Tapköprü Forest Enterprise, in press.

(1) Kafkas University, Artvin Facultry of Forestry, 080000 Artvin, TURKEY.
(2) Karadeniz Technical University, Faculty of Forestry, 61080 Trabzon, TURKEY.